Engineering Materials Handbook

Engineering Materials Handbook

Editor: Sally Renwick

NY RESEARCH
P R E S S

New York

Published by NY Research Press
118-35 Queens Blvd., Suite 400,
Forest Hills, NY 11375, USA
www.nyresearchpress.com

Engineering Materials Handbook
Edited by Sally Renwick

International Standard Book Number: 978-1-63238-630-4 (Hardback)

Cataloging-in-Publication Data

Engineering materials handbook / edited by Sally Renwick.
    p. cm.
Includes bibliographical references and index.
ISBN 978-1-63238-630-4
1. Materials. 2. Engineering. I. Renwick, Sally.
TA403 .E54 2019
620.11--dc23

Contents

Preface

Materials that can be used for the construction or manufacturing of other products through an organized engineering activity are called engineering materials. Advancements in engineering materials drive the design and manufacture of new products. Some of the industrial processes surrounding such materials are processing techniques, quality analytic methods, material characterization and purification. These materials are tested for their properties and are accordingly divided into four essential categories: Metal alloys, Composite, Ceramic and Polymer. This book is compiled in such a way that it provides extensive information about the theories and practices that are applied to this field while also examining new methods and applications in close detail. This book includes some of the vital pieces of work being conducted across the world, on various topics related to engineering materials. With its detailed analyses and data, this book will prove immensely beneficial to professionals and students involved in this area at various levels.

The information contained in this book is the result of intensive hard work done by researchers in this field. All due efforts have been made to make this book serve as a complete guiding source for students and researchers. The topics in this book have been comprehensively explained to help readers understand the growing trends in the field.

I would like to thank the entire group of writers who made sincere efforts in this book and my family who supported me in my efforts of working on this book. I take this opportunity to thank all those who have been a guiding force throughout my life.

Editor

Recrystallization-based formation of uniform fine-grained austenite structure before polymorphic transition in high-strength steels for Arctic applications

Pavel Layus[1][*] [iD], Paul Kah[1], Alexander Zisman[2], Markku Pirinen[1] and Sergey Golosienko[2]

Abstract

Background: Growing demand for materials suitable for Arctic marine constructions necessitates the development of lowcost steels with excellent low-temperature properties. One way of addressing this issue is to obtain a finegrained structure by careful control of the recrystallization process.

Methods: The research methods used in this paper includes conducting experiments and making observations. The study investigates mechanisms of grain structure formation in low-carbon cold-resistant F620 steels additionally alloyed with Nb or V during hot plastic deformation.

Results: The obtained results evidence that static recrystallization is the most effective means to refine the austenite structure under conditions of industrial rolling. On this basis, a temperature-strain scheme of fractional hot rolling is proposed that provides the most refined uniform structures of bainite and/or martensite after quenching. Additionally, the results show that substitution of the alloying element V by Nb makes the structure of the hot-rolled austenite more finely dispersed.

Conclusions: New cold-resistant F620 steels for Arctic applications alloyed with Nb of 50-mm thickness were produced utilizing the research results. These grades of steels have reduced alloying content and exhibit improved properties, in particular, cold resistance down to −60 °C and high fracture toughness at −50 °C.

Keywords: Arctic conditions, Recrystallization, Static recrystallization, Dynamic recrystallization, Cold-resistance steel, High-strength steel

Background

Simultaneously high levels of strength and cold resistance are commonly obtained in shipbuilding steels by alloying with elements such as Ni, Mo, Cr, Cu, Nb, Ti, and V which add considerably to the cost of the steel (Khlusova et al. 2007). Consequently, the growing demand for Arctic steels is promoting research in and development of suitable low-cost steels and their respective technologies. At the same time, however, quality requirements are becoming still more severe. For example, the Russian Maritime Register of Shipping has imposed more stringent requirements on ductility, cold resistance, and fracture toughness of "Arc-steels" while maintaining previous high-strength requirements (Russian Maritime Register of Shipping. Rules for the Classification & Construction and Equipment of Mobile Offshore Platforms 2012). Such a complex combination of qualities is feasible in thick rolled products only when fine-grained structures are present (Calcagnotto et al. 2009; Leinonen 2004; Salvatori 2006). Refining of austenite grains is possible by utilizing recrystallization (dynamic—DRx, metadynamic—MDRx, and static—SRx) (Gorelik et al. 2005) and fragmentation (Kodzhaspirov et al. 2006) processes under specific thermomechanical conditions (Fernández et al. 2003). In addition, the structure formation over the whole cycle of hot rolling is sensitive to small additions (~0.01 %) of Nb, Ti, or V,

* Correspondence: pavel.layus@lut.fi
[1]Welding Technology Laboratory, Lappeenranta University of Technology, PL 20, 53851 Lappeenranta, Finland
Full list of author information is available at the end of the article

which hinder boundary migration and, hence, grain growth (Sha & Sun 2009; Sha et al. 2011; Jung et al. 2011). The present paper investigates the effects of temperature and strain degree on DRx and SRx kinetics in austenite of cold-resistant high-strength steel F620 presented with two micro-alloying options—with V or Nb. The chemical composition of steel F620 according to various certification societies and the one used in this study is provided in Table 1.

Methods

Dynamic recrystallization experiments

The threshold strain (εp) related to DRx start was determined with stress-strain diagrams (σ-ε) based on characteristic stress maxima. Stress-strain diagrams for the steels in this study are shown in Fig. 1. At the highest temperature, 1150 °C, both steels display maximum stress values ($\sigma \approx 90$ MPa) at lower values of strain (εp ≈ 0.3 %), as shown in Fig. 1 a by the red dotted line. Here, Nb augments εp by about 0.005 %, slightly hindering DRx though not stopping the process. When the temperature decreases down to 1050 °C, the Nb effect becomes most pronounced, since, unlike steel F620 alloyed with V (εp $\approx$ 0.45 %), there are no signs of austenite DRx in F620 alloyed with Nb up to true strain of 0.9 %. At the lowest temperature, 950 °C, the DRx process ceases in both steels. The diagrams in Fig. 1 also show that the general hardening effect due to Nb becomes stronger when the deformation temperature is reduced.

In order to form a sufficiently uniform refined structure, the small austenite grains inherent in DRx should cover a high enough volume fraction. However, even at $T > 1050$ °C, this is possible only if single strains exceed 0.9 %, which is unrealistic in industrial rolling. Therefore, austenite formation during DRx remains incomplete, and hence, an unwanted non-uniform structure forms (coexistence of fine and coarse grains) that reduces the ductility of the steel. As evident in Fig. 1, this characteristic previously known with the traditional steel F620 alloyed with V (Zisman et al. 2012) remains valid for its modification alloyed with Nb. Therefore, the growth of threshold εp due to Nb proves to have an additional positive effect because it more reliably excludes incomplete DRx and related grain size variation. The grain size was measured according to ISO 643: 2003 standard.

Static recrystallization experiments

To investigate SRx of austenite, the stress relaxation method (Zisman et al. 2012; Perttula & Karjalainen 1998) was implemented on a Gleeble 3800 simulator. The selected degrees of 0.25 and 0.10 of single strains correspond, respectively, to the maximum thickness reduction during rough rolling and the average reduction in finish rolling of the investigated steels. Distinct

changes in the softening rate dσ/dt allow determination of the incubation period of SRx and the time span of its primary stage (Zisman et al. 2012; Perttula & Karjalainen 1998). Stress relaxation curves for austenite of F620 alloyed with V and Nb, correspondingly, are indicated in Figs. 2 and 3 by labels "V" and "Nb." When the incubation period and duration of primary SRx could be derived from the diagrams, these time parameters are indicated by light and dark circles, respectively; stars indicate the time when the SRx degree reached 80 %. Following, the SRx degree is evaluated as (Zisman et al. 2012)

$$P(t) = (\sigma\text{max}-\sigma t)/(\sigma\text{max}-\sigma\text{min}), \qquad (1)$$

where σmax and σmin are the maximum and minimum stress magnitudes during the SRx, respectively, and σt is the stress magnitude at a current time t.

Stress relaxation diagrams after $\varepsilon = 0.25$ % (Fig. 2) show that during technological pauses (several seconds or tens of seconds) austenite SRx in both steels becomes complete only at $T \geq 1000$ °C. At $T = 950$ °C, this process remains uncompleted in the steel micro-alloyed by Nb and at lower temperatures in both steels. After $\varepsilon = 0.10$ % (Fig. 3), SRx becomes much slower and becomes complete in the abovementioned pauses only at relatively high temperatures of $T \geq 1100$ °C. Earlier termination of SRx due to Nb addition accelerates the growth of the dislocation density and hence extends the range of austenite fragmentation (subgrain structure evolution) to higher temperatures (Kodzhaspirov et al. 2006; Rybin 1986). Nevertheless, previous SRx still remains significant, since the fragmentation rate increases with the grains' refinement (Rybin 1986).

Results and discussion

Samples of both F620 steels alloyed with V or Nb were subjected to two treatments on the Gleeble 3800 simulator. The treatments included five sequential strains, separated by pauses, in a temperature range of 950 to 1150 °C. In the first treatment, the pauses were not regulated; in the second treatment, the pause durations, increasing with the temperature reduction, were selected to provide complete primary SRx in each case. The true strains, strain rate, and following cooling rate were 0.15 %, 1 s^{-1} and 30 °C/s, respectively.

As can be seen from Figs. 4a–c and 5a–c, the most dispersed structure and therefore the highest hardness (Figs. 4d and 5d) are provided by the treatment with regulated pauses selected based on the austenite SRx kinetics of F620 steels alloyed with V or Nb.

Rolled products of up to 50-mm thickness were fabricated from a new lower-alloyed cold-resistant steel, F620 (alloyed with Nb), with a rolling scheme based on the above findings. The average tensile strength was 760 MPa (700–890 MPa required by specification)

Table 1 Chemical composition of F620, wt. %

Steel grade	C	Mn	Si	P	S	Cu	Ni	Mo	Cr	Al	Nb	V	Ti	V + Nb + Ti	N	B
F620 (BV[a])	0.18	1.6	0.55	0.025	0.02	1.5	2	1	2	0.015	0.06	0.1	0.2	–	0.02	0.06
F620W (DNV[b])	0.18	1.6	0.1–0.55	0.025	0.025	0.35	0.8	0.08	0.2	0.02	0.02–0.05	0.05–0.1	0.007–0.02	0.12	0.02	–
F620W (Lloyd[c])	0.18	1.6	0.55	0.025	0.025	–	–	–	–	0–0.015	0.02–0.05	0.03–0.1	0.02	0.12	0.02	–
F620W (RMRS[d])	0.18	1.6	0.55	0.025	0.025	–	–	–	–	–	–	–	–	–	0.02	–
F620W with V (TU[e])	0.08–0.10	0.3–0.6	0.17–0.37	–	–	Cu + Ni:2.2–2.9		Mo + Cr:0.55–1.05		–	–	0.01–0.03	–	–	–	–
F620W with Nb (TU[e])	0.08–0.10	0.3–0.6	0.17–0.37	–	–	Cu + Ni:2.2–2.9		Mo + Cr:0.55–1.05		–	0.01–0.03	–	–	–	–	–

If the range is not specified, the maximum allowable value is given
[a]Bureau Veritas
[b]Det Norske Veritas
[c]Lloyd's Register
[d]Russian Maritime Register of Shipping
[e]TU 5.961-11571-95 (chemical composition used in this study)

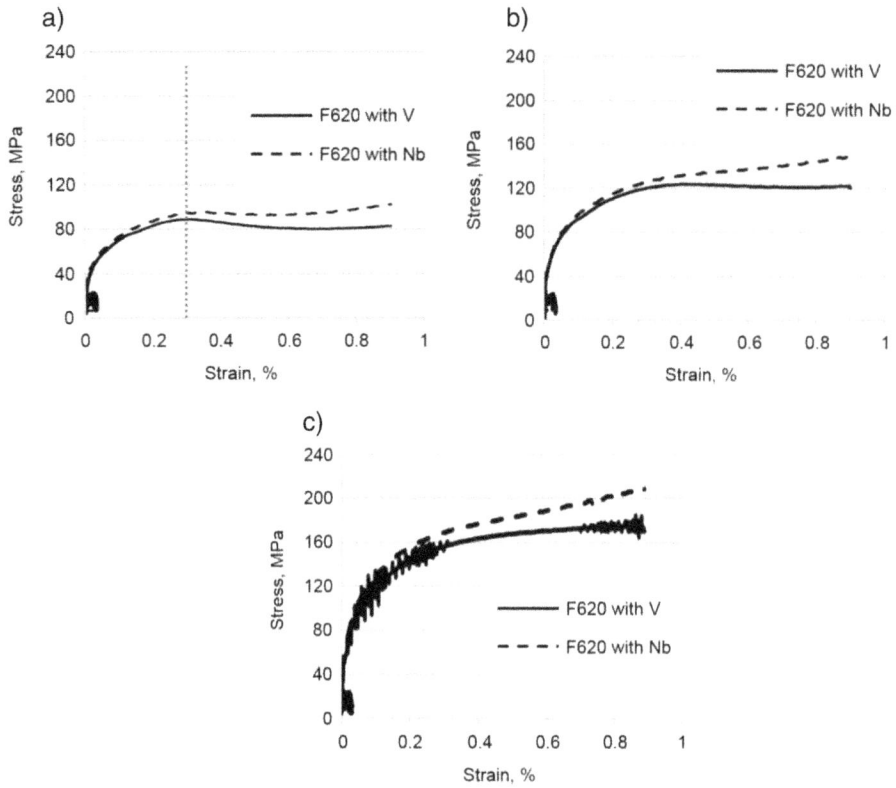

Fig. 1 Loading stress-strain diagrams of F620 with V and Nb steels at **a** $T = 1150\ ^\circ\text{C}$, **b** $T = 1050\ ^\circ\text{C}$, and **c** $T = 950\ ^\circ\text{C}$

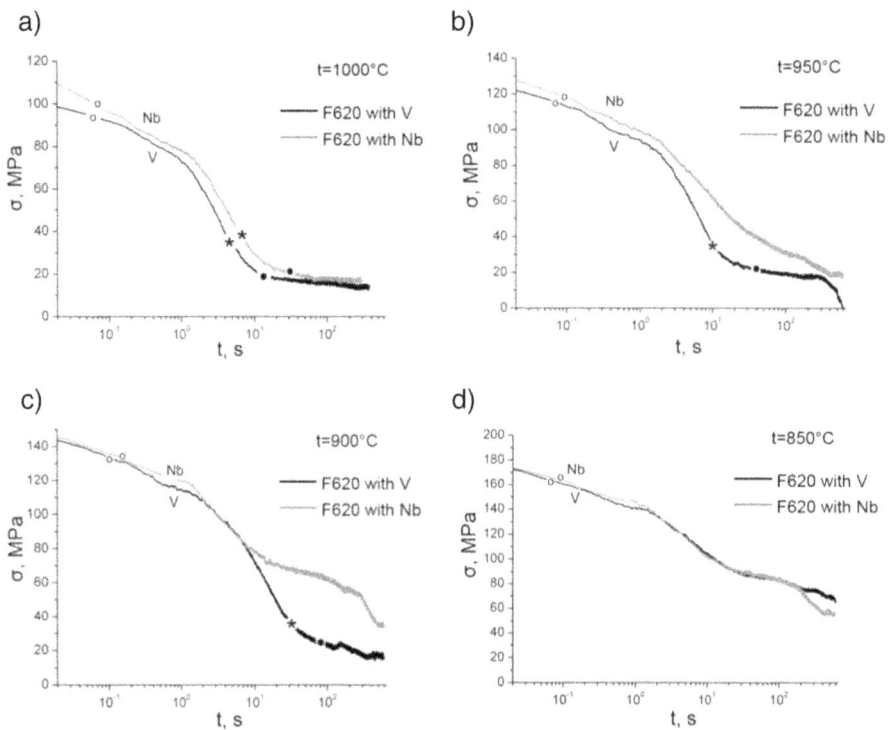

Fig. 2 Stress relaxation of F620 steel alloyed with V and Nb after true strain of 0.25 % at temperatures of **a** 1000 ℃, **b** 950 ℃, **c** 900 ℃, and **d** 850 ℃

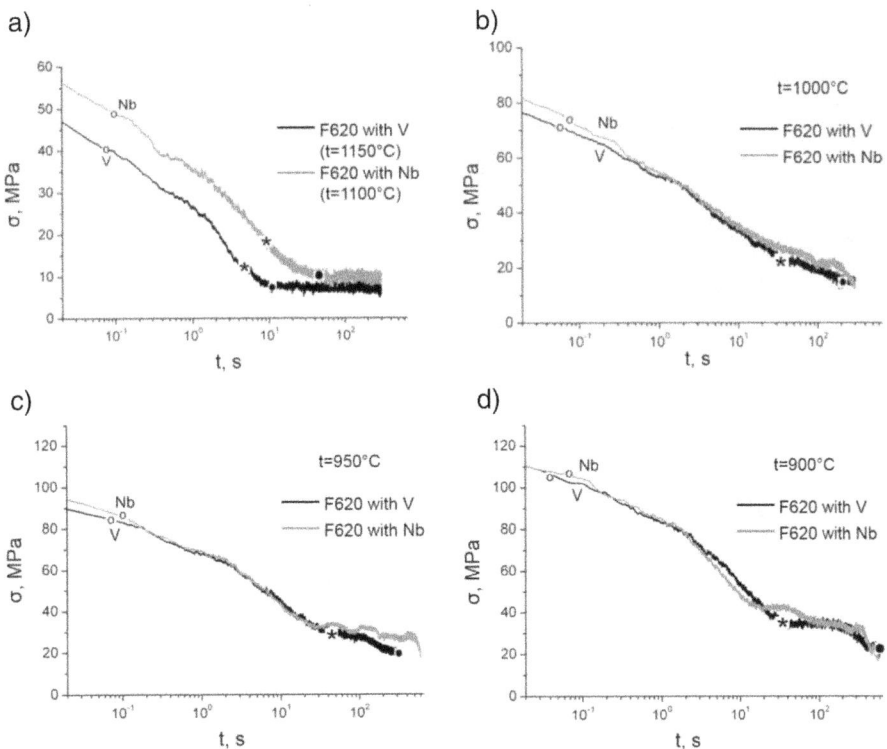

Fig. 3 Stress relaxation of F620 steel alloyed with V and Nb after true strain of 0.10 % at temperatures of **a** 1100 and 1150 °C, **b** 1000 °C, **c** 950 °C, and **d** 900 °C

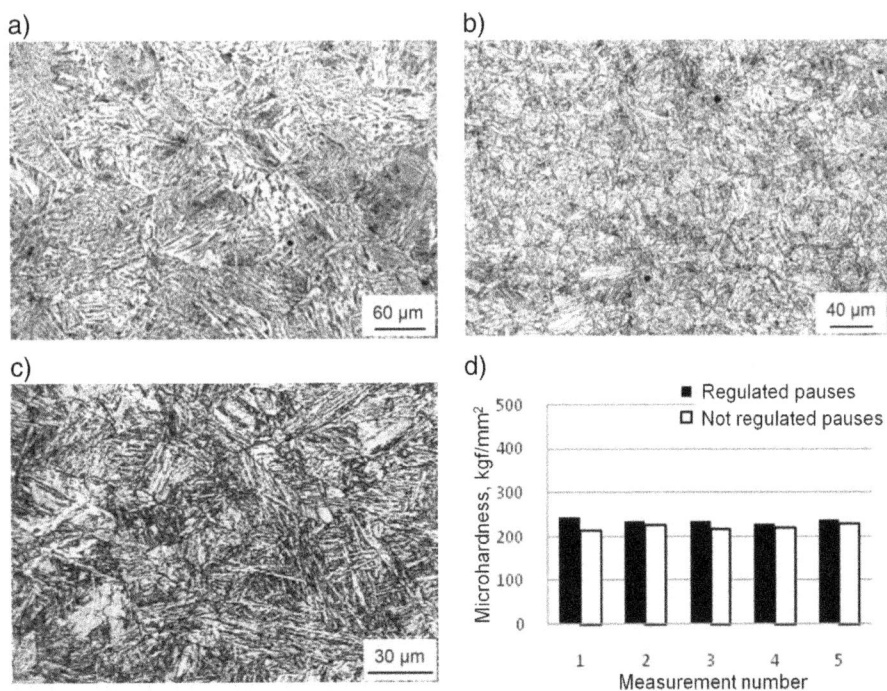

Fig. 4 Microstructure F620 steel alloyed with V: **a** transformed from austenite exposed for 60 s at $T = 1200$ °C, $D_\gamma = 120$ μm; **b** austenite deformed with pauses regulated, $D_\gamma = 17$ μm; **c** not regulated, $D_\gamma = 90$ μm; and **d** microhardness

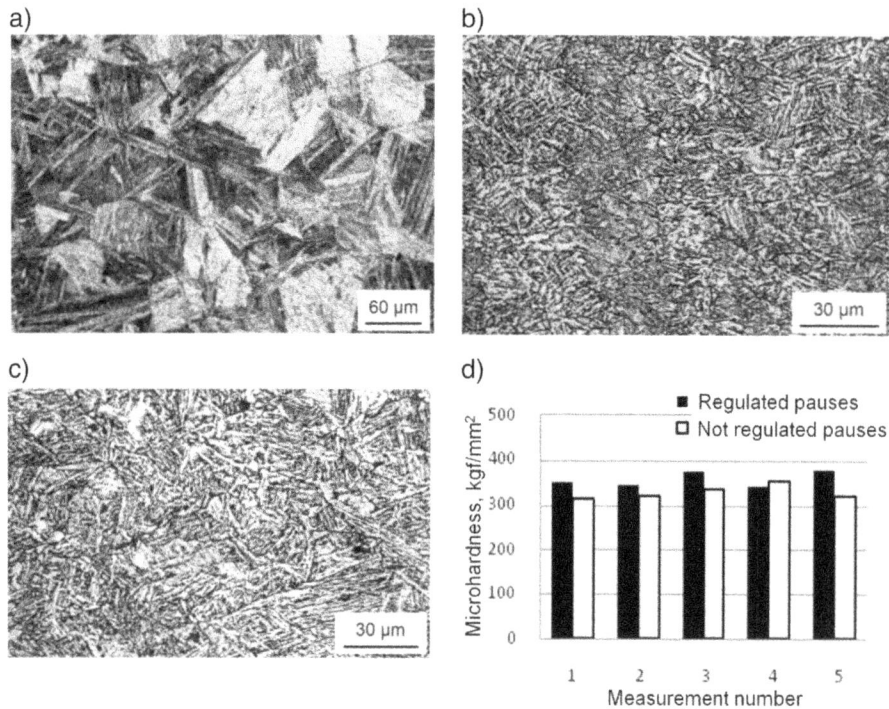

Fig. 5 Microstructure of F620 steel alloyed with Nb: **a** transformed from austenite exposed for 60 s at $T = 1200$ °C, $D_\gamma = 100$ μm; **b** austenite deformed with pauses regulated, $D_\gamma = 15$ μm; **c** not regulated, $D_\gamma = 50$ μm; and **d** microhardness

and the yield stress 700 MPa (≥620 MPa required by specification). The alloying level, as reflected by the toughness parameter Pcm, was 0.22 %. Apart from the required strength, the rolled products of Arc-steel showed high ductility ($\delta_5 \geq 20$ %), good toughness, and cold resistance down to −60 °C (Figs. 5 and 6). All tests were performed on whole thickness samples. The requirements for steels are provided according to Russian Maritime Register of Shipping (RMRS) rules. The quality characteristics of the lower-alloyed cold-resistant steel are the result of the finely dispersed structures of bainite-martensite mixture obtained by γ-α transition from the fine-grained austenite. The testing results of F620 steels alloyed with V or Nb demonstrate the effectiveness of the new procedure.

Conclusions

The outcomes and conclusions of the conducted research are:

1. During rough rolling ($T > 1050$ °C for F620 steel alloyed with V and $T > 1100$ °C for Nb-alloyed steel), incomplete DRx can occur, resulting in non-uniformity (grain size) of the austenite structure. To prevent this unwanted effect, it is necessary to regulate successive deformations.

2. A uniform fine-grained austenite structure may be obtained during rough rolling if primary SRx is completed within each pause; to this end, it is necessary to gradually increase the durations of the pause between successive deformations in the range 1150–950 °C for

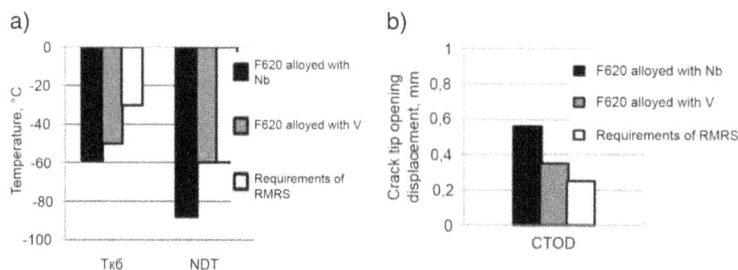

Fig. 6 Properties of F620 steel. **a** Parameters of cold resistance (characteristic temperatures Тк6 and NDT); **b** fracture toughness (CTOD at −40 °C)

F620 V-alloyed steel and 1150–1050 °C for F620 Nb-alloyed steel.

3. Substitution of V by Nb in Cr-Ni-Mo steels makes the structure of the hot-rolled austenite more finely dispersed as a result of a number of effects: (a) limitation on grain growth during the metal heating before rolling; (b) prevention of DRx, which may be only partial (incomplete) for technological reasons and can result in structural non-uniformity; (c) prevention of growth of new grains formed by primary SRx; and (d) extension to higher temperatures of the range of austenite fragmentation, forming new interfaces within grains.

4. Rolled products of 50-mm thickness fabricated from a new cold-resistant Nb-alloyed F620 steel with reduced alloying and made following the discussed procedure display improved product characteristics, in particular, cold-resistance down to –60 °C and high fracture toughness at –50 °C. The steel is recommended for Arctic applications.

Competing interests
The authors declare that they have no competing interests.

Authors' contributions
PL has drafted the paper, prepared data and provided explanations for the experimental results. PK shared expert advice and helped to complete the paper with European perspective on the study. AZ conducted the experiments and shared expert advice on the paper and proposed conclusions of the paper. MP managed the research project and took part in the discussions and conclusion formulating. SG participated in the experiments and supported the paper development. All authors read and approved the final manuscript.

Acknowledgements
This research was supported by an ENPI project Arctic Materials Technology Development.

Author details
[1]Welding Technology Laboratory, Lappeenranta University of Technology, PL 20, 53851 Lappeenranta, Finland. [2]Central Research Institute of Structural Materials Prometey, Saint-Petersburg, Russia.

References
Khlusova, EI, Kruglova, AA, & Orlov, VV (2007). The effect of chemical composition, heat and strain treatment on the size of austenite grains in low-carbon steels (in Russian). *Metall heat Treat Met., 12*, 3–8.

Russian Maritime Register of Shipping (2012). Rules for the classification, construction and equipment of mobile offshore platforms. *2*, 445.

Calcagnotto M, Ponge D, Adachi Y, Raabe D (2009) Effect of grain refinement on strength and ductility in dual-phase steels [Internet]. In: *Proceedings of the 2nd International Symposium on Steel Science (ISSS 2009)*. Kyoto: Iron and Steel Institute of Japan (ISIJ); pp. 195–198. Available from: http://edoc.mpg.de/439324.

Leinonen, JI (2004). Superior properties of ultra-fine-grained steels. *Acta Polytechnica, 44*(3), 37–40.

Salvatori, I (2006). Ultrafine grained steels by advanced thermomechanical processes and severe plastic deformations. *La Metall Ital., 5*, 41–47.

Gorelik, SS, Dobatkin, SV, & Kaputkina, LM (2005). *Recrystallization of metals and alloys*. Moscow: MISIS.

Kodzhaspirov, GE, Rudskoy, AI, & Rybin, VV (2006). *Physical bases and sustainable manufacturing techniques for plastic deformation (in Russian)*. Saint Petersburg, Russia: Nauka.

Fernández, AI, Uranga, P, & López, B (2003). Dynamic recrystallization behavior covering a wide austenite grain size range in Nb and Nb–Ti microalloyed steels. *Materials Science and Engineering A, 361*, 367–376.

Sha, Q, & Sun, Z (2009). Grain growth behavior of coarse-grained austenite in a Nb–V–Ti microalloyed steel. *Materials Science and Engineering A, 523*, 77–84.

Sha, Q, Huang, G, Guan, J, Ma, X, & Li, D (2011). A new route for identification of precipitates on austenite grain boundary in an Nb-V-Ti microalloyed steel. *Journal of Iron and Steel Research, International [Internet], 18*(8), 53–57. Available from: http://dx.doi.org/10.1016/S1006-706X(11)60104-0.

Jung, J, Park, J, Kim, J, & Lee, Y (2011). Carbide precipitation kinetics in austenite of a Nb–Ti–V microalloyed steel. *Materials Science and Engineering: A [Internet], 528*(16-17), 5529–5535. Available from: http://dx.doi.org/10.1016/j.msea.2011.03.086.

Zisman, AA, Soshina, TV, & Khlusova, EI (2012). Studies of austenite recrystallization in hot rolled steel 09HN2MD by stress relaxation (in Russian). *Progress in Materials Science, 2*, 16–24.

Perttula, JS, & Karjalainen, LP (1998). Recrystallisation rates in austenite measured by double compression and stress relaxation methods. *Materials Science and Technology, 14*(7), 626–630.

Rybin, VV (1986). *Large plastic deformation and fracture*. Moscow: Metallurgia.

2

MHD squeeze flow and heat transfer of a nanofluid between parallel disks with variable fluid properties and transpiration

K. Vajravelu[1*], K. V. Prasad[2], Chiu-On Ng[3] and Hanumesh Vaidya[2]

Abstract

Background: The purpose of the study is to investigate the effects of variable fluid properties, the velocity slip and the temperature slip on the time-dependent MHD squeezing flow of nanofluids between two parallel disks with transpiration.

Methods: The boundary layer approximation and the small magnetic Reynolds number assumptions are used. The non-linear governing equations with appropriate boundary conditions are initially cast into dimensionless form by using similarity transformations and then the resulting equations are solved analytically via Optimal Homotopy Analysis Method (OHAM). A detailed parametric analysis is carried out through plots and tables to explore the effects of various physical parameters on the velocity temperature and nanoparticles concentration fields.

Results: The velocity distribution profiles for transpiration (suction/blowing) are parabolic in nature. In general, at the central region, these profiles exhibit the cross-flow behavior and also exhibit the dual behavior with the increase in the pertinent parameters. The temperature distribution reduces in the case of suction whereas the reverse trend is observed in the case of injection.

Conclusion: The effects of temperature dependent thermophysical properties are significant on the flow field. For higher values of the fluid viscosity parameter, the velocity field increases near the walls. However, the transpiration effects are dominant and exhibit the cross-flow behavior as well as the dual behavior. The temperature and the concentration fields are respectively the increasing functions of the variable thermal conductivity and the variable species diffusivity parameters.

Keywords: Squeeze flow, Nanofluid, Wall slip, OHAM, Transpiration

Background

Squeeze flow has promising applications in engineering and industrial processes such as bio-mechanics, flow through arteries, food processing, polymer processing, compression, injection modeling, and mechanical, industrial, and chemical engineering. Generally, squeeze flows are generated in many hydro-dynamical machines and tools where vertical velocities or normal stresses are applied due to moving boundary. Stefan (1874) initiated the study by considering the squeeze flow of a Newtonian fluid with lubrication approximation. Leider and Bird (1974) and Hamza (1988) extended the pioneering work of Stefan (1874) for flow between two parallel disks. Domairry and Aziz (2009) performed a comprehensive analysis of an electrically conducting fluid between two parallel disks of which the upper disk is impermeable and the lower disk is permeable. Joneidi et al. (2010) analyzed the effects of suction/injection and squeeze Reynolds number on the magnetohydrodynamic flow between two parallel disks. Hayat et al. (2012) extended the work of Domairry and Aziz (2009) to second-grade fluid using HAM homotopy analysis method. Further, Hussain et al. (2012) examined the unsteady MHD flow and heat transfer of a viscous fluid squeezed between two parallel disks. Furthermore, Shaban et al. (2013) reported the time-dependent squeeze MHD flow between two parallel disks via a new hybrid method based on the Tau method and the homotopy analysis method.

In recent years, manufacturing industries have begun to choose the fluids based on the heat transmission property: This has considerable implications in the performance of many devices such as in air-conditioning, electronic,

* Correspondence: kuppalapalle.vajravelu@ucf.edu
[1]Department of Mathematics, University of Central Florida, Orlando, FL 32816, USA
Full list of author information is available at the end of the article

chemical, and power. The traditional fluids such as water, ethylene glycol, mineral oils have limited heat transfer ability. Choi (1995) coined the word "nanofluid" for the fluids with suspended nanometer sized (10^{-9} nm) particles called nanoparticles, which were dispersed in traditional fluids. The distinctive nature of the nanofluid, higher thermal conductivity at low nanoparticle aggregation, strong temperature-dependent thermal conductivity, and non-linear increase in thermal conductivities are more useful in industrial and engineering applications which include nuclear reaction cooling, geothermal power extraction, automobile fuels, radiator cooling, cooling of electronic devices, smart fluids, and in bio- and pharmaceutical industry. Motivated by the aforementioned applications, numerous researchers have engaged in the discussion of flows of nanofluids through different geometries (see Vajravelu et al. 2011; Makinde and Aziz 2011; Bachok et al. 2012; Safaei et al. 2014; Safaei et al. 2014; Goodarzi et al. 2014; Togun et al. 2014; Prasad et al. 2016). The MHD squeezing flow of nanofluid between parallel disks is studied by Hashmi et al. (2012). Recently, Das et al. (2016) presented a numerical study on the squeeze flow of a nanofluid between two parallel disks in the presence of a magnetic field considering slip effects. Mohyud-Din et al. (2016) considered both velocity and temperature slip effects on squeeze MHD flow of a nanofluid between parallel disks.

All the above studies analyzed the characteristics of nanofluid squeeze flows assuming constant thermo-physical properties. However, several researchers (see Lai and Kulacki (1990), Vajravelu et al. (2013), Prasad et al. (2016)) have shown that the thermo-physical properties of the ambient fluid may change with temperature, especially the fluid viscosity and the fluid thermal conductivity. For lubricating fluids, heat generated by internal friction and the corresponding rise in the temperature affects the physical properties of the nanofluid, and hence the properties of the fluid are no longer assumed to be constant. The increase in temperature leads to an increase in the transport phenomena and so the heat transfer at the wall is also affected. The experimental results show that even a very low volume fraction of nanoparticles can significantly affect the thermo-physical properties of a nanofluid (see Vajjha and Das (2012)). Therefore, to predict the flow, heat, and mass transfer rates, it is necessary to take the variable fluid properties into account. Thus, the purpose here is to investigate the effects of variable fluid properties, the velocity slip, and the temperature slip on the time-dependent MHD squeezing flow of nanofluids between two parallel disks with transpiration. The governing partial differential equations are transformed into a set of ordinary differential equations: The non-linear coupled systems of equations have been solved for various values of sundry parameters via an efficient analytical method, optimal homotopy analysis method (OHAM) (for details see Liao (2003), Fan and You (2013)). The analysis revels that the fluid flow is appreciably influenced by the physical parameters. It is expected that the results presented here will not only provide useful information for industrial applications but also complement the existing literature.

Mathematical formulation

Let us consider the MHD incompressible flow of an electrically conducting nanofluid squeezed between two parallel disks separated by a variable distance $h(t) = H(1 - \alpha t)^{1/2}$ where α is a parameter having the dimension of inverse time. The upper disk at $z = h(t)$ is moving with the velocity $dz/dt = -(1/2)\left(\alpha H/\sqrt{(1-\alpha t)}\right)$ towards or away from the stationary lower disk at $z = 0$. Mathematically, $w = \partial h/\partial t$ represents squeezing of the upper disk with the lower disk. The axial coordinate is denoted by z and the radial coordinate by r. u and w are the radial and axial velocities, respectively. A magnetic field of strength $B(t) = B_0(1 - \alpha t)^{-1/2}$ is applied

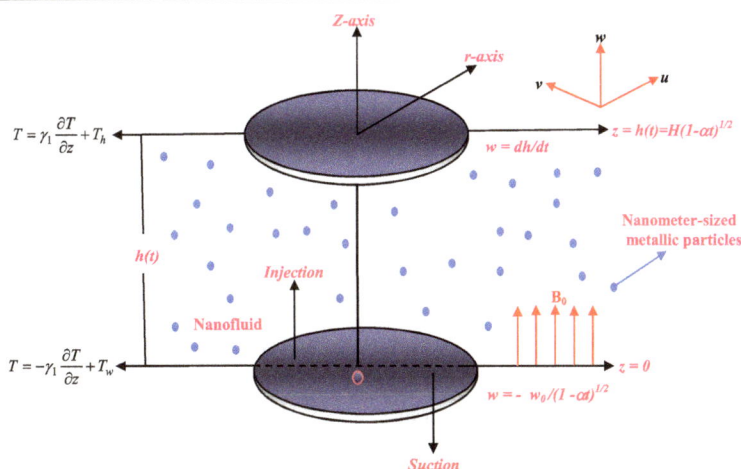

Fig. 1 Geometry of the problem

perpendicular to the disks. On the basis of low Reynolds number assumption, the induced magnetic field is neglected. Here, T_w and C_w denote the temperature and nanoparticles concentration at the lower disk while the temperature and concentration at the upper disk are T_h and C_h, respectively. The physical configuration is represented in Fig. 1. Here, the cylindrical polar coordinates are chosen for the system. Due to the rotational symmetry of the flow, the azimuthal component of the velocity vanishes identically. Under these assumptions, the governing equations for unsteady two-dimensional flow and heat transfer of a viscous nanofluid are (see Hashmi et al. (2012) and Das et al. (2016)):

$$\frac{\partial u}{\partial r} + \frac{u}{r} + \frac{\partial w}{\partial z} = 0, \tag{1}$$

$$\frac{\partial u}{\partial t} + u\frac{\partial u}{\partial r} + w\frac{\partial u}{\partial z} = -\frac{1}{\rho_f}\frac{\partial p}{\partial r} + \frac{1}{\rho_f}\left(\begin{array}{c} \frac{\partial}{\partial r}\left(\mu(T)\frac{\partial u}{\partial r}\right) + \frac{\partial}{\partial z}\left(\mu(T)\frac{\partial u}{\partial z}\right) \\ +\mu(T)\left(\frac{1}{r}\frac{\partial u}{\partial r} - \frac{u}{r^2}\right) \end{array} \right)$$
$$-\frac{\sigma B_0^2(t)}{\rho_f}u, \tag{2}$$

$$\frac{\partial w}{\partial t} + u\frac{\partial w}{\partial r} + w\frac{\partial w}{\partial z} = -\frac{1}{\rho_f}\frac{\partial p}{\partial z}$$
$$+\frac{1}{\rho_f}\left(\frac{\partial}{\partial r}\left(\mu(T)\frac{\partial w}{\partial r}\right) + \frac{\mu(T)}{r}\frac{\partial w}{\partial r} + \frac{\partial}{\partial z}\left(\mu(T)\frac{\partial w}{\partial z}\right)\right), \tag{3}$$

$$\frac{\partial T}{\partial t} + u\frac{\partial T}{\partial r} + w\frac{\partial T}{\partial z} = \frac{1}{\rho_f C_p}\left(\begin{array}{c} \frac{\partial}{\partial r}\left(\lambda(T)\frac{\partial T}{\partial r}\right) + \frac{\partial}{\partial z}\left(\lambda(T)\frac{\partial T}{\partial z}\right) \\ +\lambda(T)\frac{1}{r}\frac{\partial T}{\partial r} \end{array} \right)$$
$$+\tau\left[\begin{array}{c} D_B(C)\left(\frac{\partial C}{\partial r}\frac{\partial T}{\partial r} + \frac{\partial C}{\partial z}\frac{\partial T}{\partial z}\right) \\ +\frac{D_T}{T_m}\left\{\left(\frac{\partial T}{\partial r}\right)^2 + \left(\frac{\partial T}{\partial z}\right)^2\right\} \end{array} \right], \tag{4}$$

$$\frac{\partial C}{\partial t} + u\frac{\partial C}{\partial r} + w\frac{\partial C}{\partial z} = \left(\begin{array}{c} \frac{\partial}{\partial r}\left(D_B(C)\frac{\partial C}{\partial r}\right) + \frac{D_B(C)}{r}\frac{\partial C}{\partial r} \\ +\frac{\partial}{\partial z}\left(D_B(C)\frac{\partial C}{\partial z}\right) \end{array} \right)$$
$$+\frac{D_T}{T_m}\left\{\frac{\partial^2 T}{\partial z^2} + \frac{\partial^2 T}{\partial r^2} + \frac{1}{r}\frac{\partial T}{\partial r}\right\}. \tag{5}$$

Here, ρ_f is the density of the fluid, σ is the electric conductivity, T is the temperature of the fluid, and C is the nanoparticle concentration. Further, μ is the coefficient of viscosity which is considered to vary as an inverse function of temperature (see Lai and Kulacki (1990), Vajravelu et al. (2013), Prasad et al. (2016)) as

$$\mu = \frac{\mu_h}{[1 + \zeta(T - T_h)]} \quad \text{i.e.,} \quad \frac{1}{\mu}$$
$$= a(T - T_r), \quad \text{where } a = \frac{\zeta}{\mu_h} \text{ and } T_r$$
$$= T_h - \frac{1}{\zeta}. \tag{6}$$

Here, both a and T_r are constants, and their values depend on the reference state and the parameter ζ, reflecting a thermal property of the fluid. In general, $a > 0$ corresponds to liquid and $a < 0$ to gasses. Also, let θ_r be the constant which is defined by $\theta_r = (T_r - T_h)/\Delta T = -1/(\zeta \Delta T)$, where $\Delta T = (T_w - T_h)$. It is worth mentioning here that for $\zeta \to 0$ i.e., $\mu = \mu_h$ (constant), $\theta_r \to \infty$. It is also important to note that θ_r is negative for liquids and positive for gasses. This is due to the fact that viscosity of a liquid usually decreases with increasing temperature while it increases for gasses. The temperature-dependent thermal conductivity $\lambda(T)$ and the species dependence on molecular diffusion of the diffusing species $D_B(C)$ (diffusion coefficient) in the fluid are assumed to vary as linear functions of temperature and nanoparticle species diffusion respectively in the following forms:

$$\frac{1}{\rho_h C_p}\lambda(T) = \lambda_h \left(= \frac{k_h}{\rho_h C_p}\right)$$
$$\times \left(1 + \varepsilon_1\left(\frac{T - T_h}{T_w - T_h}\right)\right) \text{ and } D_B(C)$$
$$= D_{Bh}\left(1 + \varepsilon_2\left(\frac{C - C_h}{C_w - C_h}\right)\right), \tag{7}$$

where ε_1 and ε_2 are the small parameters, respectively, called variable thermal conductivity parameter and the variable species diffusivity parameter. T_w and C_w are the temperature and the nanoparticle concentration at the lower disk whereas $\mu_h, k_h,$ and D_{Bh} are respectively the fluid viscosity, thermal conductivity, and the species diffusivity/Brownian diffusion coefficient of the fluid at the upper disk, T_m is the mean fluid temperature, D_T is the thermophoretic diffusion coefficient, $\tau = (\rho_\infty c_p)_p/(\rho_\infty c_p)_f$ is the ratio between the effective heat capacity of the nanoparticle material and heat capacity of the fluid. Thus, the value of τ will be therefore different for different fluids and nanoparticle materials. The appropriate boundary conditions for the problem (see Das et al. (2016)) are

$$u = -\beta_i\frac{\partial u}{\partial z}, \ w = -\frac{w_0}{\sqrt{1 - \alpha t}}, \ T = -\gamma_1\frac{\partial T}{\partial z} + T_w,$$
$$C = C_w \text{ at } z = 0,$$
$$u = \beta_i\frac{\partial u}{\partial z}, \ w = w_h\left(= \frac{dh}{dt}\right), \ T = \gamma_1\frac{\partial T}{\partial z} + T_h,$$
$$C = C_h \text{ at } z = h(t) \tag{8}$$

where β_1 is the velocity slip parameter and γ_1 is the thermal slip parameter.

Similarity equations for nanofluid model

For the physical model considered, it is observed that the nanoparticle species diffusion and thermal and momentum boundary layers exist when the surface temperature and the nanoparticle species diffusion differs from that of fluid temperature as well as nanoparticles species diffusion. Using the following similarity transformation (see Das et al. (2016))

$$\eta = \frac{z}{H\sqrt{1-\alpha t}}, \; u = \frac{\alpha r}{2(1-\alpha t)}f'(\eta), \; w$$
$$= -\frac{\alpha H}{\sqrt{1-\alpha t}}f(\eta), \; \theta = \frac{T-T_h}{T_w-T_h}, \quad \phi = \frac{C-C_h}{C_w-C_h}$$

(9)

Equations (2) and (3) reduce to

$$\left(f''\left(1-\frac{\theta}{\theta_r}\right)^{-1}\right)'' -S(\eta f''' + 3f'' -2ff''')-Mnf'' = 0.$$

(10)

Now, Eqs. (4) and (5) take the following forms with the associated boundary conditions

$$((1+\varepsilon_1\theta)\theta')' + \Pr S(2f\theta' -\eta\theta')$$
$$+ \Pr\theta'(Nb\phi'(1+\varepsilon_2\phi) + Nt\theta') = 0,$$

(11)

$$((1+\varepsilon_2\phi)\phi')' + Le\,S(2f\phi' -\eta\phi') + \frac{Nt}{Nb}\theta'' = 0, \quad (12)$$

$$\left.\begin{array}{l} f(\eta) = A, f'(\eta) + \beta f''(\eta) = 0, \; \theta(\eta) + \gamma\theta'(\eta) = 1, \; \phi(\eta) = 1 \text{ at } \eta = 0 \\ f(\eta) = \frac{1}{2}, f'(\eta) - \beta f''(\eta) = 0, \; \theta(\eta) - \gamma\theta'(\eta) = 0, \; \phi(\eta) = 0 \text{ at } \eta = 1 \end{array}\right\}$$

(13)

The non-dimensional parameters θ_r, S, Mn, $\Pr$, Nb, Nt, Le, A, ε_1, ε_2, β, and γ denote respectively the fluid viscosity parameter, squeeze parameter, magnetic parameter, Prandtl number, Brownian motion parameter, thermophoresis parameter, the Lewis number, the suction/injection parameter, the variable thermal conductivity parameter, the variable species diffusivity parameter, the velocity slip parameter, and the temperature slip parameter. They are defined as

$$S = \frac{\alpha H^2}{2\nu_\infty}, \; Mn = \frac{H^2\sigma B_0^2}{\nu_\infty}, \; \Pr = \frac{\nu_\infty}{\lambda}, \; Nb = \frac{\tau D_B(C_w-C_h)}{\nu_\infty},$$
$$Nt = \frac{\tau D_T(T_w-T_h)}{\nu_\infty T_m}, \; Le = \frac{\nu_\infty}{D_B}, A = \frac{w_0}{\alpha H}, \beta = \frac{\beta_1}{H\sqrt{1-\alpha t}},$$
$$\gamma = \frac{\gamma_1}{H\sqrt{1-\alpha t}}.$$

Physically, squeeze parameter $S > 0$ corresponds to plates moving apart and $S < 0$ to the plates moving

closer. Furthermore, transpiration (suction/injection) $A > 0$ corresponds to suction and $A < 0$ to injection at the lower stationary disk. In the absence of variable fluid properties (i. e., $\theta_r = \varepsilon_1 = \varepsilon_2 = 0$), slip velocity (i. e., $\beta = 0$), and slip temperature (i. e., $\gamma = 0$), Eqs. (10) to (13) reduce to those of Hashmi et al. (2012). Also, in the absence of variable fluid properties (i. e., $\theta_r = \varepsilon_1 = \varepsilon_2 = 0$) the set of equations with boundary conditions (14) reduces to those of Das et al. (2016). The quantities of physical interest are the skin friction coefficient C_{fr}, the Nusselt number N_{ur}, and the Sherwood number S_{hr} which are defined as follows:

$$C_{fr} = \frac{\mu_\infty\left(\frac{\partial u}{\partial z} + \frac{\partial w}{\partial r}\right)\big|_{z=h(t)}}{\rho_\infty\left(\frac{-\alpha H}{2\sqrt{1-\alpha t}}\right)^2}, N_{ur} = \frac{-H\left(\frac{\partial T}{\partial z}\right)\big|_{z=h(t)}}{(T_w-T_h)},$$

$$S_{hr} = \frac{-H\left(\frac{\partial C}{\partial z}\right)\big|_{z=h(t)}}{(C_w-C_h)}.$$

(14)

In terms of (9), these expressions reduce to

$$C_{fr} = \frac{H^2}{r^2}C_f\mathrm{Re}_r = f''(1), N_{ur} = \sqrt{(1-\alpha t)}Nu$$
$$= -\theta'(1), S_{hr} = \sqrt{(1-\alpha t)}Sh = -\phi'(1) \quad (15)$$

where $\mathrm{Re}_r = \frac{r\alpha H\sqrt{(1-\alpha t)}}{2\nu_\infty}$ is the local squeeze Reynolds number.

Methods

Semi-analytical solution: optimal homotopy analysis method (OHAM)

The governing equations are highly non-linear, coupled ODEs with variable coefficients. We use optimal homotopy analysis method (OHAM) to obtain appropriate analytic solutions for Eqs. (10)–(12) with associated boundary conditions (13). The OHAM is based on the homotopy concept from topology. In this regard, a non-linear problem is transformed into an infinite number of linear sub-problems. In the framework of OHAM, we have great freedom to choose the auxiliary linear operators and initial approximations. This is advantageous over other iterative techniques, where convergence is largely tied to a good initial approximation of the solution. The OHAM differs from other analytic approximation methods as it does not depend on small or large physical parameters. This is achieved by inclusion of an artificial "convergence control parameter," which guarantees convergence of the solution series. The OHAM has been successfully applied to a wide variety of non-linear problems (for details see Liao (2003) Fan and You (2013)). Optimal homotopy analysis method is employed to solve the system of non-linear Eqs. (10)–(12) with boundary conditions (13).

In the outline of the OHAM, the non-linear equations are decomposed into their linear parts as follows. In accordance with the boundary conditions (13), the base functions are chosen as $\{a_m \eta^m / m \geq 0\}$, then the dimensionless velocity $f(\eta)$, temperature $\theta(\eta)$, and concentration $\phi(\eta)$ can be expressed in the series form as follows

$$f(\eta) = \sum_{k=0}^{\infty} a_k \eta^k, \quad \theta(\eta) = \sum_{k=0}^{\infty} b_k \eta^k \text{ and } \phi(\eta) = \sum_{n=1}^{\infty} c_k \eta^k$$

where a_k, b_k, and c_k are the coefficients. According to the rule of solution expression and boundary conditions (13), we assume the following:

Initial guesses for dimensionless velocity $f(\eta)$, temperature $\theta(\eta)$, and concentration $\phi(\eta)$ as

$$\left.\begin{array}{l} f_0(\eta) = A - \eta \dfrac{(3-6A)(\beta-1/2)\beta}{2\beta(2-3\beta)-1/2} + \eta^2 \dfrac{(3-6A)(\beta-1/2)}{2(2\beta(2-3\beta)-1/2)} \\ \qquad + \dfrac{\eta^3}{6}\left[(3-6A) - \dfrac{3(3-6A(\beta-1/2))}{2\beta(2-3\beta)-1/2} + \dfrac{6(3-6A)(\beta-1/2)\beta}{2\beta(2-3\beta)-1/2}\right], \\ \theta_0(\eta) = \dfrac{1-\gamma}{1-2\gamma} - \dfrac{\eta}{1-2\gamma} \text{ and } \phi_0(\eta) = 1 - \eta \end{array}\right\}$$

$$(16)$$

Linear operators L_f, L_θ, and L_ϕ as

$$L_f = \frac{d^4}{d\eta^4}, \quad L_\theta = \frac{d^2}{d\eta^2}, \text{ and } L_\phi = \frac{d^2}{d\eta^2} \quad \text{such that} \quad (17)$$

$$L_f\left[c_1 + c_2\eta + c_3\frac{\eta^2}{2} + c_4\frac{\eta^3}{6}\right] = 0, \quad L_\theta[c_5 + c_6\eta]$$
$$= 0 \text{ and } L_\phi[c_7 + c_8\eta] = 0$$

where c_i's ($i = 1, 2, 3, 4, 5, 6, 7, 8$) are arbitrary constants and $H_f(\eta) = H_\theta(\eta) = H_\phi(\eta) = 1$ are auxiliary functions.

Let us consider so called zeroth order deformation approximations as

$$(1-q)L_f\left[\hat{f}(\eta,q) - f_0(\eta)\right] = qH_f(\eta)\widehat{h}_f N_f\left[\hat{f}(\eta,q), \hat{\theta}(\eta,q)\right],$$
$$(18)$$

$$(1-q)L_\theta\left[\hat{\theta}(\eta,q) - \theta_0(\eta)\right] = qH_\theta(\eta)\widehat{h}_\theta N_\theta\left[\hat{\theta}(\eta,q), \hat{f}(\eta,q), \hat{\phi}(\eta,q)\right],$$
$$(19)$$

$$(1-q)L_\phi\left[\hat{\phi}(\eta,q) - \phi_0(\eta)\right] = qH_\phi(\eta)\widehat{h}_\phi N_\phi\left[\hat{\phi}(\eta,q), \hat{f}(\eta,q), \hat{\theta}(\eta,q)\right],$$
$$(20)$$

with conditions

$$\hat{f}(0,q) = A, \hat{f}'(0,q) + \beta\hat{f}''(0,q) = 0, \hat{f}(1,q) = \frac{1}{2}, \hat{f}'(1,q) - \beta\hat{f}''(1,q) = 0;$$
$$\hat{\theta}(0,q) + \gamma\hat{\theta}'(0,\eta) = 1, \hat{\theta}(1,q) - \gamma\hat{\theta}'(1,\eta) = 0; \quad \hat{\phi}(0,q) = 1, \hat{\phi}(1,q) = 0.$$

where $q \in [0,1]$ is an embedding parameter and $(\widehat{h}_f, \widehat{h}_\theta, \widehat{h}_\phi) \neq 0$ are the convergence control parameters. Here, N_f, N_θ, and N_ϕ are non-linear operators defined as

$$N_f = \left(1 - \frac{\hat{\theta}(\eta,q)}{\theta_r}\right)^2 \hat{f}^{iv}(\eta,q) + \left(1 - \frac{\hat{\theta}(\eta,q)}{\theta_r}\right)\frac{2\hat{f}'''(\eta,q)\hat{\theta}'(\eta,q)}{\theta_r}$$
$$+ \left(1 - \frac{\hat{\theta}(\eta,q)}{\theta_r}\right)\frac{\hat{f}''(\eta,q)\hat{\theta}''(\eta,q)}{\theta_r} + \frac{2\hat{f}''(\eta,q)\hat{\theta}'^2(\eta,q)}{\theta_r^2}$$
$$- \left(1 - \frac{\hat{\theta}(\eta,q)}{\theta_r}\right)^3 S\left(\eta\hat{f}'''(\eta,q) + 3\hat{f}''(\eta,q) - 2\hat{f}(\eta,q)\hat{f}''(\eta,q)\right)$$
$$- Mn^2\left(1 - \frac{\hat{\theta}(\eta,q)}{\theta_r}\right)^3 \hat{f}'(\eta,q),$$

$$N_\theta = \left((1 + \varepsilon_1\hat{\theta}(\eta,q))\hat{\theta}'(\eta,q)\right)' + \Pr S\left(2\hat{f}(\eta,q)\hat{\theta}'(\eta,q) - \eta\hat{\theta}'(\eta,q)\right)$$
$$+ \Pr Nb\hat{\theta}'(\eta,q)\hat{\phi}'(\eta,q) + \Pr Nb\varepsilon_2\hat{\phi}(\eta,q)\hat{\theta}'(\eta,q)\hat{\phi}'(\eta,q)$$
$$+ \Pr Nt\hat{\theta}'^2(\eta,q),$$

$$N_\phi = \left((1 + \varepsilon_2\hat{\phi}(\eta,q))\hat{\phi}'(\eta,q)\right)'$$
$$+ LeS\left(2\hat{f}(\eta,q)\hat{\phi}(\eta,q) - \eta\hat{\phi}(\eta,q)\right)$$
$$+ \frac{Nt}{Nb}\hat{\theta}''(\eta,q).$$

From Eqs. (18) to (20), at $q = 0$, we have

$$L_f\left[\hat{f}(\eta,0) - f_0(\eta)\right] = 0, \quad L_\theta\left[\hat{\theta}(\eta,0) - \theta_0(\eta)\right]$$
$$= 0, \text{ and } L_\phi\left[\hat{\phi}(\eta,0) - \phi_0(\eta)\right] = 0$$

which implies that $\hat{f}(\eta,0) = f_0(\eta)$, $\hat{\theta}(\eta,0) = \theta_0(\eta)$, and $\hat{\phi}(\eta,0) = \phi_0(\eta)$ respectively, whereas at $q = 1$, we have

$$N_f\left[\hat{f}(\eta,1), \hat{\theta}(\eta,1)\right] = 0, N_\theta\left[\hat{\theta}(\eta,1), \hat{f}(\eta,1), \hat{\phi}(\eta,1)\right] =$$
$$0, \text{ and } N_\phi\left[\hat{\phi}(\eta,1), \hat{f}(\eta,1), \hat{\theta}(\eta,1)\right] = 0 \quad \text{which implies}$$

that $\hat{f}(\eta,1) = f(\eta)$, $\hat{\theta}(\eta,1) = \theta(\eta)$, and $\hat{\phi}(\eta,1) = \phi(\eta)$ respectively.

Hence, by defining

$$f_m(\eta) = \frac{1}{m!}\frac{d^m f(\eta,q)}{d\eta^m}\bigg|_{q=0}, \quad \theta_m(\eta) = \frac{1}{m!}\frac{d^m \theta(\eta,q)}{d\eta^m}\bigg|_{q=0},$$

$$\phi_m(\eta) = \frac{1}{m!}\frac{d^m \phi(\eta,q)}{d\eta^m}\bigg|_{q=0}$$

we expand $\hat{f}(\eta,q)$, $\hat{\theta}(\eta,q)$, and $\hat{\phi}(\eta,q)$ by means of Taylor's series as

$$\hat{f}(\eta,q) = f_0(\eta) + \sum_{m=1}^{\infty} f_m(\eta)q^m,$$

$$\hat{\theta}(\eta,q) = \theta_0(\eta) + \sum_{m=1}^{\infty} \theta_m(\eta)q^m, \text{ and } \hat{\phi}(\eta,q) = \phi_0(\eta)$$

$$+ \sum_{m=1}^{\infty} \phi_m(\eta)q^m.$$

$$(21)$$

If series (21) converges at $q = 1$, we get the homotopy series solutions

$$f(\eta) = f_0(\eta) + \sum_{m=1}^{\infty} f_m(\eta), \quad \theta(\eta) = \theta_0(\eta) + \sum_{m=1}^{\infty} \theta_m(\eta), \quad \text{and}$$

$$\phi(\eta) = \phi_0(\eta) + \sum_{m=1}^{\infty} \phi_m(\eta).$$

$$(22)$$

Optimal convergence control parameter

It should be noted that $f(\eta)$, $\theta(\eta)$, and $\phi(\eta)$ in Eq. (22) contain the unknown convergence control parameters $\widehat{h}_f, \widehat{h}_\theta$, and $\widehat{h}_\phi$ which can be used to adjust and control the convergence region and the rate of convergence of the homotopy series solution. The mth order deformation equations and the conditions are as follows:

$$L_f\left[f_m(\eta) - \chi_m f_{m-1}(\eta)\right] = H_f(\eta)\widehat{h}_f R_m^f(\eta),$$
$$L_\theta\left[\theta_m(\eta) - \chi_m \theta_{m-1}(\eta)\right] = H_\theta(\eta)\widehat{h}_\theta R_m^\theta(\eta),$$
$$L_\phi\left[\phi_m(\eta) - \chi_m \phi_{m-1}(\eta)\right] = H_\phi(\eta)\widehat{h}_\phi R_m^\phi(\eta),$$

with $f_m(0) = 0$, $f'_m(0) + \beta f''_m(0) = 0$, $f_m(1) = 0, f'_m(1) - \beta f''_m(1) = 0$, $\theta_m(0) + \gamma \theta'(0) = 0$, $\theta_m(1) - \gamma \theta'(1) = 0$, $\phi_m(0) = 0$, $\phi_m(1) = 0$, where

$$R_m^\theta = \theta''_{m-1} + \varepsilon_1\left(\sum_{k=0}^{m-1}\theta''_{m-1-k}\theta_k + \sum_{k=0}^{m-1}\theta'_{m-1-k}\theta'_k\right)$$

$$+ \Pr S\left(2\sum_{k=0}^{m-1}\theta'_{m-1-k}f_k - \eta\theta'_{m-1}\right)$$

$$+ \Pr Nb\sum_{k=0}^{m-1}\theta'_{m-1-k}\phi'_k + \Pr Nb\varepsilon_2\sum_{k=0}^{m-1}\phi'_{m-1-k}\sum_{j=0}^{k}\theta'_{k-j}\phi_j$$

$$+ \Pr Nt\sum_{k=0}^{m-1}\theta'_{m-1-k}\theta'_k$$

$$R_m^\phi = \phi''_{m-1}(\eta) + \varepsilon_2\left(\sum_{k=0}^{m-1}\phi''_{m-1-k}\phi_k + \sum_{k=0}^{m-1}\phi'_{m-1-k}\phi'_k\right)$$

$$+ LeS\left(2\sum_{k=0}^{m-1}\phi'_{m-1-k}f_k - \eta\phi'_{m-1}\right) + \frac{Nt}{Nb}\theta''_{m-1}$$

and

$$X_m = \begin{cases} 0, m \leq 1 \\ 1, m > 1 \end{cases}.$$

Error analysis

Now we evaluate the error and minimize over $\widehat{h}_f, \widehat{h}_\theta$, and $\widehat{h}_\phi$ in order to obtain the optimal values for $\widehat{h}_f, \widehat{h}_\theta$, and $\widehat{h}_\phi$ and the least possible error. In the process of error analysis, two different methods are employed, namely, exact residual error and average residual error. For different order approximation, CPU time required

$$R_m^f = f^{iv}_{m-1} + \left(\frac{1}{\theta_r^2}\right)\sum_{k=0}^{m-1}f^{iv}_{m-1-k}\sum_{j=0}^{k}\theta_{k-j}\theta_j - \left(\frac{2}{\theta_r}\right)\sum_{k=0}^{m-1}f^{iv}_{m-1-k}\theta_k + \left(\frac{2}{\theta_r}\right)\sum_{k=0}^{m-1}f'''_{m-1-k}\theta'_k - \left(\frac{2}{\theta_r^2}\right)\sum_{k=0}^{m-1}f'''_{m-1-k}\sum_{j=0}^{k}\theta_{k-j}\theta_j$$

$$+ \left(\frac{1}{\theta_r}\right)\sum_{k=0}^{m-1}f''_{m-1-k}\theta''_k - \left(\frac{1}{\theta_r^2}\right)\sum_{k=0}^{m-1}f''_{m-1-k}\sum_{j=0}^{k}\theta''_{k-j}\theta_j + \left(\frac{2}{\theta_r^2}\right)\sum_{k=0}^{m-1}f''_{m-1-k}\sum_{j=0}^{k}\theta'_{k-j}\theta'_j - S\eta f'''_{m-1}$$

$$+ \left(\frac{S\eta}{\theta_r^3}\right)\sum_{k=0}^{m-1}f'''_{m-1-k}\sum_{j=0}^{k}\theta_{k-j}\sum_{i=0}^{j}\theta_{j-i}\theta_i + \left(\frac{3S\eta}{\theta_r}\right)\sum_{k=0}^{m-1}f''_{m-1-k}\theta_k - \left(\frac{3S\eta}{\theta_r^2}\right)\sum_{k=0}^{m-1}f''_{m-1-k}\sum_{j=0}^{k}\theta_{k-j}\theta_j - 3Sf''_{m-1}$$

$$+ \left(\frac{3S}{\theta_r^3}\right)\sum_{k=0}^{m-1}f''_{m-1-k}\sum_{j=0}^{k}\theta_{k-j}\sum_{i=0}^{j}\theta_{j-i}\theta_i + \left(\frac{9S}{\theta_r}\right)\sum_{k=0}^{m-1}f''_{m-1-k}\theta_k - \left(\frac{9S}{\theta_r^2}\right)\sum_{k=0}^{m-1}f''_{m-1-k}\sum_{j=0}^{k}\theta_{k-j}\theta_j + 2S\sum_{k=0}^{m-1}f'''_{m-1-k}f_k$$

$$- \left(\frac{2S}{\theta_r^3}\right)\sum_{k=0}^{m-1}f'''_{m-1-k}\sum_{j=0}^{k}f_{k-j}\sum_{i=0}^{j}\theta_{j-i}\sum_{l=0}^{i}\theta_{i-l}\theta_l - \left(\frac{6S}{\theta_r}\right)\sum_{k=0}^{m-1}f'''_{m-1-k}\sum_{j=0}^{k}f_{k-j}\theta_j$$

$$+ \left(\frac{6S}{\theta_r^2}\right)\sum_{k=0}^{m-1}f'''_{m-1-k}\sum_{j=0}^{k}f_{k-j}\sum_{i=0}^{j}\theta_{j-i}\theta_i - M^2f''_{m-1} + \left(\frac{M^2}{\theta_r^3}\right)\sum_{k=0}^{m-1}f''_{m-1-k}\sum_{j=0}^{k}\theta_{k-j}\sum_{i=0}^{j}\theta_{j-i}\theta_i$$

$$+ \left(\frac{3M^2}{\theta_r}\right)\sum_{k=0}^{m-1}f''_{m-1-k}\theta_k - \left(\frac{3M^2}{\theta_r^2}\right)\sum_{k=0}^{m-1}f''_{m-1-k}\sum_{j=0}^{k}\theta_{k-j}\theta_j$$

for evaluation of $f'(1)$ is observed. As for the CPU time is concerned, the average residual error needs less time compared to that of the exact residual error for increasing values of m (for details, see Table 1). At the mth order deformation equation, the exact residual errors are given by

$$\hat{E}_m^{\,f} = \int_0^1 \left(N_1 \left[\sum_{n=0}^m f_n(\eta) \right] \right)^2 d\eta$$

$$\hat{E}_m^{\,\theta} = \int_0^1 \left(N_2 \left[\sum_{n=0}^m \theta_n(\eta) \right] \right)^2 d\eta$$

$$\hat{E}_m^{\,\phi} = \int_0^1 \left(N_3 \left[\sum_{n=0}^m \phi_n(\eta) \right] \right)^2 d\eta.$$

But in practice the evaluation of $\hat{E}_m^f$, $\hat{E}_m^\theta$, and $\hat{E}_m^\phi$ is much time-consuming so instead of exact residual errors, we use average residual errors defined as

$$E_m^{\,f} = \frac{1}{M+1} \sum_{k=0}^M \left(N_f \left[\sum_{n=0}^m f_n(\eta_k) \right] \right)^2,$$

$$E_m^{\,\theta} = \frac{1}{M+1} \sum_{k=0}^M \left(N_\theta \left[\sum_{n=0}^m \theta_n(\eta_k) \right] \right)^2,$$

$$E_m^{\,\phi} = \frac{1}{M+1} \sum_{k=0}^M \left(N_\phi \left[\sum_{n=0}^m \phi_n(\eta_k) \right] \right)^2, \text{ and}$$

$$E_m^t = E_m^{\,f} + E_m^{\,\theta} + E_m^{\,\phi}$$

where E_m^t is the total squared residual error, $\eta_k = k\Delta\eta = \frac{k}{M}$; $k = 0, 1, 2, \ldots M$. Now the error functions E_m^f, E_m^θ, and E_m^ϕ are minimized over $\hat{h}_f, \hat{h}_\theta$, and $\hat{h}_\phi$ to obtain the optimal values. Evidently, $\lim_{m\to\infty} E_m^{\,f} = 0$, $\lim_{m\to\infty} E_m^{\,\theta} = 0$, $\lim_{m\to\infty} E_m^{\,\phi} = 0$ correspond to a convergent series solution. Average residual errors for f, θ, and ϕ are obtained by considering the optimal values of $\hat{h}_f(-0.821699)$, $\hat{h}_\theta$ (-0.919399), and $\hat{h}_\phi(-0.918880)$ in Table 3, which has been obtained by minimizing the squared residual errors of f, θ, and ϕ at the approximation $m = 6$ as shown in Table 2. It is observed from Table 3 that

CPU time increases consistently as the approximation increases and with the increase of order of approximation, the average residual error for f, θ, and ϕ decreases continuously. Hence, the optimal value of f, θ, and ϕ helps in the fastest convergence of solution series. Substituting these optimal values of $\hat{h}_f, \hat{h}_\theta$, and $\hat{h}_\phi$ in Eq. (22), we get the approximate solutions of Eqs. (10) to (12) satisfying the boundary conditions (13).

Results and discussion

The system of Eqs. (10) to (12) is highly non-linear coupled ordinary differential equations. The system of equations with the boundary conditions (13) is solved analytically via efficient OHAM. In order to validate the method and to judge the accuracy of the analysis, the values of the skin friction, local Nusselt number, and the local Sherwood number are compared with the previously published results for special cases in which the thermo-physical fluid properties are neglected (see Tables 4, 5, and 6). It can be found from Tables 4, 5, and 6 that the present results agree very well with those of Hashmi et al. (2012) and Das et al. (2016).

On employing the above analytical schemes, the system of equations are solved for several sets of values of the pertinent parameters, namely, the fluid viscosity parameter θ_r, the squeeze parameter S, the magnetic parameter Mn, the Prandtl number Pr, the thermophoresis parameter Nt, the Brownian motion parameter Nb, the variable thermal conductivity parameter ε_1, the variable species diffusivity parameter ε_2, the Lewis number Le, the suction/injection parameter A, velocity slip parameter β, and the temperature slip parameter γ. In order to get clear insight into the effects of these parameters on flow characteristics, the obtained results are analyzed graphically in Figs. 2, 3, and 5. It may be note that, $\theta_r < 0$ for liquids and $\theta_r > 0$ for gasses. This is due to the fact that viscosity of a liquid decreases with increasing temperature while it increases for gasses. Hence, in this steady, we considered negative values of θ_r. The computed numerical values for $f'(1), \theta'(1)$, and $\phi'(1)$ are tabulated in Table 7.

Velocity field

The effect of θ_r, Mn, S, and β on the velocity distribution $f'(\eta)$ for transpiration (suction/blowing) is elucidated in Fig. 2a–d. Here, the profiles are parabolic in nature. In general, at the central region, these profiles exhibit the cross-flow behavior and also exhibit the dual behavior with the increase in the pertinent parameters. From Fig. 2a it is clear that, for higher values of θ_r, the velocity profile increases near the walls where the transpiration effects are dominant when $\eta \le 0.44$, $\eta \le 0.46$, and $\eta \le 0.49$. However, for $\eta > 0.44$, $\eta > 0.46$, and $\eta > 0.49$,

Table 1 Comparison of $f'(1)$ and CPU time (s) incurred to evaluate the mth order approximation by exact residual error and average residual error when $M = \text{Pr} = 1, S = Nb = Nt = 0.5$, $Le = 2, A = 0.01, \gamma = \varepsilon_1 = \varepsilon_2 = 0.1, \theta_r = -5.0$, and $\beta = 0.01$.

Order m	Using exact residual error		Using average residual error	
	$-f'(1)$	CPU time (s)	$-f'(1)$	CPU time (s)
1	2.87099	9.83	2.87063	0.61
2	2.86492	26.86	2.86492	2.14
3	2.84464	73.93	2.84498	5.94
4	2.84315	1234.25	2.84332	14.43

Table 2 Values of convergence control parameters h_f, h_θ, and h_φ and the corresponding total residual errors E_m^t for different orders of approximation m with $S = M = 0.1$, $\mathrm{Pr} = 0.72$, $\gamma = 0.1$, $\mathrm{Nb} = \mathrm{Nt} = 0.1$, $\mathrm{Le} = 1$, $A = 0.01$, $\varepsilon_1 = \varepsilon_2 = 0.1$, $\theta_r = -10.0$, $\beta = 0.1$.

m	$-h_f$	$-h_\theta$	$-h_\phi$	E_m^t	CPU time (s)
1	0.855974	0.913900	0.213443	2.51×10^{-1}	1.14
2	0.914521	0.894988	0.120164	1.08×10^{-2}	5.34
3	0.905082	0.873693	0.917802	2.98×10^{-4}	14.92
4	0.886267	0.835341	0.896250	5.54×10^{-6}	38.95
5	0.858155	0.913425	0.876598	2.40×10^{-7}	72.10
6	0.821699	0.919399	0.918880	3.42×10^{-8}	74.52

the velocity profiles decreases as $\theta_r \to 0$. This may be attributed to the fact that, for a given base fluid (air or water), when ζ is fixed, lesser θ_r implies higher temperature difference between the wall and the ambient fluid. Hence, the results explicitly manifest that θ_r is the indicator of the variation of fluid viscosity with temperature which has a substantial effect on $f(\eta)$ and hence on $f'(1)$ (see Table 7 for details). The effect of Mn on $f(\eta)$ is demonstrated in Fig. 2b. It is observed that, for the higher values of Mn, the fluid velocity increase and thereby decrease the thickness of the momentum boundary layer near the wall region, for absence of suction/injection parameter and $0.23 \geq \eta \geq 0.67$ for suction) However, the reverse trend is observed in the central region. This dual behavior of the flow is due to the mass conservation constraint (see Lawal and Kalyon (1998)). Figure 2c depicts the behavior of S on $f'(\eta)$ It is seen that $f'(\eta)$ decreases with an increase in S at the central region for suction $(0.32 \leq \eta \leq 0.74)$ and the reverse trend is recorded in the case of injection. Physically, in the case of suction, there will be an increase in $h(t)$ which will cause it to decrease in velocity and the increase in squeezing parameter. Further, when $\eta \leq 0.32$ or $\eta \leq 0.74$, that is the neighborhood of 0.32 and 0.74, there will be a corresponding decrease in $h(t)$ which in turn will increase in both fluid velocity and squeezing parameter. The effect of β on $f'(\eta)$ is shown in Fig. 2d. The behavior of $f'(\eta)$ at the surface of the disks is zero when $\beta = 0$ (no slip at the wall). For the suction flow, an increase in β strengthens the $f'(\eta)$ near the wall, that is for $\eta \geq 0.16$ and $\eta \geq 0.82$, while at the central region, rising β results in weakening of $f'(\eta)$. However, for the case of injection, the effect of $f'(\eta)$ is opposite to that accounted for suction flow. Physically, in the case of injection, an increase in $f'(\eta)$ near the wall region with increasing β gives rise to a decrease in velocity gradient at there. However, when the mass flow rate is kept constant, an increase in the fluid velocity near the wall region will be remunerated by an analogous fall in the fluid velocity near the mid region so that mass conservation limitation will not be dishonored (see ref. (Lawal and Kalyon 1998)).

Temperature distribution and nanoparticle volume distribution

The effects of S, γ, ε_1, Pr, and Nt on the temperature distribution $\theta(\eta)$ for transpiration are presented in Fig. 3a–c. Figure 3a illustrates the effect of S on $\theta(\eta)$. For larger S, the temperature distribution reduces in the case of suction whereas the reverse trend is observed in the case of injection. Higher values of S indicate a decrease in the kinematic viscosity which depends on the velocity and distance between disks. The effect of γ on $\theta(\eta)$ is depicted in Fig. 3b. It is observed that in the absence of thermal slip (when $\gamma = 0$); the temperature of the fluid and temperature of the disks' surfaces remain same (here, 0 for the lower disk and 1 for the upper disk). These results are found to

Table 3 Individual average residual error with $S = M = 0.1$, $\mathrm{Pr} = 0.72$, $\gamma = 0.1$, $\mathrm{Nb} = \mathrm{Nt} = 0.1$, $\mathrm{Le} = 1$, $A = 0.01$, $\varepsilon_1 = \varepsilon_2 = 0.1$, $\theta_r = -10.0$, $\beta = 0.1$.

m	E_m^f	E_m^θ	E_m^ϕ	CPU time (s)
1	1.85×10^{-1}	4.24×10^{-3}	1.09×10^{-1}	0.17
2	1.02×10^{-2}	1.15×10^{-4}	7.63×10^{-3}	1.33
3	5.44×10^{-4}	4.05×10^{-6}	2.06×10^{-4}	4.84
4	1.94×10^{-5}	1.82×10^{-7}	8.31×10^{-6}	13.63
5	3.64×10^{-7}	7.83×10^{-9}	4.67×10^{-7}	33.22
6	4.13×10^{-9}	2.58×10^{-10}	2.68×10^{-8}	71.22
7	2.55×10^{-9}	7.23×10^{-12}	1.01×10^{-9}	139.98
8	5.32×10^{-10}	1.62×10^{-13}	3.31×10^{-11}	280.03
9	6.70×10^{-11}	5.94×10^{-15}	8.03×10^{-13}	454.01
10	6.44×10^{-12}	3.05×10^{-16}	2.73×10^{-14}	743.12

Table 4 Comparison of the values of the of skin friction coefficient for various values of Mn and S with $A = 2$, $Pr = Le = Nb = 1$, $\theta_r \rightarrow \infty$, $\varepsilon_1 = \varepsilon_2 = \beta = \gamma = Nt = 0$.

Mn	S	Hashmi et al. (Hashmi et al. 2012)	Present Study			CPU time in seconds
			$-f'(1)$	$-h_f$	E^f_{10}	
0	1	7.53316579	7.533166134027173	0.9677134841537185	$1.734689146770528 \times 10^{-7}$	12.0019918
2		8.26387231	8.263872005406636	0.9704415447020044	$1.803753433567724 \times 10^{-7}$	11.4906528
3		9.09732572	9.097325884047699	0.9714084422981315	$2.017660259031556 \times 10^{-7}$	11.3115338
5		11.3492890	11.349079068376334	0.9671141942979257	$3.832529114797568 \times 10^{-7}$	12.2261407
1	0.1	8.97552394	8.975523936918258	0.6586335149025293	$3.901356871103989 \times 10^{-6}$	11.6837806
	0.5	8.34924578	8.34924579267211	0.8848994230900776	$5.175581204948535 \times 10^{-7}$	11.4095983
	1	7.72194601	7.721945958968551	0.9686906697036372	$1.753316413825429 \times 10^{-7}$	11.1624343
	2	6.94077326	6.940264738765628	0.7930138508564271	$1.820417866243763 \times 10^{-3}$	11.1854494

be identical to the results of Das et al. (2016) when $\gamma = 0$. In the case of suction, temperature of the fluid decreases as γ increases when $\eta \leq 0.74$ and for injection, the trend is reverse when $\eta \geq 0.34$. The influence of the Pr on $\theta(\eta)$ is displayed in Fig. 3c. It is worth mentioning here that, from the experimental studies it has been noted that at 20 °C, the Prandtl number for air is 0.72; at 300 °C, the Prandtl number for water is 1.09; at 40 °C, the Prandtl number for ammonia is 2.0; and at 417 °C, the Prandtl number for molten salt is 5.09 (see Tables 4, 5, and 6). Physically, Pr < < 1 means high thermal conductivity but low viscosity, while Pr >> 1 corresponding to high-viscosity oils (see Kothandaraman and Subramanyan (2014)). Rapid increase in temperature profile is recorded for decreasing Pr, while the thermal boundary layer thickness decreases with increasing values of Pr (as evident from Table 7). This is due to fluids with higher values of Pr possess a large heat capacity and hence intensifies the heat transfer which leads to a fall in the rate of heat transfer; in addition to this, the assumption of temperature-dependent thermal conductivity suggests a reduction in the magnitude of the velocity by a

quantity $\frac{\partial}{\partial r}\left(\lambda(T)\frac{\partial T}{\partial r}\right) + \frac{\partial}{\partial z}\left(\lambda(T)\frac{\partial T}{\partial z}\right) + \lambda(T)\frac{1}{r}\frac{\partial T}{\partial r}$ which can be seen in Eq. (4). Therefore, the rate of cooling is much faster for the coolant material having small values for the thermal conductivity. Opposite behavior of $\theta(\eta)$ is observed in the case of ε_1. Figure 3d reflects the effects of suction and injection for increasing values of Nt and Nb on $\theta(\eta)$. The increase in both Nt and Nb leads increase in $\theta(\eta)$ for suction and for injection opposite nature is observed. The behavior of concentration profile (nanoparticle volume distribution) $\phi(\eta)$ is demonstrated in Fig. 4a, b. Figure 4a shows the effect of Nt and Nb on $\phi(\eta)$. Both Nt and Nb exerts opposite effects on $\phi(\eta)$. Increase in Nt results in a decrease in $\phi(\eta)$, this results is due to a larger mass flux at the lower disk. The effect of Le and ε_2 on $\phi(\eta)$ is discussed in Fig. 4b. The increase in ε_2 results in an increase in concentration and the opposite results are found in the case of Le. Figure 5a–c presents the three-dimensional plot of the velocity components. These plots exhibit similar results as those of $f(\eta)$.

In Table 7, we present the results of $f'(1)$, $\theta'(1)$, and $\phi'(1)$ for several sets of values of the physical

Table 5 Comparison of the values of the local Nusselt number for various values of Nb and Nt with $A = 2$, $Mn = Pr = S = Le = 1$, $\theta_r \rightarrow \infty$, $\varepsilon_1 = \varepsilon_2 = \beta = \gamma = 0$

Nb	Nt	Hashmi et al. (2012)	Das et al.(2016)	Present Study			CPU time of the system in seconds
				$-\theta'(1)$	$-h_\theta$	E^θ_{15}	
0.1	0.1	0.52628540	0.526285397692707	0.5262854795861431	0.908516	$1.412748795457583 \times 10^{-11}$	54.439
0.5		0.63433253	0.526285397692707	0.6343325523545158	0.914084	$1.52497172481667 \times 10^{-13}$	54.525
1.0		0.78636385	0.634332530012476	0.7863638216341705	0.928943	$9.805935149902197 \times 10^{-13}$	54.009
1.5		0.95569955	0.955699547716439	0.9556995220292246	0.918411	$7.022421405151564 \times 10^{-12}$	54.625
1.5	0.5	1.17682119	1.176821184883260	1.1768208205076784	0.956179	$2.358155300119884 \times 10^{-10}$	53.940
	1	1.48581207	1.485811936635642	1.4858197859422315	1.017842	$1.217973748574313 \times 10^{-8}$	53.585
	1.5	1.82305276	1.823053529110100	1.8231897567384212	0.953137	$1.724206185011289 \times 10^{-6}$	54.439
	2	2.17915991	2.179227931099257	2.1772753199939645	0.817733	$7.893665960448466 \times 10^{-4}$	57.834

Table 6 Comparison of the values of the local Sherwood number for various values of Nb and Nt with $A = 2$, $Mn = Pr = S = Le = 1$, $\theta_r \rightarrow \infty$, $\varepsilon_1 = \varepsilon_2 = \beta = \gamma = 0$.

Nb	Nt	Hashmi et al. (2012)	Das et al.(2016)	Present study			CPU time of the system in seconds
				$-\phi'(1)$	$-h_\phi$	E_{15}^ϕ	
0.1	0.1	0.86604666	0.866046665141329	0.8660467842530996	0.936748	$2.426680292036448 \times 10^{-10}$	54.439
0.5		0.53012814	0.530128143896286	0.5301281821145534	0.928890	$9.992908438598528 \times 10^{-13}$	54.525
1.0		0.48603919	0.486039186120291	0.48603915346137855	0.891607	$4.403963236292989 \times 10^{-13}$	54.009
1.5		0.46986157	0.469861566526085	0.4698615495700593	0.884253	$3.657227369041722 \times 10^{-13}$	54.625
1.5	0.5	0.40180718	0.401807177532398	0.40180541220583815	0.908420	$2.420566500000295 \times 10^{-10}$	53.940
	1	0.12619334	0.126193335885242	0.12617433001495504	0.949699	$2.832474925670632 \times 10^{-8}$	53.585
	1.5	0.39083080	0.390839865635371	0.3900777099943474	0.865141	$6.281273402797455 \times 10^{-6}$	54.439
	2	1.16777723	1.16800852390237	1.1587904533044613	0.912246	$3.5954596655078387 \times 10^{-4}$	57.834

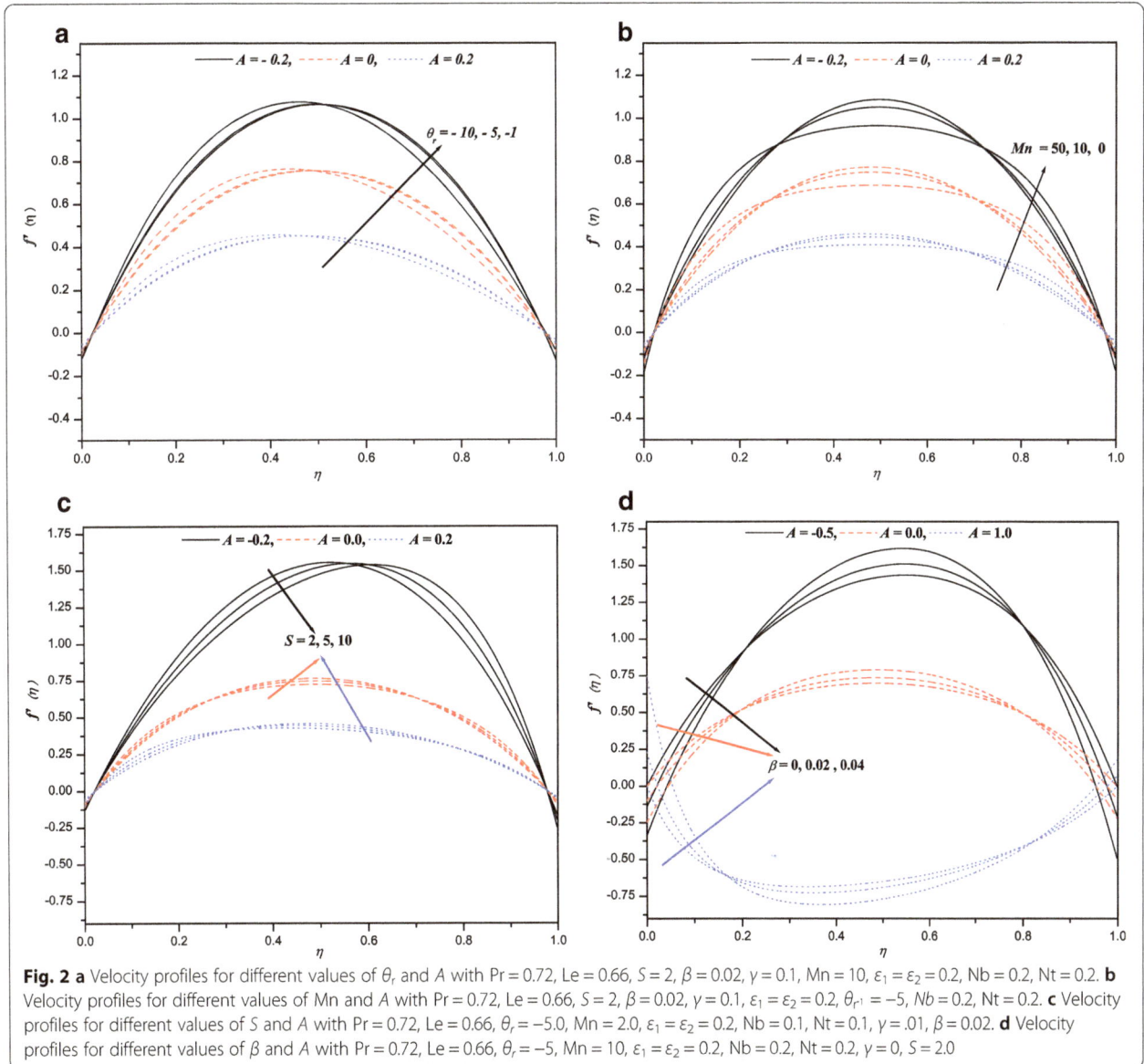

Fig. 2 a Velocity profiles for different values of θ_r and A with $Pr = 0.72$, $Le = 0.66$, $S = 2$, $\beta = 0.02$, $\gamma = 0.1$, $Mn = 10$, $\varepsilon_1 = \varepsilon_2 = 0.2$, $Nb = 0.2$, $Nt = 0.2$. **b** Velocity profiles for different values of Mn and A with $Pr = 0.72$, $Le = 0.66$, $S = 2$, $\beta = 0.02$, $\gamma = 0.1$, $\varepsilon_1 = \varepsilon_2 = 0.2$, $\theta_{r^1} = -5$, $Nb = 0.2$, $Nt = 0.2$. **c** Velocity profiles for different values of S and A with $Pr = 0.72$, $Le = 0.66$, $\theta_r = -5.0$, $Mn = 2.0$, $\varepsilon_1 = \varepsilon_2 = 0.2$, $Nb = 0.1$, $Nt = 0.1$, $\gamma = .01$, $\beta = 0.02$. **d** Velocity profiles for different values of β and A with $Pr = 0.72$, $Le = 0.66$, $\theta_r = -5$, $Mn = 10$, $\varepsilon_1 = \varepsilon_2 = 0.2$, $Nb = 0.2$, $Nt = 0.2$, $\gamma = 0$, $S = 2.0$

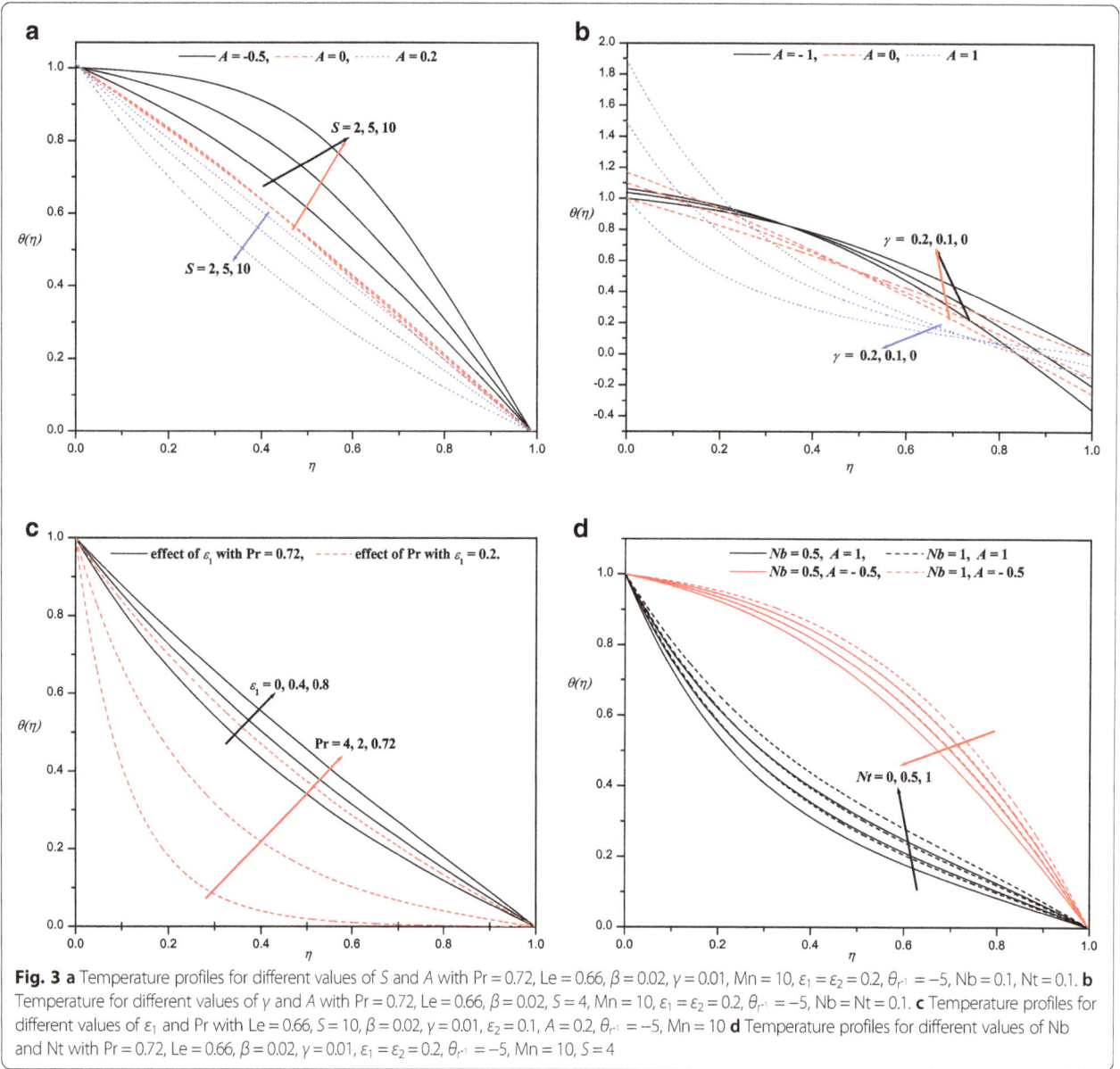

Fig. 3 a Temperature profiles for different values of S and A with Pr = 0.72, Le = 0.66, β = 0.02, γ = 0.01, Mn = 10, $\varepsilon_1 = \varepsilon_2$ = 0.2, θ_{r^1} = −5, Nb = 0.1, Nt = 0.1. **b** Temperature for different values of γ and A with Pr = 0.72, Le = 0.66, β = 0.02, S = 4, Mn = 10, $\varepsilon_1 = \varepsilon_2$ = 0.2, θ_{r^1} = −5, Nb = Nt = 0.1. **c** Temperature profiles for different values of ε_1 and Pr with Le = 0.66, S = 10, β = 0.02, γ = 0.01, ε_2 = 0.1, A = 0.2, θ_{r^1} = −5, Mn = 10 **d** Temperature profiles for different values of Nb and Nt with Pr = 0.72, Le = 0.66, β = 0.02, γ = 0.01, $\varepsilon_1 = \varepsilon_2$ = 0.2, θ_{r^1} = −5, Mn = 10, S = 4

parameters. Increase in the magnetic parameter reduces the skin friction but increases the Nusselt number as well as the Sherwood number. Furthermore, an increase in the variable thermal conductivity parameter and the variable species diffusivity parameter results in an increase in the Nusselt number and the Sherwood number.

Conclusions

In this paper, an analytical technique is used to solve the mathematical model of MHD squeeze flow of nanofluid between parallel disks. This analysis gives unified results for the parameters Mn, S, Nt, and Nb, from which one can obtain the results for the special cases of Hashmi et al.

(2012) and Das et al. (2016). The effects of temperature-dependent thermo-physical properties are significant on the flow field. For higher values of the fluid viscosity parameter, the velocity field increases near the walls. However, the transpiration effects are dominant and exhibit the cross-flow behavior as well as the dual behavior. The temperature and the concentration fields are respectively the increasing functions of the variable thermal conductivity and the variable species diffusivity parameters.

Nomenclature

A suction/injection parameter
a constant in Eq. (6)
C nanoparticle volumetric fraction

Table 7 Values of skin friction, Nusselt number, and Sherwood number for different physical parameters with $\gamma = \beta = 0.1$, $Nt = Nb = 0.1$, $S = 0.1$, $A = 0.01$.

Pr	Le	ε_2	ε_1	θ_r	Mn	$-f''(1)$	$-h_f$	E^f_{10}	$-\theta'(1)$	$-h_\theta$	E^θ_{10}	$-\phi'(1)$	$-h_\phi$	E^ϕ_{10}	CPU time
6.2	1	0.1	0.1	−10	0.5	6.93724	0.521876	1.85×10^{-5}	2.56836	0.795575	3.93×10^{-6}	−0.20702	0.816672	6.56×10^{-5}	452.31
					1	7.16163	0.529837	1.90×10^{-5}	2.56728	0.795370	3.87×10^{-6}	−0.20606	0.816681	6.49×10^{-5}	417.61
					1.5	7.54951	0.540578	2.20×10^{-5}	2.56541	0.794987	3.77×10^{-6}	−0.20439	0.816682	6.37×10^{-5}	466.91
					2	8.12409	0.551879	2.94×10^{-5}	2.56262	0.794530	3.16×10^{-6}	−0.20191	0.816656	6.21×10^{-5}	415.70
					2.5	8.92326	0.561811	4.11×10^{-5}	2.55874	0.793334	3.42×10^{-6}	−0.19845	0.816579	5.99×10^{-5}	417.18
6.2	1	0.1	0.1	−10	0.1	3.23547	0.539759	3.88×10^{-6}	2.61368	0.800566	4.60×10^{-6}	−0.24769	0.820601	7.36×10^{-5}	445.34
				−5		6.37108	0.796477	1.40×10^{-4}	2.57160	0.796836	4.08×10^{-6}	−0.20998	0.817371	6.64×10^{-5}	404.28
				−1		3.66322	0.344926	4.32×10^{-2}	2.59901	0.400879	1.80×10^{-4}	−0.23434	0.538266	3.87×10^{-4}	377.96
6.2	1	0.1	0	−10	0.1	6.88538	0.83426	2.12×10^{-6}	4.37052	0.753998	2.89×10^{-6}	−0.05358	0.838724	5.27×10^{-6}	413.24
			0.2			6.84778	0.563282	1.20×10^{-5}	3.29095	0.747359	7.96×10^{-5}	−0.38818	0.772511	2.83×10^{-4}	416.86
			0.4			6.80107	0.690673	6.36×10^{-4}	2.76736	0.689505	4.11×10^{-4}	−0.84408	0.694220	7.80×10^{-4}	416.80
			0.8			6.66595	0.605364	0.602742	2.40018	0.386541	4.20245	1.00077	0.003287	1.63×10^{-4}	536.38
6.2	1	0	0.1	−10	0.1	6.87282	0.817156	2.97×10^{-5}	2.53070	0.787063	1.65×10^{-6}	−0.22122	0.827749	4.33×10^{-5}	407.73
		0.2				6.86217	0.528059	1.83×10^{-5}	2.60736	0.795988	8.99×10^{-6}	−0.19379	0.805701	1.08×10^{-4}	461.73
		0.4				6.85327	0.780916	1.10×10^{-4}	2.68649	0.785656	3.00×10^{-5}	−0.16792	0.782738	3.30×10^{-4}	384.13
		0.8				6.83347	0.580338	2.35×10^{-5}	2.85208	0.739950	1.48×10^{-4}	−0.12115	0.716450	3.26×10^{-3}	416.60
6.2	1	0.1	0.1	−10	0.1	6.86664	0.519571	1.84×10^{-5}	2.56923	0.795649	3.95×10^{-6}	−0.20845	0.816398	6.51×10^{-5}	429.74
	5.0					6.86634	0.522037	1.71×10^{-5}	2.57158	0.795701	3.95×10^{-6}	−0.21346	0.815193	6.23×10^{-5}	444.22
	10.0					6.86703	0.806771	4.15×10^{-5}	2.57445	0.795678	3.96×10^{-6}	−0.21969	0.813673	5.88×10^{-5}	435.78
	25.0					6.86487	0.535357	1.18×10^{-5}	2.58273	0.795091	4.05×10^{-6}	−0.23816	0.809001	4.96×10^{-5}	436.47
0.72	1	0.1	0.1	−10	0.1	7.02711	0.910736	4.87×10^{-15}	1.43830	0.913630	2.82×10^{-16}	0.86420	0.918855	2.80×10^{-14}	380.47
1.09						7.01781	0.896013	8.61×10^{-14}	1.49661	0.908669	2.18×10^{-14}	0.80733	0.905835	6.45×10^{-13}	391.07
2.0						6.99425	0.869605	2.11×10^{-11}	1.64904	0.830756	2.05×10^{-11}	0.65997	0.877078	1.20×10^{-10}	848.27
5.09						6.90493	0.826361	2.89×10^{-6}	2.28346	0.796751	2.40×10^{-7}	0.05966	0.832669	4.16×10^{-6}	552.53
6.2						6.86798	0.809634	4.61×10^{-5}	2.56870	0.795627	3.95×10^{-6}	−0.20732	0.816668	6.58×10^{-5}	534.81

C_w nanoparticle concentration at the lower disk

T temperature (K)

T_r constant in Eq. (6)

T_w temperature at the lower disk (K)

T_∞ ambient temperature (K)

T_m mean fluid temperature (K)

ΔT temperature difference (K)

D_T thermophoretic diffusion coefficient (kg/ms K)

$D_{B\infty}$ Brownian diffusion coefficient (kg/ms)

B_0 uniform magnetic field (Tesla)

C_p specific heat at constant pressure (J/kg K)

C_{fr} skin friction

$h(t)$ variable distance $= H(1 - \alpha t)^{1/2}$

$K(T)$ temperature dependent thermal conductivity (W/m K)

K_∞ thermal conductivity of the fluid far away from the sheet (W/m K)

Le Lewis number

Mn magnetic parameter

Nb Brownian motion parameter

Nt thermophoresis parameter

N_{ur} reduced Nusselt number

Pr Prandtl number

Re_x local squeeze Reynolds number

S_{hr} reduced Sherwood number

S squeezing parameter

r, u, and w radial and axial velocities (m/s)

Greek symbols

α characteristic parameter

ζ constant defined in equation (6)

ν_∞ kinematic viscosity away from the sheet (kg/m³)

ρ_∞ constant fluid density (kg/m³)

ρ_f density of the fluid (kg/m³)

σ electric conductivity

ε_1 variable thermal conductivity parameter

ε_2 variable species diffusivity parameter (m²/s)

η similarity variables

a

b

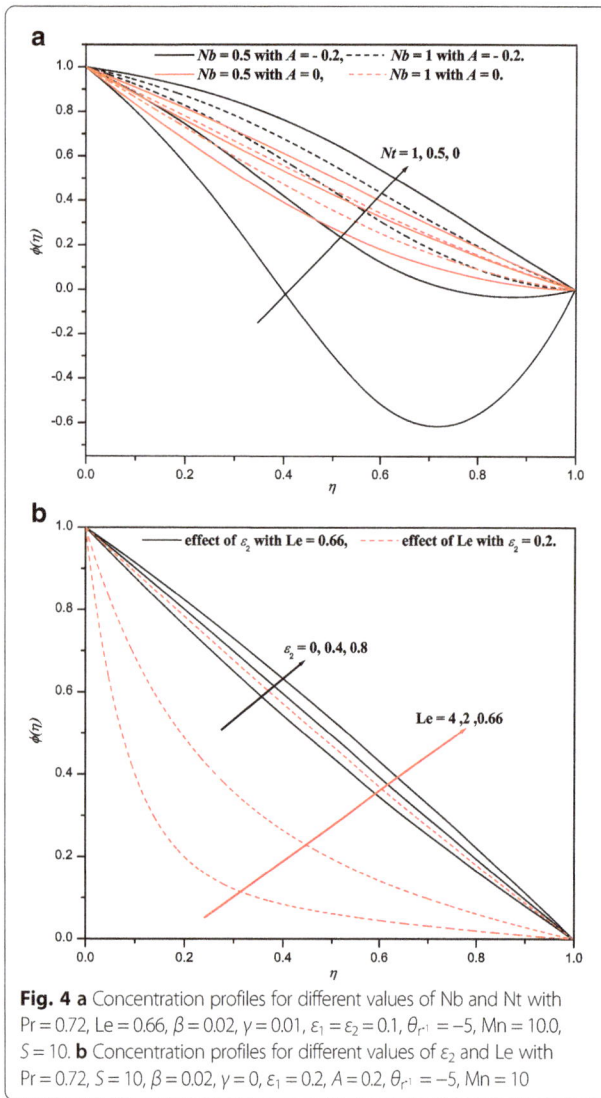

Fig. 4 a Concentration profiles for different values of Nb and Nt with Pr = 0.72, Le = 0.66, β = 0.02, γ = 0.01, ε_1 = ε_2 = 0.1, $\theta_{r'}$ = −5, Mn = 10.0, S = 10. **b** Concentration profiles for different values of ε_2 and Le with Pr = 0.72, S = 10, β = 0.02, γ = 0, ε_1 = 0.2, A = 0.2, $\theta_{r'}$ = −5, Mn = 10

a

b

c

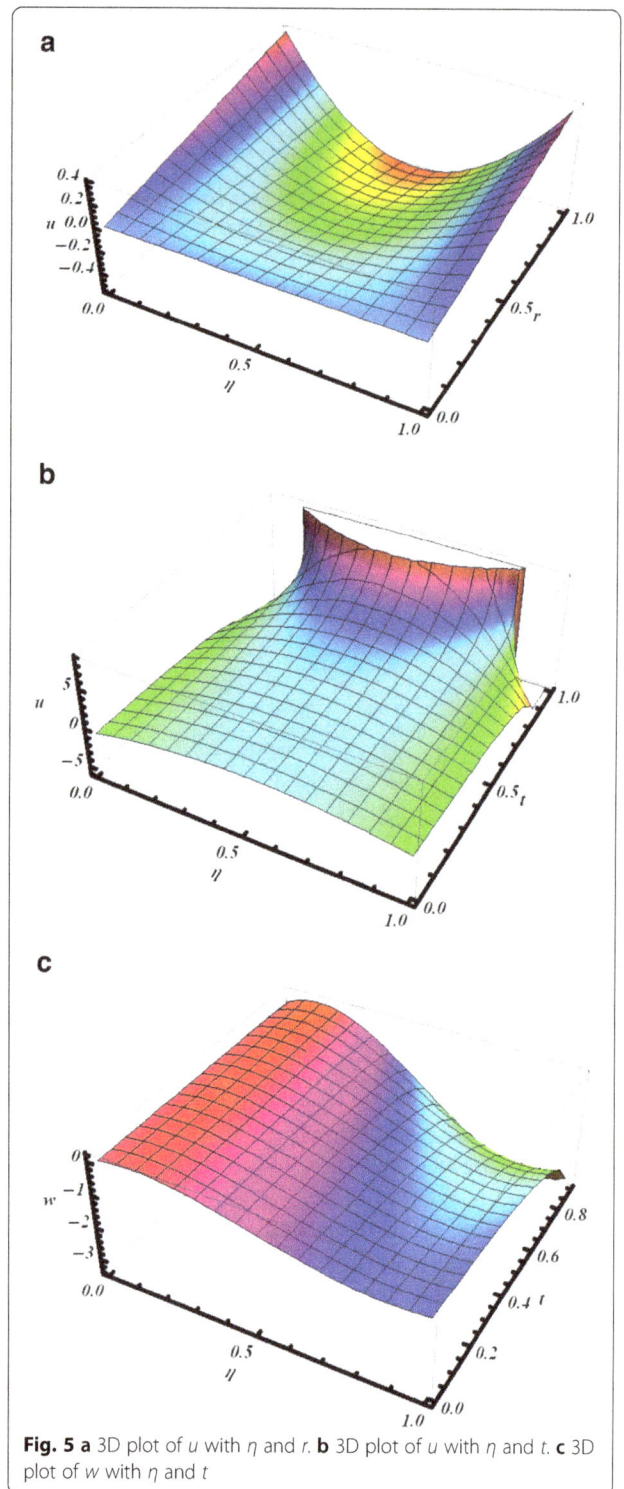

Fig. 5 a 3D plot of u with η and r. **b** 3D plot of u with η and t. **c** 3D plot of w with η and t

θ dimensionless temperature

ϕ dimensionless concentration

θ_r fluid viscosity parameter

μ dynamic viscosity (Pa s)

μ_h fluid viscosity of the fluid at the upper disk (Pa s)

k_h thermal conductivity of the fluid at the upper disk

D_{Bh} species diffusivity/Brownian diffusion coefficient of the fluid at the upper disk

μ_∞ constant value of dynamic viscosity (Pa s)

τ ratio between the effective heat capacity of the nanoparticle material and heat capacity of the fluid

τ_w wall shear stress

ψ stream function

β velocity slip parameter

γ thermal slip parameter

Subscripts

∞ condition at infinity

w condition at the wall

$'$ differentiation with respect to η

Authors' contributions
All authors read and approved the final manuscript.

Competing interests
The authors declare that they have no competing interests.

Author details
[1]Department of Mathematics, University of Central Florida, Orlando, FL 32816, USA. [2]Department of Mathematics, VSK University, Vinayaka Nagar, Ballari 583 105, Karnataka, India. [3]Department of Mechanical Engineering, The University of Hong Kong, Pokfulam, Hong Kong.

References
M.J. Stefan: Versuch über die scheinbare Adhäsion, Sitzungsber. – Abt. II, Österr. Akad. Wiss., Math.- Naturwiss. K, **69** (1874) 713–721.

Leider, P. J., & Bird, R. B. (1974). Squeezing flow between parallel disks. I. Theoretical analysis. *Industrial and Engineering Chemistry Fundamentals, 13*, 336–341.

Hamza, E. A. (1988). The magnetohydrodynamic squeeze film. *Journal of Tribology, 110*, 375–377.

Domairry, G., & Aziz, A. (2009). Approximate analysis of MHD squeeze flow between two parallel disks with suction or injection by homotopy perturbation method. *Mathematical Problems in Engineering, 2009*, 603916.

Joneidi, A. A., Domairry, G., & Babaelahi, M. (2010). Effect of mass transfer on a flow in the magnetohydrodynamic squeeze film between two parallel discs with one porous disk. *Chemical Engineering Communications, 198*, 299–311.

Hayat, T., Yousaf, A., Mustafa, M., & Obaidat, S. (2012). MHD squeezing flow of second-grade fluid between two parallel disks. *International Journal for Numerical Methods in Fluids, 69*, 399–410.

Hussain, A., Mohyud-Din, S. T., & Cheema, T. A. (2012). Analytical and numerical approaches to squeezing flow and heat transfer between two parallel disks with velocity slip and temperature jump. *Chinese Physics Letters, 29*, 114705.

Shaban, M., Shivanian, E., & Abbasbandy, S. (2013). Analyzing magneto-hydrodynamic squeezing flow between two parallel disks with suction or injection by a new hybrid method based on the Tau method and the homotopy analysis method. *European Physical Journal Plus, 128*, 1–10.

S.U.S. Choi, J.A. Eastman: Enhancing thermal conductivity of fluids with nanoparticles. Proceedings of the 1995 ASME Int. Mech. Eng. Cong. and Exposition, San Francisco, *USA, ASME FED* 231/MD 66:99,105, 1995.

Vajravelu, K., Prasad, K. V., Lee, J., Lee, C., Pop, I., & Van Gorder, R. A. (2011). Convective heat transfer in the flow of viscous Ag-water and Cu-water nanofluids over a stretching surface. *International Journal of Thermal Sciences, 50*, 843–851.

Makinde, O. D., & Aziz, A. (2011). Boundary layer flow of a nanofluid past a stretching sheet with a convective boundary condition. *International Journal of Thermal Sciences, 50*, 1326–1332.

Bachok, N., Ishak, A., & Pop, I. (2012). Unsteady boundary-layer flow and heat transfer of a nanofluid over a permeable stretching/shrinking sheet. *International Journal of Heat and Mass Transfer, 55*, 2102–2109.

Safaei, M. R., Togun, H., Vafai, K., Kazi, S. N., & Badarudin, A. (2014). Investigation of heat transfer enhancement in a forward-facing contracting channel using FMWCNT nanofluids. *Numerical Heat Transfer Part A, 66*, 1321–1340.

Safaei, M. R., Mahian, O., Garoosi, F., Hooman, K., Karimipour, A., Kazi, S. N., & Gharehkhani, S. (2014). Investigation of micro and nano-sized particle erosion in a 90° pipe bend using a two-phase discrete phase model. *Scientific World Journal, 2014*, 740578.

Goodarzi, M., Safaei, M. R., Vafai, K., Ahmadi, G., Dahari, M., Kazi, S. N., & Jomhari, N. (2014). Investigation of nanofluid mixed convection in a shallow cavity using a two-phase mixture model. *International Journal of Thermal Sciences, 75*, 204–220.

Togun, H., Safaei, M. R., Sadri, R., Kazi, S. N., Badarudin, A., Hooman, K., & Sadeghinezhad, E. (2014). Numerical simulation of laminar to turbulent nanofluid flow and heat transfer over a backward-facing step. *Applied Mathematics and Computation, 239*, 153–170.

Prasad, K. V., Vajravelu, K., Shivakumara, I. S., Vaidya, H., & Basha, N. Z. (2016). Flow and heat transfer of a Casson nanofluid over a nonlinear stretching sheet. *Journal of Nanofluids, 5*, 743–752.

Hashmi, M. M., Hayat, T., & Alsaedi, A. (2012). On the analytic solutions for squeezing flow of nanofluid between parallel disks. *Nonlinear Anal Model Control, 17*, 418–430.

Das, K., Jana, S., & Acharya, N. (2016). Slip effects on squeezing flow of nanofluid between two parallel discs. *International Journal of Applied Mechanics and Engineering, 21*, 5–20.

S.T. Mohyud-Din, S.I. Khan, B. Bin-Mohsin: Velocity and temperature slip effects on squeezing flow of nanofluid between parallel disks in the presence of mixed convection, Neural Comput. Applic., (2016). DOI: 10.1007/s00521-2016-2329-1.

Lai, F. C., & Kulacki, F. A. (1990). The effect of variable viscosity on convective heat transfer along a vertical surface in a saturated porous medium. *International Journal of Heat and Mass Transfer, 33*, 1028–1031.

Vajravelu, K., Prasad, K. V., & Ng, C. O. (2013). Unsteady convective boundary layer flow of a viscous fluid at a vertical surface with variable fluid properties. *Nonlinear Anal Real World Applications, 14*, 455–464.

Prasad, K. V., Vajravelu, K., & Vaidya, H. (2016). MHD Casson nanofluid flow and heat transfer at a stretching sheet with variable thickness. *Journal of Nanofluids, 5*, 423–435.

Vajjha, R. S., & Das, D. K. (2012). A review and analysis on influence of temperature and concentration of nanofluids on thermophysical properties, heat transfer and pumping power. *International Journal of Heat and Mass Transfer, 55*, 4063–4078.

Liao, S. (2003). *Beyond perturbation: introduction to homotopy analysis method.* London: Chapman & Hall/CRC Press.

Fan, T., & You, X. (2013). Optimal homotopy analysis method for nonlinear differential equations in the boundary layer. *Numerical Algorithms, 62*, 337–354.

Lawal, A., & Kalyon, D. M. (1998). Squeezing flaw of viscoplastic fluids subject to wall slip. *Polymer Engineering and Science, 38*, 1793–1804.

C P Kothandaraman, S Subramanyan. Heat and mass transfer data book, *New Age International (P) Ltd., Publishers*, 2014.

3

Experimental study of forces and energies during shearing of steel sheet with angled tools

E. Gustafsson[1]* ⓘ, L. Karlsson[1] and M. Oldenburg[2]

Abstract

Background: Shearing is a fast and inexpensive method to cut sheet metal that has been used since the beginning of the industrialism. Consequently, published experimental studies of shearing can be found from over a century back in time. Recent studies, however, are due to the availability of low-cost digital computation power, mostly based on finite element simulations that guarantees quick results. Still, for validation of models and simulations, accurate experimental data is a requisite. When applicable, 2D models are in general desirable over 3D models because of advantages like low computation time and easy model formulation. Shearing of sheet metal with parallel tools is successfully modeled in 2D with a plane strain approximation, but with angled tools, the approximation is less obvious.

Methods: Plane strain approximations for shearing with angled tools were evaluated by shear experiments of high accuracy. Tool angle, tool clearance, and clamping of the sheet were varied in the experiments.

Results: The results showed that the measured forces in shearing with angled tools can be approximately calculated using force measurements from shearing with parallel tools. Shearing energy was introduced as a quantifiable measure of suitable tool clearance range.

Conclusions: The effects of the shearing parameters on forces were in agreement with previous studies. Based on the agreement between calculations and experiments, analysis based on a plane strain assumption is considered applicable for angled tools with a small (up to 2°) rake angle.

Keywords: Sheet metal, Experiment, Shearing, Force, Clearance, Angle

Background

Shearing is the process where sheet metal is mechanically cut between two tools as shown in Fig. 1. A thorough review on various aspects of cutting in the engineering, physical, and biological sciences is provided by Atkins (2009), who, among other topics, discusses shearing of ductile sheets and plates, and guillotining, i.e., perform a progressive cut using an angled tool. Some examples of more specific studies of sheet metal shearing are given in the sequel of this section.

Experiments on sheet metal shearing have been of interest since the industrial revolution. Measurements of forces on various materials, including carbon steels, were performed in the early twentieth century by Izod (1906).

Similar experiments, i.e., symmetric double shearing with parallel tools, were done by Chang and Swift (1950) in a study of varied clearance evaluated on required force and produced sheared surfaces. Further development of shearing experiments were done by Gustafsson et al. (2014) to include measurement of the force that strives to separate the two tools and the clearance changes during shearing.

Rotational symmetric blanking has several features in common with sheet metal shearing, for instance, forces and sheared surface geometries were studied by Crane (1927); and tool wear was examined by Hambli et al. (2003).

The geometry of the sheared surface is a topic addressed in several studies: the sheared surface was partitioned into various regions and the mechanisms that was active in these regions was identified by Atkins (1981); the influence of tool sharpness on the surface geometry was studied by Suliman (2001); the surface geometry was also

*Correspondence: egu@du.se
[1]Dalarna University, SE-791 88 Falun, Sweden
Full list of author information is available at the end of the article

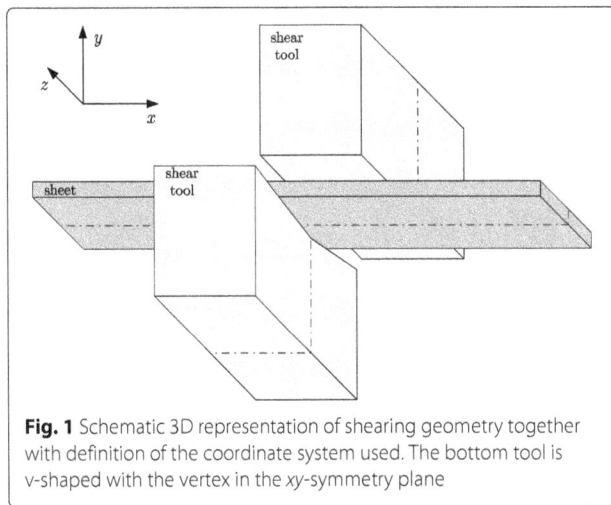

Fig. 1 Schematic 3D representation of shearing geometry together with definition of the coordinate system used. The bottom tool is v-shaped with the vertex in the *xy*-symmetry plane

examined by Hilditch and Hodgson (2005) who concluded that the extent of the regions defined by Atkins (1981) depends on the shearing geometry as well as on the material properties.

Further analyses of sheared samples, in addition to surface geometry, include the hardness distribution on the sheared surface by Weaver and Weinmann (1985) and an evaluation of the post-shearing strain from microstructure examination of the deformed area by Wu et al. (2012).

A number of articles cover the post-shearing deformation properties of sheet metal, and to name a few of these are as follows: crack formation in post-shearing bending deformations was studied by Weaver and Weinmann (1985); problems with cracks in the sheared surface in combination with a subsequent rolling operation were studied by Hubert et al. (2010); and effects of rake angle on hardness and post-shearing stretch-flange formability were studied by Matsuno et al. (2015).

When angled tools are used to cut the sheet (see Fig. 1), the progressive contact between the sheet and the angled tool reduces the required force at the expense of an increased tool stroke, and as stated by Guimaraes (1988), an increased deformation of the sheared sheet. Rake angles (the rake angle is defined in Fig. 3) up to 2° are common in larger shearing equipment, but angles larger than 5° are seldom used due to deformations that result in curl, camber, or bow of narrow strips, as discussed by Guimaraes (1988). Tools with rake angles also introduce a force along the tools (*z*-direction in Fig. 1). Although this force is small compared to forces in the other directions, it may affect the shearing if the equipment has a low stiffness in that direction. V-shaped tools balance the *z*-directional force and are therefore favorable, compared with unsymmetrical tools.

Shearing experiments are time consuming and requires specialized equipment. Studies involving a large number of parameters and coupled effects are therefore preferably

performed by finite element (FE) based simulations. Accurate experimental data is still a prerequisite to validate such simulations. There is, however, a shortage of accurate experimental data to validate such simulations, as noticed by, for example, Saanouni et al. (2010). One purpose of this work is therefore, at least to some extent, to fill this gap and provide accurate experimental data of forces from shearing with both parallel and inclined tools.

A simplified analytical model of forces during shearing with angled tools was developed by Atkins (1990) in guillotining of copper plates. In that model, the force can be computed provided the materials shear yield stress, work hardening index, and fracture toughness in the plane of cut are known. The latter two parameters are, however, often not so readily available as the yield stress. Therefore, in this paper, a simple model is developed, based on the assumption of plane strain, in which the forces in shearing with angled tools are calculated from forces measured with parallel tools. The main purpose of this work is to compare the forces predicted by the model, with forces measured with inclined tools. Forces were accurately measured, both with parallel and angled tools, using an experimental set-up with a high stiffness and a stable tool clearance providing reliable reproducible results, as previously demonstrated by Gustafsson et al. (2014). If the model can predict the forces in shearing with angled tools, this implies that a 2D plane strain FE analysis is applicable to obtain forces in shearing with angled tools. Furthermore, the presented experimental data can also be used to verify numerical models of sheet metal shearing.

Simplified shearing model

In this section, a simple model for calculating the forces during shearing with angled tools is presented. The model assumes that plane strain prevails (in the *xy*-plane); forces from measurements on parallel tools was used as input to the model.

Definition of shearing arrangement

Some important shearing parameters and boundary conditions are defined in Fig. 2, which shows the arrangement when parallel tools are used. Shearing of wide sheet strips generally uses a rake angle, either by rotation of one tool around the *x*-axis or by a v-shaped tool as the bottom tool in Fig. 1. The rake angle, θ, is defined in Fig. 3.

Approximation of shearing forces

When shearing sheets of a given thickness and strength using parallel tools (the rake angle $\theta = 0$) the force, $F^0(u)$ (superscript denotes $\theta = 0$), acting on the tool is proportional to the sheet width 2w. That is, $F^0(u) = 2w \cdot P(u)$, where $P(u)$ is the force per unit sheet width and u is the magnitude of the penetration of the tool into the sheet in the *y*-direction. During shearing, the tool is in

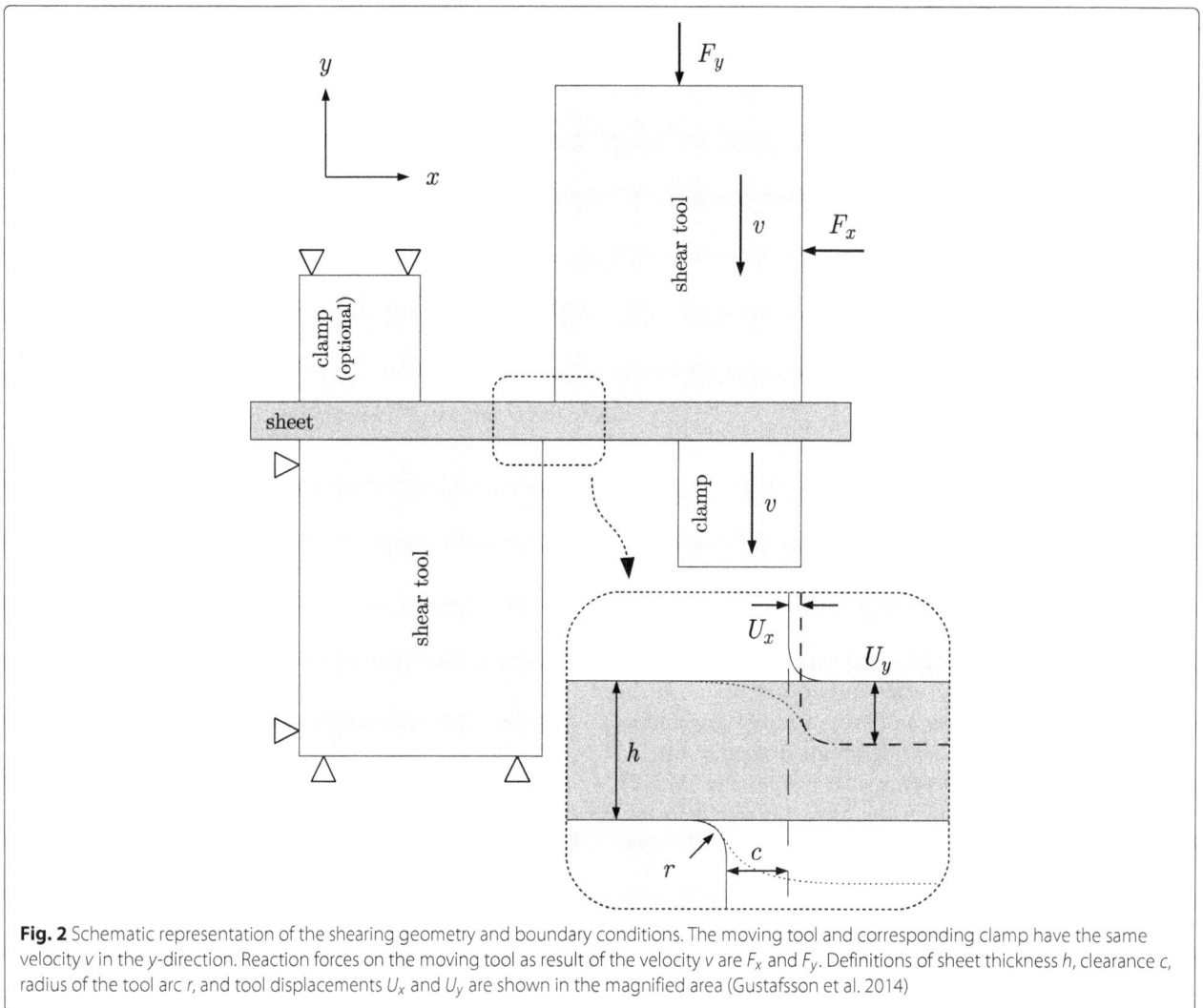

Fig. 2 Schematic representation of the shearing geometry and boundary conditions. The moving tool and corresponding clamp have the same velocity v in the y-direction. Reaction forces on the moving tool as result of the velocity v are F_x and F_y. Definitions of sheet thickness h, clearance c, radius of the tool arc r, and tool displacements U_x and U_y are shown in the magnified area (Gustafsson et al. 2014)

contact with the sheet over the entire sheet width, with u constant over the width and $u = U$, where U is the magnitude of the tool displacement in the y-direction. The tool displacement increases from zero at shearing start to $U = u_f$ at fracture of the sheet.

If instead, a rake angle $\theta \neq 0$ is used, the contact between the tool and the sheet is confined to a short region along the sheet width. At any moment, the local tool penetration, u, will therefore vary along the sheet width with $u \neq U$. In this contact region, a plastic zone of length $z_{AB} = z_B - z_A$ is formed in the sheet, where z_A and z_B are the z-coordinates of the boundaries A and B, respectively, as shown in Fig. 3.

If $P(u)$ is assumed to be the same for parallel and angled tools, then $P(u) = F^0(u)/(2w)$ can be obtained from measurements of tool force on parallel tools. Thus, the tool force, $F^\theta(u)$, at rake angles $\theta \neq 0$ can be computed by integrating $P(u)$ along the plastic zone. That is, due to the symmetric set-up shown in Fig. 3, $F^\theta(U) =$

$2 \int_{z_A}^{z_B} P(u)dz$, since $P(u) = 0$ outside the plastic region. Using the geometric relation $du = dz \tan \theta$, the integral can be rewritten as $F^\theta(U) = (2/\tan\theta) \int_{u_A}^{u_B} P(u)du$, where u_A and u_B are the magnitudes of the local penetration of the tool at the boundaries of the plastic zone. The limits of integration, either z_A and z_B or u_A and u_B, are functions of the global tool displacement, U, which determines the size of the plastic zone formed in the contact region. It is therefore natural to consider the shearing process in terms of the plastic zone formed in the sheet.

We will regard the entire shearing process to consist of the following three stages (see also Fig. 3): (i) formation of a plastic zone whose length gradually increases to a constant value z_{AB}^{max}; (ii) the plastic zone of length z_{AB}^{max} propagates along the sheet width; and (iii) the plastic zone front has reached the middle of the sheet (at the vertex of the v-shaped tool) and the zone size gradually decreases to zero, followed by final fracture of the sheet. These three

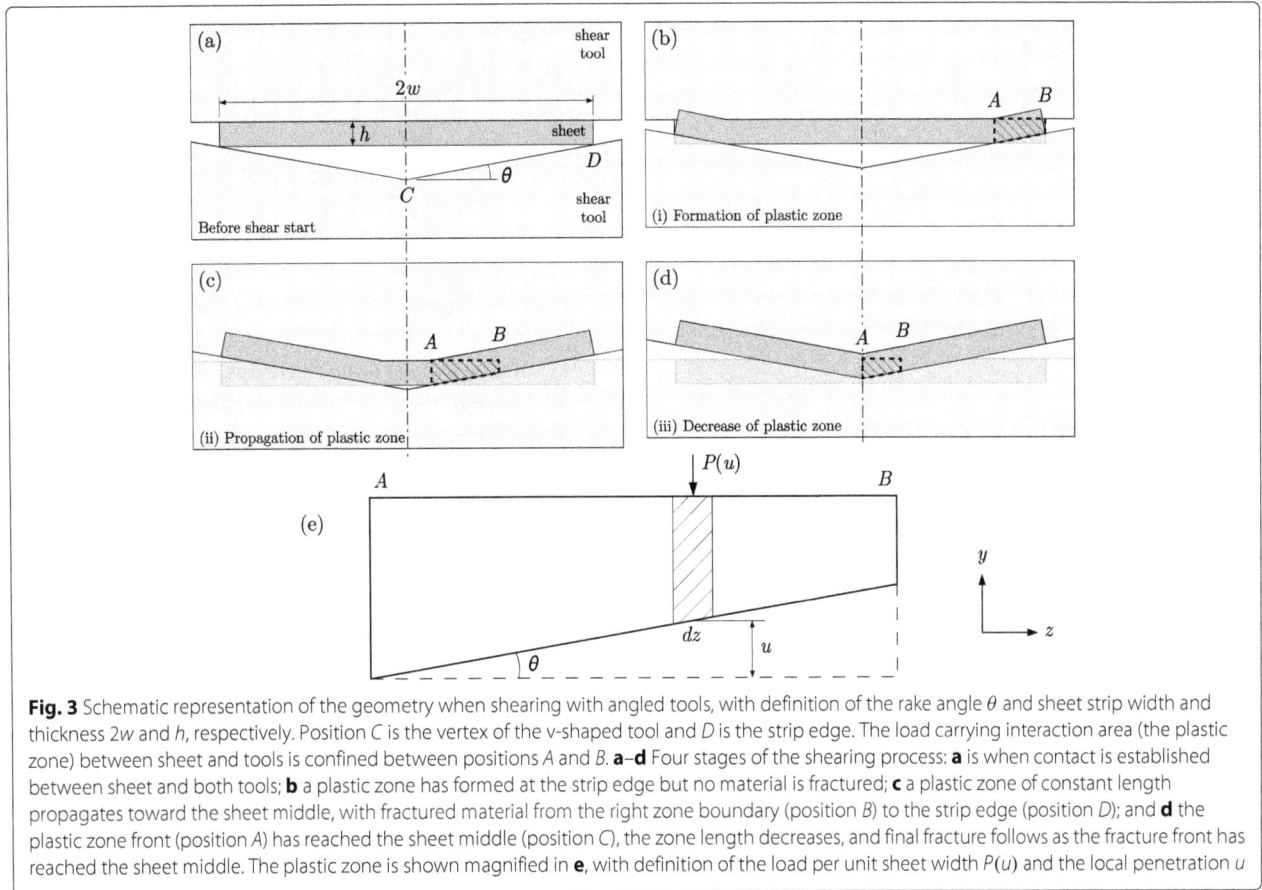

Fig. 3 Schematic representation of the geometry when shearing with angled tools, with definition of the rake angle θ and sheet strip width and thickness $2w$ and h, respectively. Position C is the vertex of the v-shaped tool and D is the strip edge. The load carrying interaction area (the plastic zone) between sheet and tools is confined between positions A and B. **a–d** Four stages of the shearing process: **a** is when contact is established between sheet and both tools; **b** a plastic zone has formed at the strip edge but no material is fractured; **c** a plastic zone of constant length propagates toward the sheet middle, with fractured material from the right zone boundary (position B) to the strip edge (position D); and **d** the plastic zone front (position A) has reached the sheet middle (position C), the zone length decreases, and final fracture follows as the fracture front has reached the sheet middle. The plastic zone is shown magnified in **e**, with definition of the load per unit sheet width $P(u)$ and the local penetration u

stages are in turn considered in detail below. Using basic geometrical relations, the stages can be formulated as intervals of the global tool displacement, U, and similarly for the limits of integration, namely (z_A, z_B) or (u_A, u_B).

Formation of plastic zone: $(0 < U < u_f)$. The initial contact between the tool and the sheet occurs at the sheet edge (position D in Fig. 3). At this moment $z_{AB} = 0$, since positions A and B both coincide with D at the edge of the sheet. As the tool is pressed into the sheet, z_{AB} will increase to a constant value $z_{AB}^{\max}$ that is dependent on the material properties and shearing parameters. The plastic zone boundaries are

$$z_A = w - U/\tan\theta, \qquad z_B = w,$$

and the corresponding limits of integration with respect to u are

$$u_A = 0, \qquad u_B = U.$$

Propagation of plastic zone: $(u_f < U < w\tan\theta)$. A zone of plastically deformed material of length $z_{AB}^{\max}$ is translated at constant force along the sheet in the z-direction. The sheet is fractured to the right of the plastic zone marked in Fig. 3 $(z > z_B)$ and is undeformed to the left of the zone $(z < z_A)$. The boundary between undeformed

and plastically deformed material is hence at $z = z_A$, and similarly, $z = z_B$ is the boundary between plastically deformed and fractured material. Now,

$$z_A = w - U/\tan\theta, \qquad z_B = w - \left(U - u_f\right)/\tan\theta,$$
$$u_A = 0, \qquad u_B = u_f.$$

Decrease of plastic zone: $(w\tan\theta < U < w\tan\theta + u_f)$. As the front of the plastic zone has reached the sheet middle (positions A and C in Fig. 3 coincide), the length z_{AB} of the zone starts to decrease and the load decreases. Eventually, the entire sheet is fractured when the fracture front has reached the sheet middle (positions B and C coincide) so that $z_{AB} = 0$. Here,

$$z_A = 0, \qquad z_B = w - \left(U - u_f\right)/\tan\theta,$$
$$u_A = U - w\tan\theta, \qquad u_B = u_f.$$

The total tool force, $F^\theta(U)$, as a function of the global tool displacement, U, can hence be expressed as three integrals:

$$F^{\theta}(U) = \begin{cases} \frac{2}{\tan\theta}\int_{0}^{U}P(u)du & : 0 < U < u_f \\ \frac{2}{\tan\theta}\int_{0}^{u_f}P(u)du & : u_f \leq U \leq w\tan\theta \\ \frac{2}{\tan\theta}\int_{U-w\tan\theta}^{u_f}P(u)du & : w\tan\theta < U < w\tan\theta + u_f \end{cases},$$

(1)

corresponding to the three stages with respect to the plastic zone. The above integrals are evaluated as the relevant area under the curve of $P(u) = F^0(u)/(2w)$ versus U, which may be obtained experimentally from measurements on shearing with parallel tools.

Experimental set-up

Shearing experiments were performed with a slightly modified version of the previously developed procedure and experimental set-up described by Gustafsson et al. (2014) and schematically shown in Fig. 4. Two simultaneous and symmetrical shears were applied for clearance stability and internal balancing of forces in the x-direction. Thus, forces, guides, and their inevitable friction losses were avoided and high accuracy force measurements were hence possible. Sheet metal strips were sheared between each pair of tools and clamps were available on both sides of the sheet. The forces, F_x and F_y were measured with strain gauges attached to the experimental set-up.

Studies of the rake angle were made possible with v-shaped tools that ensured symmetry and zero net force in the z-direction and symmetric force distributions in the x- and y-directions. Still, in order to verify this symmetry, the optical tool tracking used by Gustafsson et al. (2014) was substituted with displacement transducers that allowed measurements of the tool rotation and/or bending around the x- and y-axes. The displacement transducers improved tool tracking abilities and simplified the data processing, as displacements from the resistive transducers were sampled together with data from the strain gauges on a common time base. All signals were sampled at 5 kHz. Linear calibration of the displacement transducers was done by means of gauge blocks, and the transducers were stable during the shearing experiments; the strain gauges used in force calculations were shunt calibrated before each experiment.

Displacements transducers

Displacement transducers were positioned on the tools according to Fig. 5, to record tool displacements in the x-direction (transducers A, B, and D) and y-direction (transducers C and E). The transducers were free to move on the tool surface and merely recorded translations in the indicated directions. Rotations of the tools around the x-, y-, and z-directions, as well as translations of the tool edge, were calculated from the transducer recordings as shown below, assuming rigid tools and small displacements.

Let U_x^A, U_x^B, and U_x^D denote the tool displacements in the x-direction recorded by transducers A, B, and D,

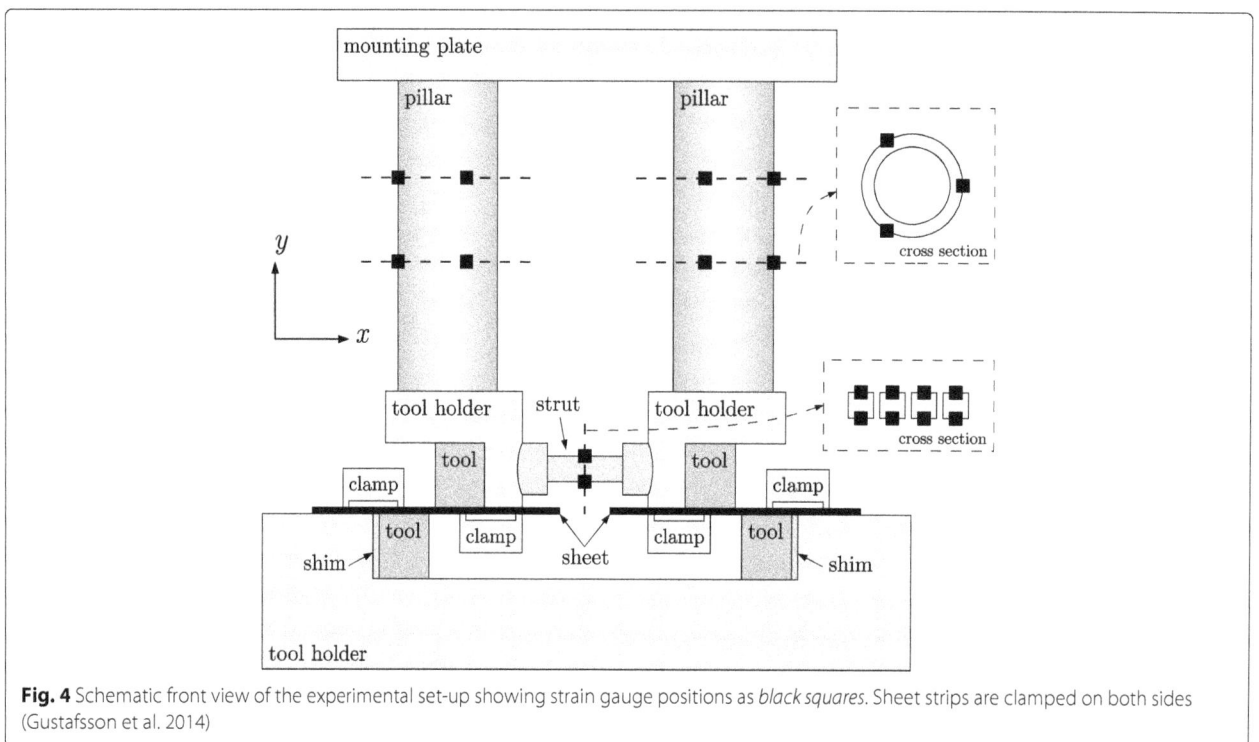

Fig. 4 Schematic front view of the experimental set-up showing strain gauge positions as *black squares*. Sheet strips are clamped on both sides (Gustafsson et al. 2014)

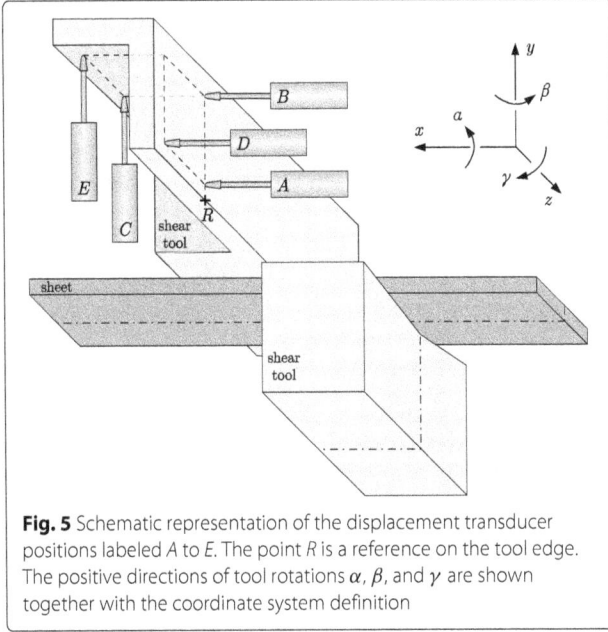

Fig. 5 Schematic representation of the displacement transducer positions labeled A to E. The point R is a reference on the tool edge. The positive directions of tool rotations α, β, and γ are shown together with the coordinate system definition

Methods

Material

A cold rolled and continuously annealed martensitic steel with a minimum yield strength of 950 MPa, a tensile strength between 1200 and 1400 MPa and a minimum elongation, A_{80}, of 3 % was used in the study. Samples with dimensions of 140 × 150 mm (140 mm perpendicular to the rolling direction) were laser cut from a 1250 mm (perpendicular to the rolling direction) wide sheet. Samples along the edges of the sheet were discarded due to thickness variations. All sheared sheet samples had a thickness between 2.04 and 2.05 mm. The samples were sheared perpendicular to the rolling direction in all experiments.

Shearing experiments

Shearing experiments were performed at two rake angles, 1° and 2°, and with parallel tools. Two clamping configurations, one side clamped and both sides clamped (see also Fig. 2), were used together with parallel tools while the experiments with rake angles were performed only with one side clamped. Tool clearances of 0.05h, 0.10h, 0.15h, and 0.20h, where h is the sheet thickness, were evaluated for all combinations of rake angle and clamping configuration. With parallel tools, the studied clearance range was extended to larger clearances as shown in Table 1. Three experiments were performed for each configuration in Table 1.

Tool displacements and forces were continuously measured during shearing. Shearing energies, W, could therefore be calculated by integration of $dW = FdU$ from shearing start to fracture. The integration was done numerically by approximating the area under the load-displacement curve by a set of rectangles and summation of the rectangle areas. Only the F_y-$|U_y|$-curve was considered since the displacement, U_x, was negligible compared to U_y and F_x was much smaller than F_y.

respectively. Similarly, U_y^C and U_y^E are the tool displacements in the y-direction recorded by transducers C and E, respectively. The tool rotations around the x-, y-, and z-axes are denoted α, β, and γ, respectively. For small angles, $\alpha \approx \tan \alpha$, and similarly for β and γ, then

$$\alpha \approx \frac{U_y^E - U_y^C}{z_{EC}}, \tag{2}$$

$$\beta \approx \frac{U_x^A - U_x^D}{z_{AD}}, \tag{3}$$

$$\gamma \approx \frac{U_x^A - U_x^B}{y_{AB}}, \tag{4}$$

where z_{EC} is the distance in the z-direction between E and C and similarly for z_{AD} and y_{AB}. The positive directions of the rotations are shown in Fig. 5.

Consider a point, R, on the tool edge with the same z-coordinate as transducers A, B, and C, and let U_x^R and U_y^R denote the displacement of R in the x- and y-direction. Assuming negligible rotations α and β, then, as a first approximation, the displacements of the tool edge (point R) are

$$U_x^R \approx U_x^A + \gamma \left(y_{RA} - U_y^R \right), \tag{5}$$

$$U_y^R \approx U_y^C - \gamma x_{RC}, \tag{6}$$

where x_{RC} is the distance in the x-direction between R and C at shearing start, and similarly for y_{RA}. In the following, U_x^R and U_y^R are simply referred to as U_x and U_y. These displacements are, respectively, the change of the tool clearance and the global tool displacement during shearing.

Table 1 Matrix of clearances and rake angles used in the experimental study

Clearance	Parallel tools, 1 clamp	Parallel tools, 2 clamps	1° rake angle, 1 clamp	2° rake angle, 1 clamp
0.05h	x	x	x	x
0.10h	x	x	x	x
0.15h	x	x	x	x
0.20h	x	x	x	x
0.25h	x			
0.30h		x		
0.40h		x		
0.50h		x		

The sheared surfaces were ocular examined, although not quantitatively characterized as done by Gustafsson et al. (2016). Primarily, the fracture surfaces were checked for excessive roughness. Results from the ocular examinations are discussed when relevant but are not presented in a separate section.

Tensile and compression testing

One purpose of this work is to present experimental shearing results that allow for validation of numerical models of the shearing process. In such models, material isotropy is often assumed. To check whether this assumption holds for the material investigated, the degree of anisotropy was assessed from flow characteristics in three directions of the sheet sample.

Material characterization was conducted by means of uniaxial tensile and compression tests. Tensile test samples were prepared in the x-direction (sheet rolling direction) and the z-direction (in sheet plane direction that is transverse to the rolling direction), and compression test samples were prepared in the y-direction (sheet thickness direction). The tensile tests were conducted at 0.02 mm s^{-1} on specimens with a 5.2 mm waist and an active gauge length of 50 mm. The compression tests followed the procedure described by Gustafsson et al. (2014) on cylindrical samples, 3.2 mm in diameter, with a length of 6.15 mm obtained by stacking three layers of sheet. Each flow stress curve in Fig. 6 is representative for a series of three tests per direction. The flow stresses showed some anisotropy with lower flow stress in the rolling direction compared with the other two measured directions.

Results

Results from the shearing experiments in terms of forces, F_x and F_y, and clearance changes, U_x, are presented in

Figs. 7, 8, 9, and 10 as functions of tool displacement, U_y. The parameters U_x and U_y were calculated from displacement transducers according to Eqs. (5) and (6), respectively. Only results from one of the two symmetric sides of the experimental set-up (see Fig. 4) are shown since variations in F_y and U_x between the sides were less than 1 %. Among the three experiments performed for each configuration, the difference in force was less than 1% and U_y varied a few percent at the point of fracture. Therefore, only one out of the three curves for each configuration is shown in Figs. 7, 8, 9, and 10.

For most of the experiments, the tool rotation γ (Eq. (4)) was between −1 and −2 mrad. The rotations α and β (Eqs. (2) and (3), respectively) were between −0.5 and 0.5 mrad, except when shearing with 0.20h clearance and 1° rake angle (where α almost reached 3 mrad). Thus, α and β were considered negligible, as assumed in the derivation of Eqs. (5) and (6).

Clearance changes, U_x, were, with exception of 0.20h clearance at 1° rake angle, always below 20 μm and in most cases below 10 μm. The rapid change in clearance at the

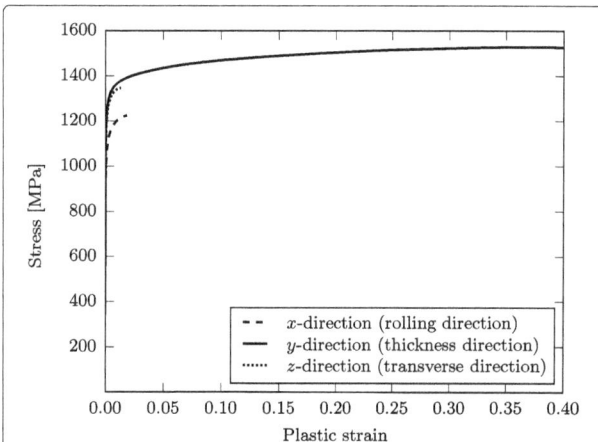

Fig. 6 Flow stress curves for the sheared material in three directions obtained from tensile (x- and z-direction) and compression (y-direction) test data

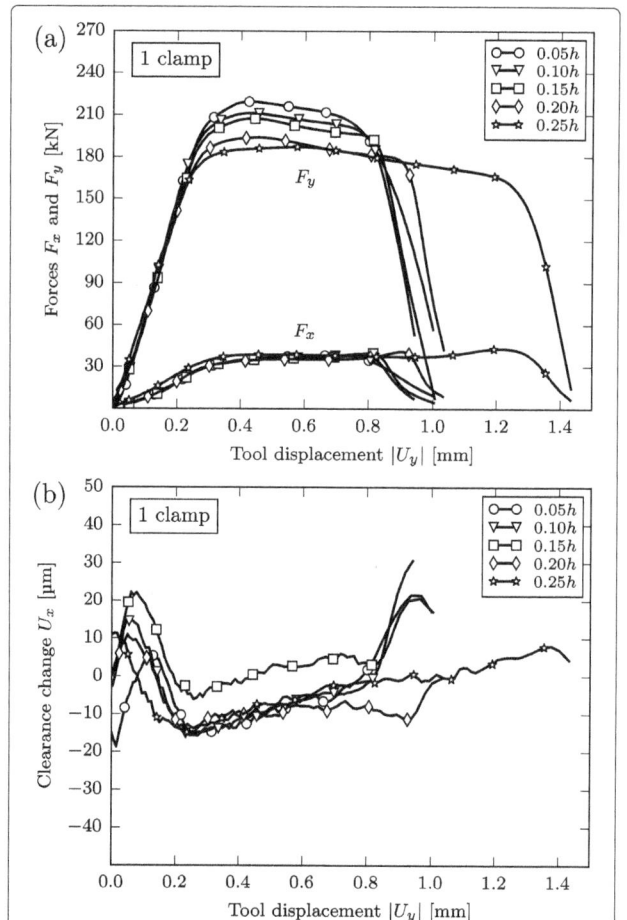

Fig. 7 Measured forces (**a**) and clearance changes (**b**) from shearing with parallel tools and one clamp. Each graph contains results from experiments with various clearances

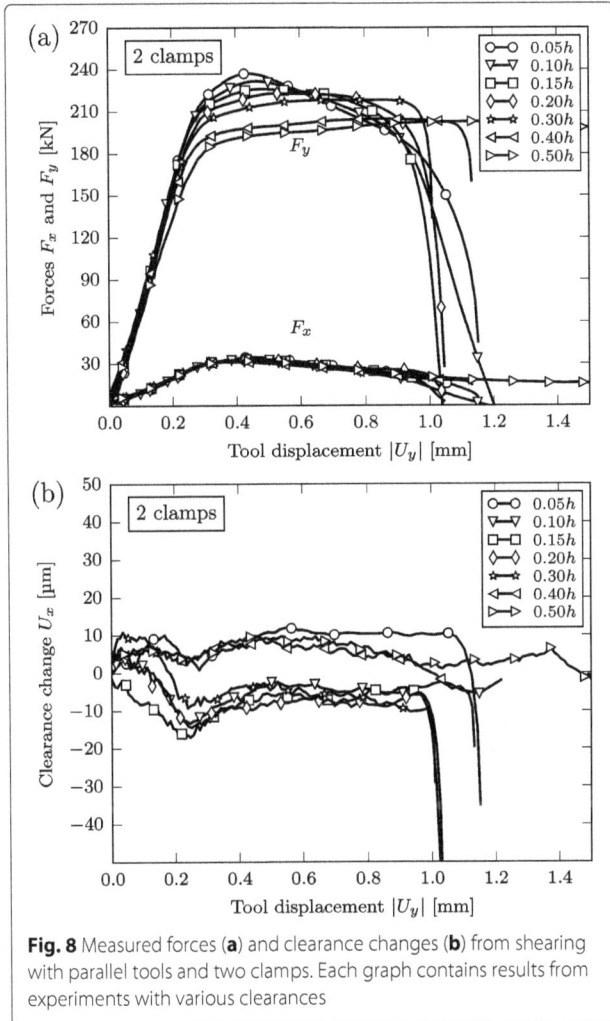

Fig. 8 Measured forces (**a**) and clearance changes (**b**) from shearing with parallel tools and two clamps. Each graph contains results from experiments with various clearances

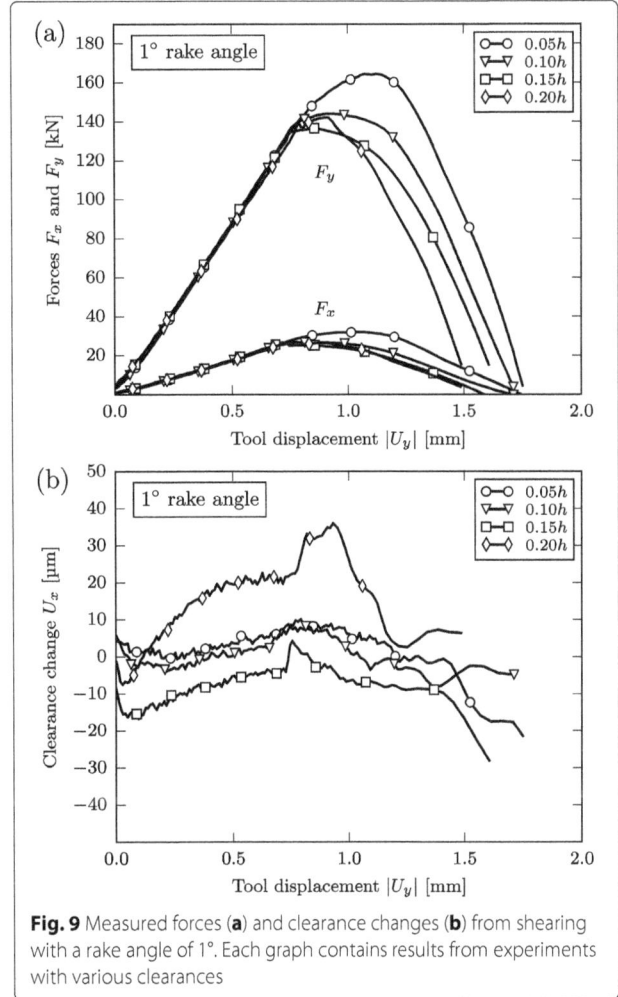

Fig. 9 Measured forces (**a**) and clearance changes (**b**) from shearing with a rake angle of 1°. Each graph contains results from experiments with various clearances

end of some curves, most pronounced in Figs. 8b and 10b, is an effect of unsymmetrical fracture between the two simultaneous shears. Shearing with v-shaped tools and clearances of 0.20h and larger caused a stepwise fracture and an oscillating rotation of the tool around the x-axis. The effect increased with increased clearance and made shearing with v-shaped tools and clearances over 0.20h useless. Slight effects of the stepwise fracture, in form of local peaks in force and clearance, were seen on the combination of 0.20h clearance and 1° rake angle at around $|U_y| = 0.9$ mm as shown in Fig. 9.

For parallel tools (Figs. 7 and 8), the variation in force among the tested clearances and clamping configurations followed the same trends as observed by Gustafsson et al. (2016). The force, F_x, was almost independent of the clearance and was slightly smaller when shearing with two clamps (Fig. 8) compared to one clamp (Fig. 7). In terms of maximum force, the values were 40 and 35 kN for shearing with one and two clamps, respectively, but the force for tool displacements $|U_y| > 0.6$ mm was considerably

smaller with two clamps. With one clamp, the maximum F_y decreased with increased clearance (Fig. 7a); the same was observed with two clamps for $|U_y| < 0.6$ mm (Fig. 8a). For the largest clearances used (0.25h, one clamp; 0.50h, two clamps), the tool displacement, $|U_y|$, required to initiate fracture was exceptionally large (compared to the required $|U_y|$ at smaller clearances) and resulted in rough sheared surfaces, as also observed by Gustafsson et al. (2016). Satisfactory results were restricted to clearances up to 0.20h for shearing with one clamp and up to 0.40h with two clamps. Larger clearances required abnormal tool displacements and produced rough surfaces. The smoothest surface, however, was obtained with the largest clearance that yielded satisfactory results, namely 0.20h and 0.40h for one and two clamps, respectively.

With angled tools (Figs. 9 and 10), both the maximum of F_x and F_y decreased with increased clearance but appeared to stabilize at a clearance of 0.15h. The stepwise fracture described earlier resulted in some bumps on the curve of F_y with 0.20h clearance (Fig. 9a). For 1° rake

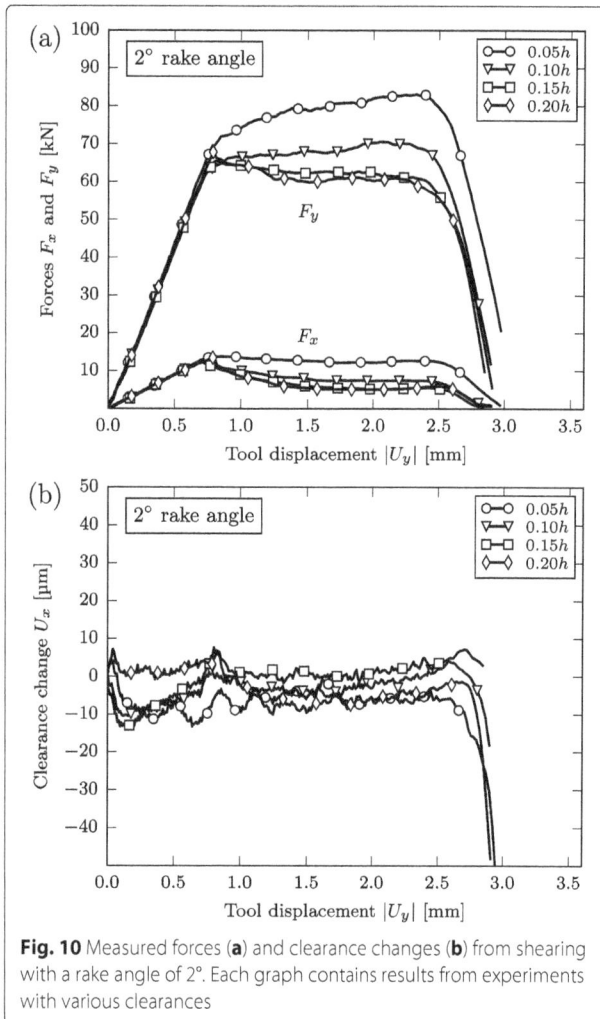

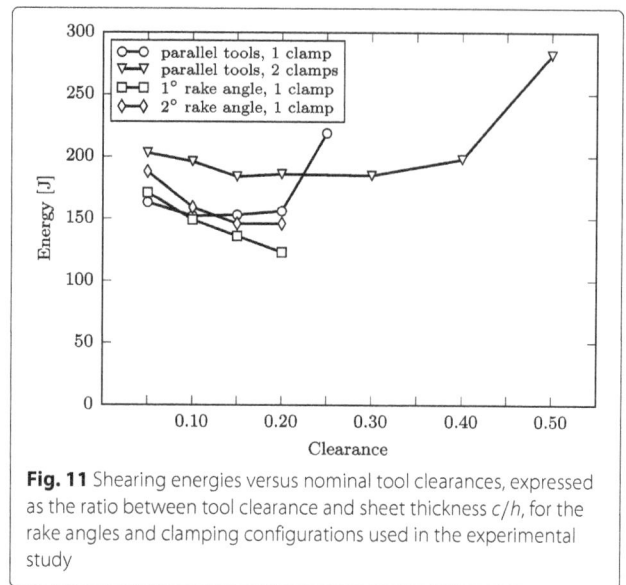

Fig. 11 Shearing energies versus nominal tool clearances, expressed as the ratio between tool clearance and sheet thickness c/h, for the rake angles and clamping configurations used in the experimental study

Fig. 10 Measured forces (**a**) and clearance changes (**b**) from shearing with a rake angle of 2°. Each graph contains results from experiments with various clearances

angle, the force-displacement curves had a fairly sharp maximum, explained as the two propagating plastic fronts almost met in the sheet middle before fracture started at the edges. With 2° rake angle, the forces were more or less constant over an interval where the shear zone propagated in the sheet sample, although, F_y increased slightly for the $0.05h$ and $0.10h$ clearances (Fig. 10a). Naturally, the forces decreased with increased rake angle at the expense of an increased tool displacement: an increase of the rake angle from 1° to 2° resulted in about halved forces.

The sheared samples remained straight without signs of bending or twisting for both 1° and 2° rake angles. Local effects in terms of larger fracture zone and smaller plastically sheared zone were, however, seen on the sheared surface in a small area around the vertex of the v-shaped tool.

Calculated shearing energies are shown in Fig. 11. Shearing with parallel tools and two clamps consumed more energy than the other three configurations studied.

Comparisons between experimentally measured forces and forces calculated with Eq. (1), from data obtained with parallel tools, at 1° and 2° rake angles are shown in Figs. 12 and 13. To calculate the force component $F_y = F^\theta(U)$, the force per length unit, $P(u)$, was obtained from the y-directional force component at parallel tools. Similarly, F_x was also calculated from Eq. (1) by letting $P(u)$ consist of the x-directional force per length unit. In Figs. 12 and 13, the lower bound for F_y corresponds to calculations from shearing with parallel tools and one clamp and the upper bound corresponds to shearing with two clamps. The bounds are the opposite for F_x although the differences in calculated F_x were small. All shearing with rake angles were performed with one clamp, but the unsheared material helps to stiffen the free end; hence, a comparison with calculations from two clamped shears seemed justified. As can be seen in Figs. 12 and 13, there was a good agreement between calculated and measured F_y before fracture start (up to the maximum on the force-displacement curves) and a fair agreement during the steady state phase with the force maximum; there was a large discrepancy between predicted and measured forces in the last phase where only partly sheared material remained.

Discussion

The comparison between measured and predicted forces mentioned in the previous paragraph should be seen as a validation of the model used to calculate forces for various rake angles. Although shearing is inevitably a 3D operation, 2D approximations are applied in FE simulations due to shorter computation times and better availability of adaptive remeshing techniques to overcome element distortion. When shearing wide strips using tools with zero rake angle, a plane strain approximation is convenient and shows good agreement with experiments, as demonstrated by Gustafsson et al. (2014). These authors hence

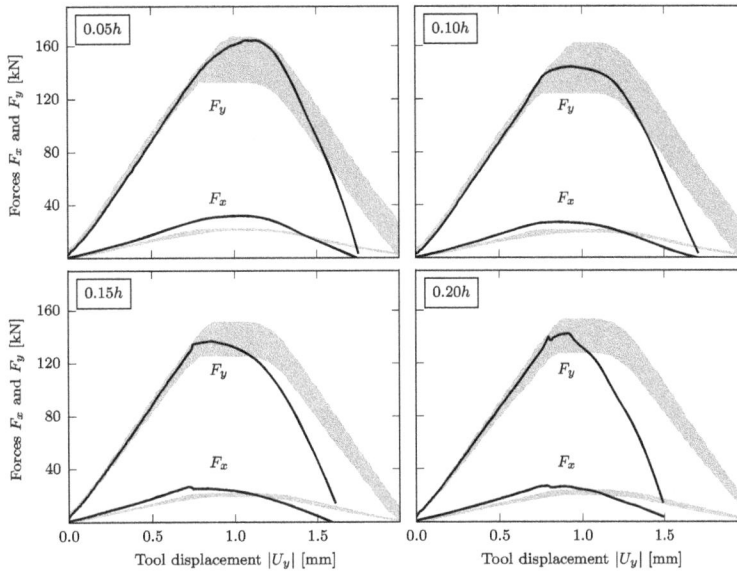

Fig. 12 Comparison of experimental and predicted force-displacement curves at 1° rake angle and clearances 0.05h, 0, 10h, 0.15h, and 0.20h. *Black lines* are experimental data and *gray areas* are calculated from measurements with parallel tools using Eq. (1). Lower and upper bounds of the gray areas for F_y represent calculations from data obtained with one clamp and two clamps, respectively; for F_x, the lower and upper bounds are reversed

showed that a 2D plain strain approximation is applicable to shearing with parallel tools. According to Oldenburg (1980), additional strains and distortions are, however, introduced with inclined tools, and thus, the validity of a plane strain assumption is not obvious in this situation. This study suggests that a 2D plane strain assumption is appropriate also for shearing with inclined tools, as it is when using a tool with zero rake angle.

According to Atkins (1990), who studied guillotining of copper plates at rake angles of 10° and 25°, the total force in the y-direction is composed of a shear component and a component required to bend the sheet to the tool inclination and where the shear component is described in terms of an effective fracture toughness. Here, with much smaller angles used, the bending component is of less importance and the force is not interpreted in terms

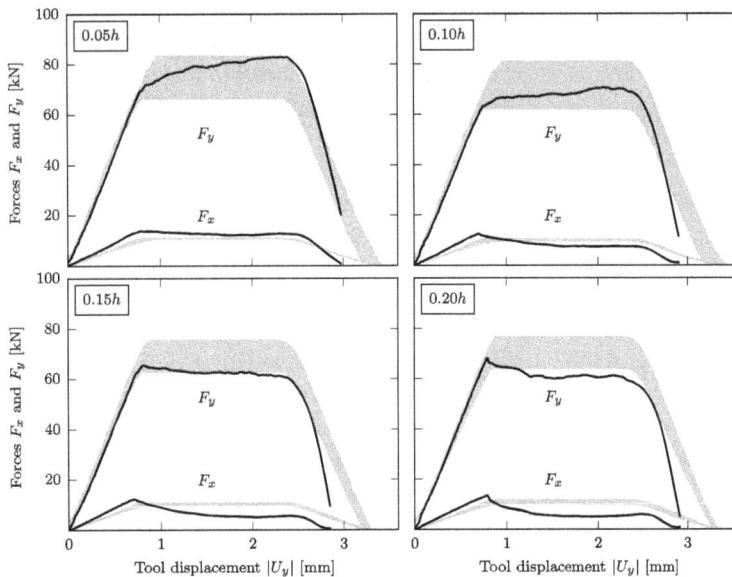

Fig. 13 Same as Fig. 12 but for 2° rake angle

of fracture toughness. A further advantage of the present model, in addition to not including the fracture toughness, is that the model includes the beginning and the end of the shearing process where the forces are not in a steady state. Consequently, the model captures forces of the entire shearing operation, from initial tool contact to final fracture, as function of the tool displacement.

There was generally a good agreement between calculated and measured forces (Figs. 12 and 13), especially for the force component F_y. The maximum F_y measured at shearing with angled tools were, for all combinations of rake angle and clearance, between the lower and upper boundary calculated from forces measured at shearing with parallel tools. The maximum F_x was constantly underestimated by between 15 and 30 % in the calculations. Considering that, in most shearing applications, it is sufficient to know the maximum forces, the differences between calculations and measurements shown in the last part of the force-displacement curves may not be critical. Due to interaction between the two approaching crack tips (and possibly also a small radius in the bottom of the v-shaped tool), the last parts of the experimentally measured force curves are not fully comparable to the calculated ones.

For the present material, no advantages in terms of forces, energies, and fracture surfaces were seen in shearing with a clearance smaller than $0.15h$, except for parallel tools and one clamp where $0.10h$ resulted in slightly less deformed free strip end. Also, no advantages were seen with clearances larger than $0.20h$, except for parallel tools and two clamps where $0.30h$ gave better results in general; even $0.40h$ clearance was beneficial with lower maximum force and a smoother fracture surface.

The stepwise fracture seen when shearing with clearances over $0.20h$ and angled tools affects the forces, tool displacements, and also the sheared surface. When the applied stress is large enough for the crack to start propagating, the sheet fractures rapidly ahead of the tool, as discussed and modeled by Atkins (2012), while elastic energy is unloaded. Since the tool velocity is much lower than the crack propagation velocity, the applied stress at the crack tip will decrease and the crack propagation will stop until additional tool displacement restores the stress. The problem may be aggravated by the slender experimental set-up and large amount of stored elastic energy together with the high-strength steel. In shearing equipment with low clearance stability, the phenomenon can also result in an oscillating tool clearance. If the stepwise fracture and unloading of elastic energy do not occur simultaneously at both shear fronts on the v-shaped tool, then, the z-directional symmetry is broken and the z-directional forces are no longer in balance. In that case, the stiffness in the z-direction is also important. Consequently, when stepwise fracture is observed, stiffness in

the x-, y-, and z-directions are all of importance for the sheared surface appearance. Stiffness in the x-direction (clearance stability) is probably still more important than in the y- and z-directions.

Shearing energy is easily quantifiable, and it is therefore tempting to use it as measure of suitable tool clearance. There was a slight minimum of the required shearing energy when shearing with parallel tools (Fig. 11). Most likely, shearing with rake angles also has an energy minimum; the energies decreases toward $0.20h$ clearance, but $0.20h$ was not large enough to show the increase in energy that likely follows at some larger clearance. The reliability of the data point at $0.20h$ in the curve for 1° rake angle is questionable because of the stepwise fracture and unstable clearance for that combination of shearing parameters. Redundant experiments on that parameter combination showed a variation in shearing energy that was not seen for other combinations. Shearing energy is a blunt tool to select a suitable clearance, but at least clearances outside of the relatively wide clearance range with an energy minimum may be excluded as suitable clearances. A low energy is, however, not necessarily always beneficial: shearing with two clamps required more energy compared to one clamp, and Gustafsson et al. (2016) showed that two clamps resulted in better sheared surface characteristics.

There are a few mechanisms that can explain the non-constant forces seen in Fig. 10a during the phase where the plastic zone travels across the sheet. The non-constant forces may depend upon varying geometric boundary conditions during the shear. In all experiments with angled tools, the sheet is clamped against the upper straight tool and free to move at the v-shaped tool, but due to unsheared material that counteracts the rotation of the free end, a F_y somewhere between those calculated from shearing with parallel tools in one and two clamped configuration can be anticipated. Further, when the shearing continues, the sheared edge of the free end slides along the upper tool further away from the shearing arc of the tool and acts as a support that constrain the free end. For the two largest clearances shown in Fig. 10a, the fracture initiation is followed by a slight drop in F_y. This can be due to a tearing effect where already sheared material inclines along the v-shaped tool and a stress concentration around the crack tip makes the sheet fracture at a smaller tool penetration once the crack has been initiated; on the other hand, the bending itself requires additional force although probably negligible for the small rake angles used here. The bending force might be of importance, however, when shearing with such large rake angles as 10° and 25° as in the study by Atkins (1990). The friction force component, μF_x, can also cause the drop in F_y. An increase of F_y, as seen for the two smallest clearances, can be due to an increasing strain rate as the tool movement is accelerating. The acceleration is not unique for the small clearances, but at small clearances, the strain rate is already higher because of a more concentrated deformation.

The maximum in F_x followed by a decrease to a more stable value, apparent in Fig. 10a, is most probably an effect of the position of the contact between the upper tool and the free strip end. Shearing exposes the free strip end to a torque that attempts to rotate the strip. When the leverage, i.e., the distance from the shearing arc of the tool to the point of contact between strip edge and tool, increases, F_x decreases.

Conclusions

Highly accurate measurements of force and tool displacement during shearing were possible to implement with a well defined and stable experimental procedure. Various rake angles and clearances were used for shearing of steel sheets. Furthermore, flow stress curves in three directions of the sheet metal were recorded in tension and compression, to describe the anisotropy of the material and to allow future models of shearing to conform to the experimental conditions and thus be validated against the measured shearing forces. Based on these experiments the following conclusions are made:

- The good agreement between calculated and measured forces when shearing with angled tools suggests that the force intensity is the same as for parallel tools. Thus, more importantly, results from plane strain finite element simulations are also applicable to shearing with rake angles.
- A suitable range of clearances based on forces, energies and sheared surfaces was $0.10h$ to $0.20h$ for shearing with one clamp (parallel and angled tools) and $0.15h$ to $0.40h$ for shearing with two clamps (parallel tools). Consequently, larger clearances could be used when shearing with two clamps than with one clamp.
- The experimental set-up had large stiffness and stability in the clearance direction (x-direction) but, by design, stored large amounts of elastic energy in the shearing direction (y-direction). When shearing with angled tools, a larger stiffness in the y-direction may reduce the effects of stepwise fracture.

Acknowledgements
This work was performed at Dalarna University within the Swedish Steel Industry Graduate School, with financial support from Högskolan Dalarna (Dalarna University), Jernkontoret (Swedish Steel Producers' Association), Länsstyrelsen i Gävleborg (County Administrative Board of Gävleborg), Region Dalarna (Regional Development Council of Dalarna), Region Gävleborg (Regional Development Council of Gävleborg), Sandvikens kommun (Municipality of Sandviken), and SSAB. Carl-Axel Norman is acknowledged for building the experimental set-up and preparing the material. SSAB provided the shearing tools and sheet metal samples.

Authors' contributions
EG conducted the experiments, collected the data and wrote the major part of the paper. MO and LK are EG's PhD supervisors, who guided and supported this work and contributed with their expertise and advices. All authors read and approved the final manuscript.

Competing interests
The authors declare that they have no competing interests.

Author details
[1]Dalarna University, SE-791 88 Falun, Sweden. [2]Luleå University of Technology, SE-971 87 Luleå, Sweden.

References
Atkins, A.G. (1981). Surfaces produced by guillotining. *Philosophical Magazine A*, *43*, 627–641.

Atkins, A.G. (1990). On the mechanics of guillotining ductile metals. *Journal of Materials Processing Technology*, *24*, 245–257.

Atkins, T. (2009). *The science and engineering of cutting*: Butterworth-Heinemann, Elsevier. ISBN: 075068531X, 9780750685313.

Atkins, T. (2012). Perforation of metal plates due to through-thickness shearing and cracking. optimum toughness/strength ratios, deformation transitions and scaling. *International Journal of Impact Engineering*, *48*, 4–14.

Chang, T.M., & Swift, H.W. (1950). Shearing of metal bars. *Journal of the institute of metals*, *78*, 119–145.

Crane, E.V. (1927). What happens in shearing metal. *Machinery*, *30*, 225–230.

Guimaraes, A. (1988). Back to basics with shearing. *Sheet Metal Industries*, *65*(1), 8–9.

Gustafsson, E., Oldenburg, M., Jansson, A. (2014). Design and validation of a sheet metal shearing experimental procedure. *Journal of Materials Processing Technology*, *214*(11), 2468–2477.

Gustafsson, E., Oldenburg, M., Jansson, A. (2016). Experimental study on the effects of clearance and clamping in steel sheet metal shearing. *Journal of Materials Processing Technology*, *229*, 172–180.

Hambli, R., Potiron, A., Kobi, A. (2003). Application of design of experiment technique for metal blanking processes optimization. *Mécanique & Industries*, *4*(3), 175–180.

Hilditch, T.B., & Hodgson, P.D. (2005). Development of the sheared edge in the trimming of steel and light metal sheet: part 1—experimental observations. *Journal of Materials Processing Technology*, *169*(2), 184–191.

Hubert, C., Dubar, L., Dubar, M., Dubois, A. (2010). Experimental simulation of strip edge cracking in steel rolling sequences. *Journal of Materials Processing Technology*, *210*(12), 1587–1597.

Izod, E.G. (1906). Behaviour of materials of construction under pure shear. *Proceedings of The Institution of Mechanical Engineers*, *70*, 5–55.

Matsuno, T., Nitta, J., Sato, K., Mizumura, M., Suehiro, M. (2015). Effect of shearing clearance and angle on stretch-flange formability evaluated by saddle-type forming test. *Journal of Materials Processing Technology*, *223*, 98–104.

Oldenburg, P.E. (1980). Study in shearing. *Machine and Tool Blue Book*, *75*, 77–90.

Saanouni, K., Belamri, N., Autesserre, P. (2010). Finite element simulation of 3d sheet metal guillotining using advanced fully coupled elastoplastic-damage constitutive equations. *Finite Elements in Analysis and Design*, *46*(7), 535–550.

Suliman, S.M.A. (2001). An experimental investigation of guillotining of aluminum alloy 5005. *Materials and Manufacturing Processes*, *16*(5), 673–689.

Weaver, H.P., & Weinmann, K.J. (1985). A study of edge characteristics of sheared and bent steel plate. *Journal of Applied Metalworking*, *3*(4), 381–390.

Wu, X., Bahmanpour, H., Schmid, K. (2012). Characterization of mechanically sheared edges of dual phase steels. *Journal of Materials Processing Technology*, *212*, 1209–1224.

4

Synthesis and characterisation of hot extruded aluminium-based MMC developed by powder metallurgy route

Sambit Kumar Mohapatra and Kalipada Maity[*]

Abstract

Background: Improvement of the mechanical and tribological properties due to extrusion can be attributed to the improved density and excellent bond strength due to high compressive stress.

Methods: To avoid the product defects mathematically contoured cosine profiled die was used for the thermomechanical treatment. Improvement of mechanical characterization like density, hardness, compression test and three-point-bend test was inquired. Two body dry sliding wear behaviour of the prepared AMCs before and after extrusion were investigated by using a pin-on-disc wear testing method by varying the variable parameters like load (N), track diameter (mm) and RPM of the counter disc.

Results: The effect of hot extrusion on mechanical and tribological characteristics of aluminium matrix composites (AMCs) developed by powder metallurgy route followed by double axial cold compaction and controlled atmospheric sintering was studied.

Conclusion: Shearing of the fine distributed graphite particles at the tribosurface acts as a solid lubricant and decreases wear rate. At higher loading and sliding velocity condition a mixed type of wear mechanism was observed.

Keywords: AMCs, Extrusion, Density, Wear, P/M

Background

Aluminium metal matrix composites are the most versatile replacement of other alloys in the sector of automotive, aerospace, defence and sports because of its high strength to weight ratio, ease and prevalence of processing techniques, excellent thermal and electrical conductivity and the ability to sustain in uncertain thermal and mechanical loading environment. This emerging area has influenced the researchers to tailor the mechanical, thermal and tribological properties of the composite using different types of reinforcements with various percentages with types of manufacturing process (Yigezu et al. 2013). The endless process of pursuance of mankind needs the material to behave substantially in critical environments.

Over the last few decades, there has been considerable attention to the evolution of Al-based MMCs developed by powder metallurgy (P/M) route of manufacturing. The main advantage of this kind of manufacturing process is the good distribution of reinforcing particles, low processing temperature and the ability to produce near net shape products with intricate designs (Min et al. 2005; Torralba et al. 2003). A number of studies have been conducted to study the effects of reinforcement of very hard metals as well as ceramics in different grades of aluminium series of powder matrix (El-Kady and Fathy 2014; Jabbari Taleghani et al. 2014; Abdollahi et al. 2014). Layers of oxide formation take place in the P/M specimen during sintering, as aluminium is highly prone to oxide formation. During thermo-mechanical treatments, the covered oxide layer breaks due to highly induced shear stress, leading to a strongly bonded microstructure and improved mechanical properties which eliminate the main drawback of AMCs (Schatt et al. 1997). The use of traditional shear-

* Correspondence: kpmaity@gmail.com
Department of Mechanical Engineering, National Institute of Technology, Rourkela, Odisha 769008, India

Table 1 Detailed composition

Specimen	Composition
Type 1	Al (92.33) + Mg (4.26) + Gr (0.85) + 1.70 Zn
Type 2	Al (92.33) + Mg (4.26) + Gr (0.85) + 1.70 Ti

faced die in extrusion causes product defects owing to the existence of higher velocity relative difference at die exit (Zhang et al. 2012a; 2012b). The use of mathematically contoured die (preferably zero, die entry and exit angle) for the MMC extrusion is highly recommendable.

Ravindran et al. (2012; 2013) investigated the effect of graphite addition, applied load, relative velocity and sliding distance on the wear behaviour of aluminium-based P/M composite. Fine graphite reinforcement acts as a solid lubricant and prevents metal to metal contact so it improves wear resistance compromising with hardness and flexural strength or fracture toughness (Baradeswaran and Perumal 2014; Suresha and Sridhara 2010). The addition of zinc with aluminium improves hot extrudability but decreases the high-temperature performances. Considering these factors, the weight percentage of reinforcements kept less in this work. Improved amount of TiC causes a marginal increase in wear rate, whereas applied load and wear rate varies linearly (Gopalakrishnan and Murugan 2012). Anilkumar et al. (2011) investigated the mechanical properties of the fly ash reinforced aluminium alloy. Improvement of mechanical properties can be achieved by adding more amount of reinforcement by compromising with ductility.

The present investigation focuses on both mechanical and tribological properties of extruded AMC synthesised by powder metallurgy route. Two types of metal reinforcements like Zn and Ti in Al + Mg + Gr matrix has been added for the comparative study. The properties of AMCs before and after thermo-mechanical treatment (extrusion) through a mathematically contoured cosine die were inquired.

Methods

Sample preparation

Aluminium, magnesium and graphite in weight percentages of 92.33, 4.26 and 0.85, respectively, were blended for the matrix composition. Two reinforcements, Zn and Ti were added in the mixture for preparing two types of sample. The compositional details of two specimens were tabulated in Table 1. The physical characterisation of the powders like size, shape and flowability was studied. Particle size and shape was analysed by SEM images.

The mixture was allowed for uniform blending in a centrifugal blender. The weight ratio of stainless steel ball to powder was maintained 10:1. At an RPM of 200 for 10 h, the mixture was allowed for blending. The flow property of the blended powders was checked by measuring apparent density and tap density of the blended compositions.

The powder was subjected to dual axial compression for the preparation of green pellets. During compaction, the powder inside the container remains in floating condition in between both of the punches. The powder was subjected to a pressure of 275 MPa with a very slow rate of rise and with a dwell period of 10 min. Green density of the prepared 10-mm-diameter pellets was measured. Green pellets were subjected to sintering in a controlled atmospheric tubular furnace in an argon atmosphere. The ramp rate of 5 °C/min was set for all the temperature rises. Dwell period of 20 min at 110 °C to remove water vapour, 30 min at 450 °C to remove lubricant (zinc stearate) and 90 min at 590 °C to form the metallic bond was set for the process.

Secondary processing

The 10-mm-diameter pellet prepared by cold compaction followed by sintering was subjected to thermo-mechanical treatment (hot extrusion) as a secondary processing. The experiment was conducted for 50% reduction of the cross-sectional area of the pellet at an

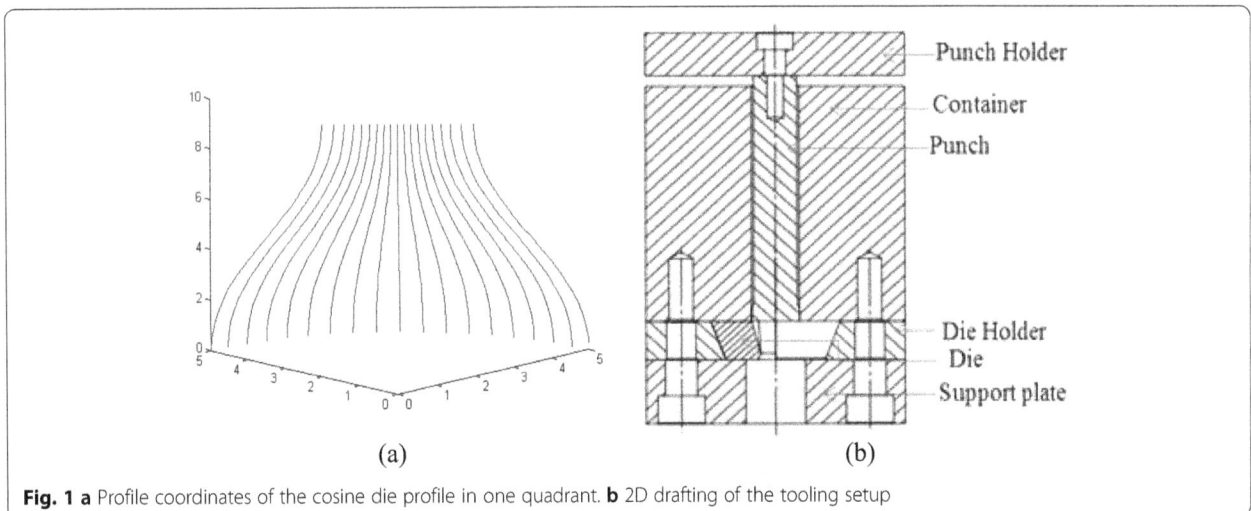

(a)   (b)

Fig. 1 a Profile coordinates of the cosine die profile in one quadrant. **b** 2D drafting of the tooling setup

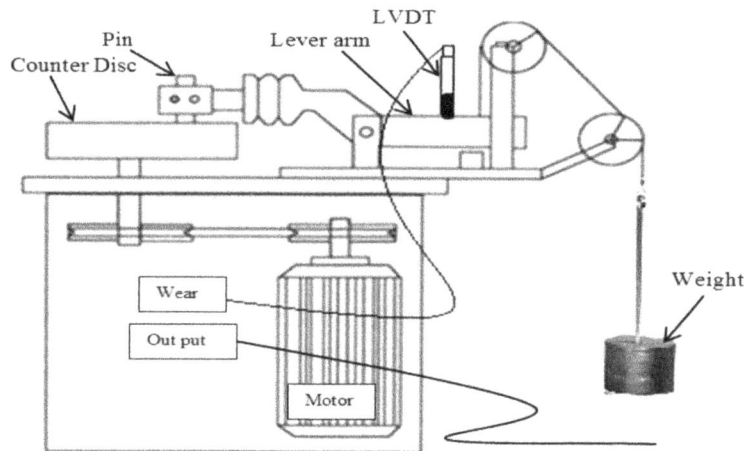

Fig. 2 Schematic layout of pin-on-disc wear testing apparatus

operating temperature of 400–450 °C with a ram rate of 3 mm/min. Extrusion through shear-faced die causes severe surface defects like surface crack and tearing in the extruded product due to velocity relative difference at the die exit. The defects are more prone in the case of extrusion of MMCs synthesised by powder metallurgy route (Prasad et al. 2001). Hence, a specially designed mathematically contoured cosine die from round to square bar extrusion was used for extrusion to avoid severe velocity relative difference. The coordinates of developed profile in one quadrant and the 2D drafting of the extrusion tooling setup are shown in Fig. 1.

Mechanical and tribological characteristics

Theoretical density, tap density, green density, sintered density and extruded density were studied by following the standard procedures. Theoretical density was measured by following the relation mentioned in Eq. 1.

$$\rho_{\text{Theoretical}} = \sum (\rho_i \times m_i) \tag{1}$$

ρ_i is the density of individual element and m_i is the mass fraction of the individual element.

The density of the solid-sintered as well as extruded pellets were measured by following Archimedes' principle. The relation for calculating the density is presented in Eq. 2.

$$\rho_{\text{sample}} = \frac{W_A \times \rho_{\text{fluid}}}{W_A - W_{\text{fluid}}} \tag{2}$$

where W_A is the mass of the sample taken at atmospheric air, W_{fluid} is the mass of the considered fluid and ρ_{fluid} is the density of the considered fluid.

Vickers micro-hardness of the sintered MMC composite was determined by dividing the applied load to the impressed area. 50 g of load were applied through a diamond pyramid having the face angle of 136° for a dwell period of 15 s to avoid spring back effects.

Three-point bend test has been performed to check the transverse rupture strength (TRS) of the sintered solid cylindrical specimen. The test was executed in UTM (universal testing machine, Instron -5979). A span of 30 mm with a compression rate of 2 mm/min at atmospheric temperature was maintained at the time of operation.

Wear testing

Dry sliding wear characteristics of the prepared aluminium MMC was studied by pin-on-disc wear testing apparatus. The pins with dimension Ø 10×25 mm were prepared with flat contact surface with smooth corners and kept stationary in the sample holder perpendicular to the counter disc. The rotating EN-31 disc of 60 HRC and average surface roughness value of 2 μm (Ra) was acted as the counter body. The counter disc and pin surface were cleaned with acetone before the experimentation. A normal load was applied on the MMC pellet through the specimen holder by a lever attachment. The line diagram of wear testing mechanism is shown in Fig. 2. The variable parameters like track diameter, normal load and RPM of the disc are well facilitated by the machine set-up, and all are needed to be fixed manually

Table 2 Variable parameters selected for experimentation

Variable parameters	Level 1	Level 2	Level 3
Wear track dia (D), (mm)	50	70	90
Normal load (L), (N)	40	60	80
RPM of counter disc (N)	200	400	600

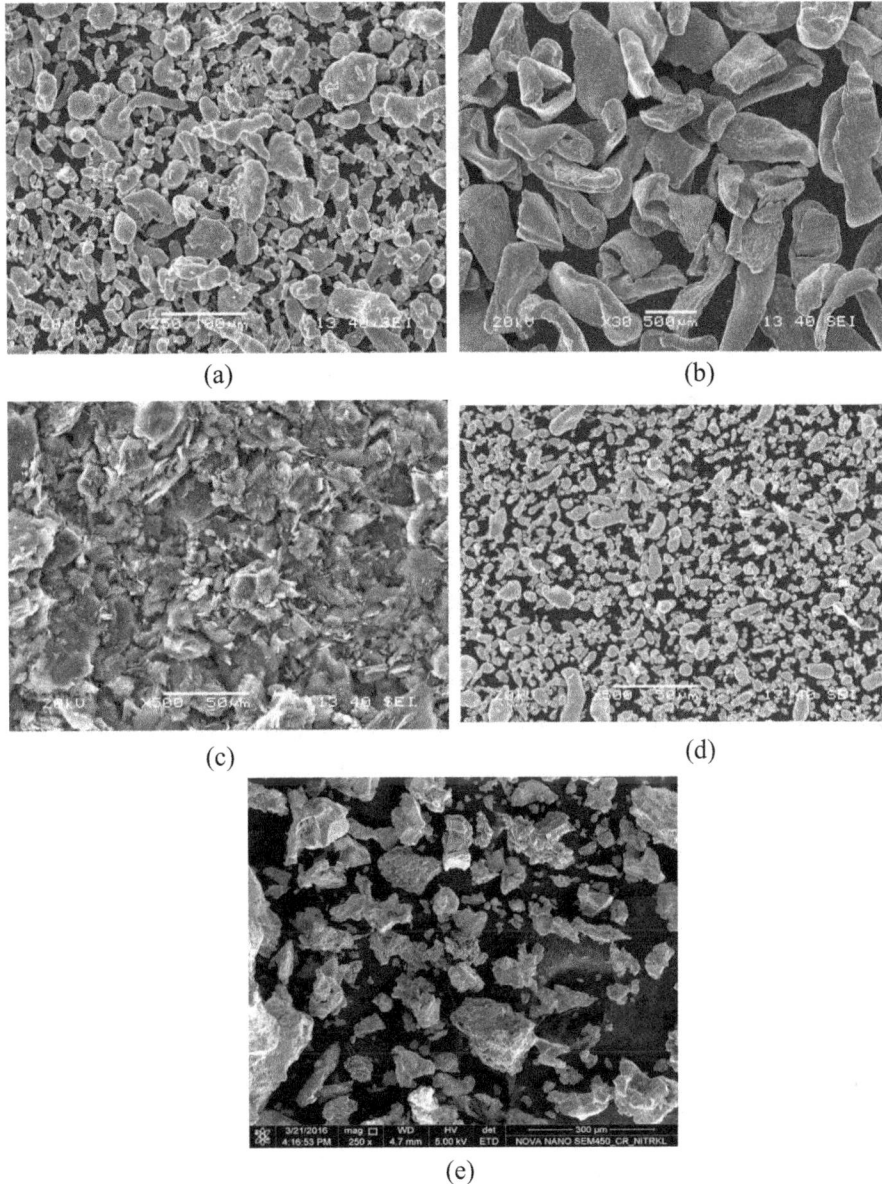

Fig. 3 SEM image of **a** aluminium, **b** magnesium, **c** graphite, **d** zinc and **e** titanium powder

anterior to experimentation. The variable parameters chosen for the wear analysis are tabulated in Table 2. Ten minutes of test duration was adopted for each experiment. With an accuracy of 0.1 mg, the MMC pin (specimen) was weighed before and after the wear operation for determining the wear loss.

Results and discussion
Physical characteristics
Flowability of the powder is directly influenced by physical properties as well as environmental conditions of the powders. The mechanical and tribological properties of the final product are directly related to flowability.

Table 3 Physical characteristics of powders

Powder	Supplier	Average size (µm)	Particle shape	Purity/assay (%)
Aluminium	Loba Chemie	45	Spherical and sub-rounded	98.0
Magnesium		140	Flakey	99.0
Graphite		20	Rounded and flakey	98.0
Zinc		22	Spherical and sub-rounded	98.0
Ti		85	Very angular and irregular	98.0

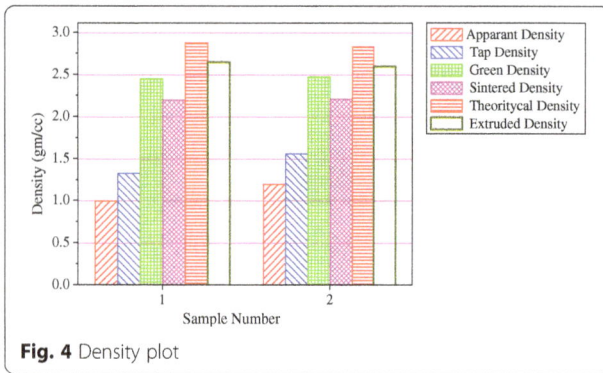

Fig. 4 Density plot

Table 4 Porosity analysis

Sample	Sample 1	Sample 2
Densification factor	−0.597	−0.742
Percentage improvement in densification (after extrusion)	20.32	17.35
Porosity before extrusion (%)	23.7	22
Porosity after extrusion (%)	8.15	8.53

$$\text{Densification parameter} = \frac{\text{(sintered density–green density)}}{\text{(theoretical density–green density)}}$$

(3)

Hence, there is a great importance for the study of physical characterisation such as size, shape and densities of the powders. Among the three most popular techniques used for powder size determination like microscopy, laser diffraction and sieve analysis, microscopy and image analysis were performed for determining the size and shape of the powders. The SEM images of all the powders are shown in Fig. 3.

Image analysis software package is employed for the analysis of particle size and shape of the powders. The detailed report is presented in Table 3.

Density analysis

Apparent density/bulk density and tap density were recorded from direct measurement. It is depicted from the graph shown in Fig. 4 that 30–35% improvement of density caused by tapping the blended powder. The green density of the pellet primarily depends on the compaction pressure. It was found a good consolidation of metallic particles even in green specimens at 275 MPa. The density was calculated by dividing the measured volume (with the accuracy of ±2%) with the measured mass (with the accuracy of 0.001 g). The non-dimensional densification parameters were calculated by the following relation illustrated in Eq. 3 (Padmavathi et al. 2011).

Positive densification parameters indicates shrinkage and negative for growth or swelling. All the samples are showing swelling behaviour at the time of sintering. The percentage improvement in densification by thermo-mechanical treatment (extrusion) is calculated by the following relation illustrated in Eq. 4.

$$\frac{\text{Percentage improvement}}{\text{in densification}} = \frac{\text{extruded density – sintered density}}{\text{sintered density}}$$

(4)

There is an increase of 15–20% of density after extrusion with 50% reduction evident in Fig. 4. The calculated values of densification factor, improvement of density and porosities are presented in Table 4. The density of sample type 1 is higher because of higher zinc density.

Compression test of sintered specimen

Compression test of all the two types of pellets was conducted by UTM (Instron-setec series) with the ram rate of 3 mm/min at room temperature. From the output results stress versus strain, curve is plotted in Fig. 5a, b. The average ultimate stress of the two samples is 409, 381 MPa for samples 1 and 2, respectively. The strength of sample type 1 is on the higher side which can be

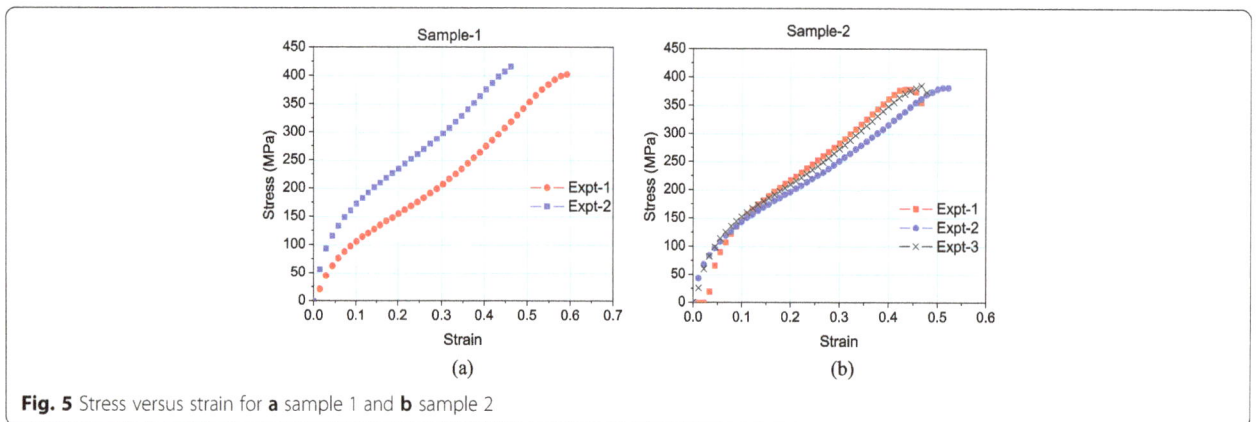

Fig. 5 Stress versus strain for **a** sample 1 and **b** sample 2

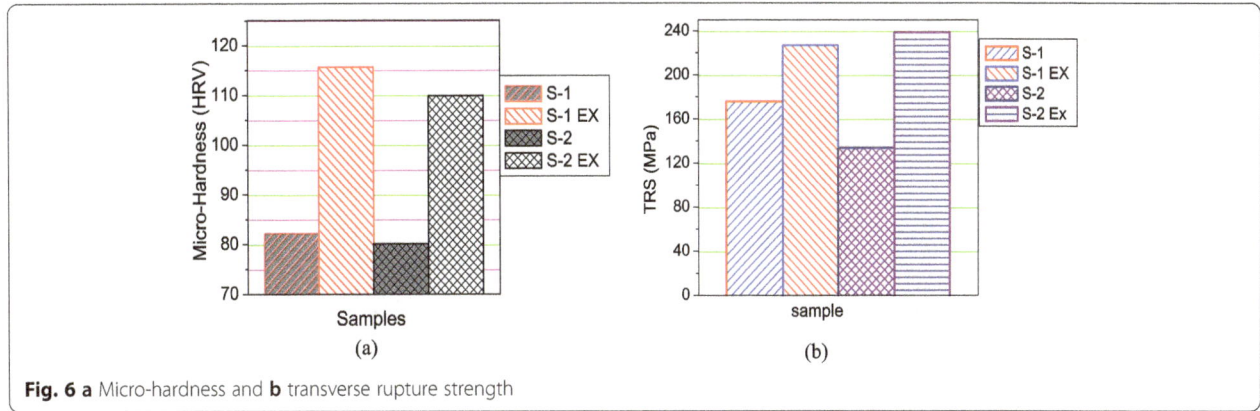

Fig. 6 a Micro-hardness and **b** transverse rupture strength

Micro-hardness

The average micro-hardness was calculated from the ten readings taken for both sintered and extruded products. Figure 6a shows there exists 35–40% of improvement of hardness in the material after extrusion prepared by the above procedure. It can also be attributed to the higher dislocation density around the reinforcement particles due to the difference of mechanical property and thermal mismatch (Ozdemir and Toparli 2003; Ramesh et al. 2011). The mismatch of the properties between the matrix and reinforcements causes the storage of large internal and thermal stress and engenders improved mechanical properties.

attributed to the excellent bond strength due to liquid phase sintering.

3-point bend test

The TRS in MPa found of the specimens was estimated for sintered as well as extruded specimen by following the relations mentioned in Eqs. 5 and 6, respectively.

$$\text{TRS} = 8Pl/\Pi D^3 \tag{5}$$

$$\text{TRS} = 3Pl/2d^3 \tag{6}$$

where P = the maximum load (N)

l = length of the sample (mm)

D = diameter of the sintered specimen (mm)

d = depth = width of the extruded square specimen (mm)

The average TRS for all sintered and extruded sample is presented in Fig. 6b. For the case of sample type 1,

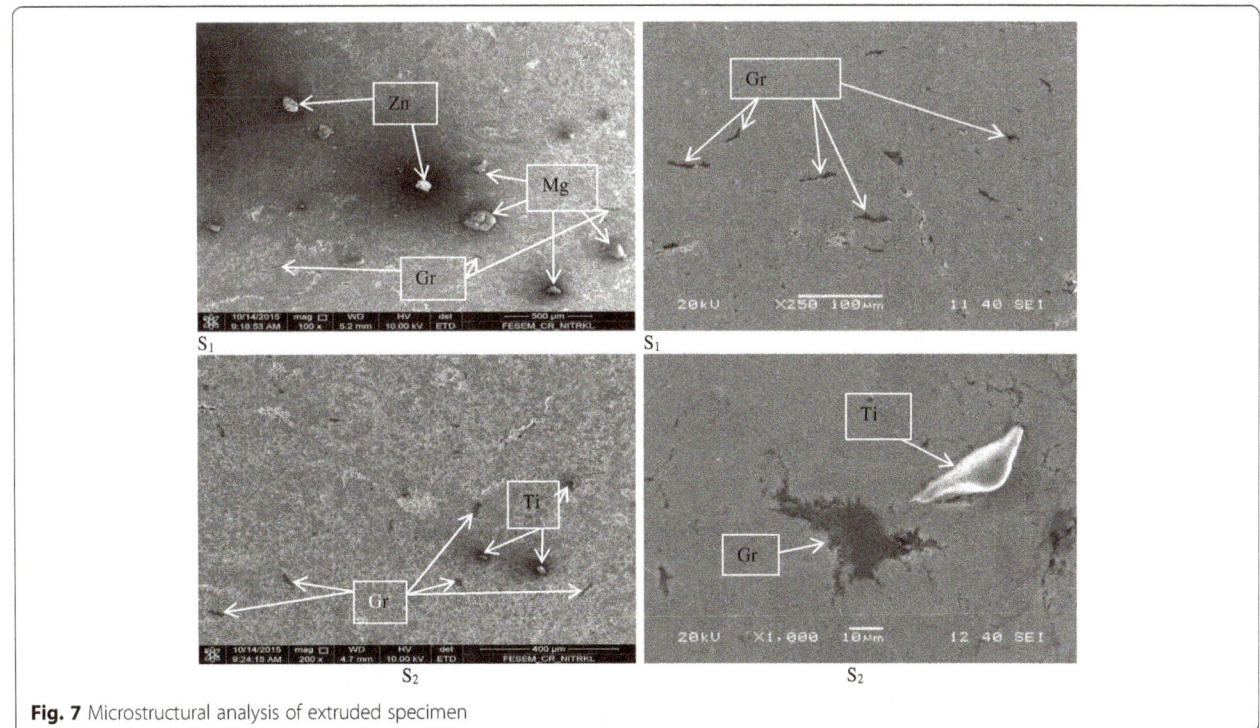

Fig. 7 Microstructural analysis of extruded specimen

presence of zinc having a melting point of 420 °C caused liquid phase sintering at a temperature of 600 °C. In the case of Ti-reinforced sample, there exists a high-stress concentration at the boundary zone of the reinforcements and causes the initiation of fracture so comparatively lesser than TRS.

Microstructural analysis

Figure 7 depicts the scanning electron micrographs of the two types of specimens after the thermo-mechanical treatment at different magnifications. A very little amount of porosity still remains in the material after extrusion which is evident from the figure below. After extrusion operation, the improved density and decreased volume of porosity is illustrated in Fig. 4. The distribution of reinforced particles in the extruded product is uniform and it also shows the good dispersion of reinforcements in matrix elements. There exists a very fine distribution of graphite and other reinforcements like Mg, Ti and Zn. There is no observation of rupture of the reinforcing particle (Ti) in sample type 2 due to the compressive pressure exhibited by extrusion. Also, there is a good bond that exists in between Ti and the matrix element due to its very angular or irregular shape.

Wear test

Considering the aforementioned three variables with three levels, an orthogonal array (Table 5) has been designed for experimentation of extruded and sintered samples. The variation of wear rates of sintered, as well as extruded specimen for two types of composites, are presented in Figs. 8 and 9. Mass loss of the pin was estimated from the recorded mass of the sample before and after wear test. The mass loss the test is defined as (Soltani et al. 2014):

$$\Delta w = (w_a - w_b) \tag{7}$$

where w_a and w_b are the mass of the sample before and after the test, respectively.

Table 5 L9 orthogonal array followed for wear analysis

Run	Load (N)	Track dia (mm)	RPM of counter disc
1	20	50	200
2	20	70	400
3	20	90	600
4	40	50	400
5	40	70	600
6	40	90	200
7	60	50	600
8	60	70	200
9	60	90	400

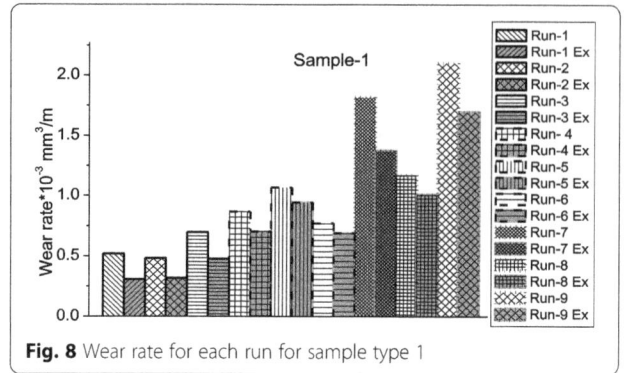

Fig. 8 Wear rate for each run for sample type 1

The volume loss of the AMCs is estimated as the following relation:

$$\text{Volume loss (mm}^3) = \text{Mass loss (g) / Density (g/mm}^3) \tag{8}$$

The volumetric wear rate W_r (mm³/Km) was estimated by following the relation

$$W_r = \text{Volume loss (mm}^3) / \text{Sliding distance (Km)} \tag{9}$$

Several researchers have reported the direct proportionality relation of hardness with wear resistance. The bond strength between the matrix and reinforcement material improves after extrusion which improves wear resistance, and it also avoids three-body abrasive wear (Ramesh et al. 1992).

Wear microscopy

SEM photomicrographs of the AMCs and extruded specimen for run-2 and 7 are shown in Fig. 10. It is observed there is lesser extent of grooving for extruded specimen than the AMCs. The grooves of the unextruded AMCs are coarse and heavy plastic deformation occuring at the groove linings. Also, granular and flake-like debris are observed around the grooves.

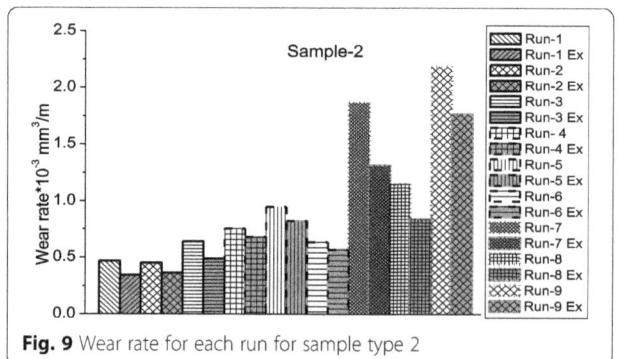

Fig. 9 Wear rate for each run for sample type 2

At higher loading conditions of 40 and 60 N with higher RPM of 600 grooving and scratching plays a predominant role over abrasion. The images showing a large amount of white particles present at the tribosurface which can be attributed to the oxidation of the surface due to frictional heating as the aluminium surface is oxide-prone.

At very high loading and high-velocity condition delamination and combination of abrasion, delamination and adhesion mechanism of wear came into the picture.

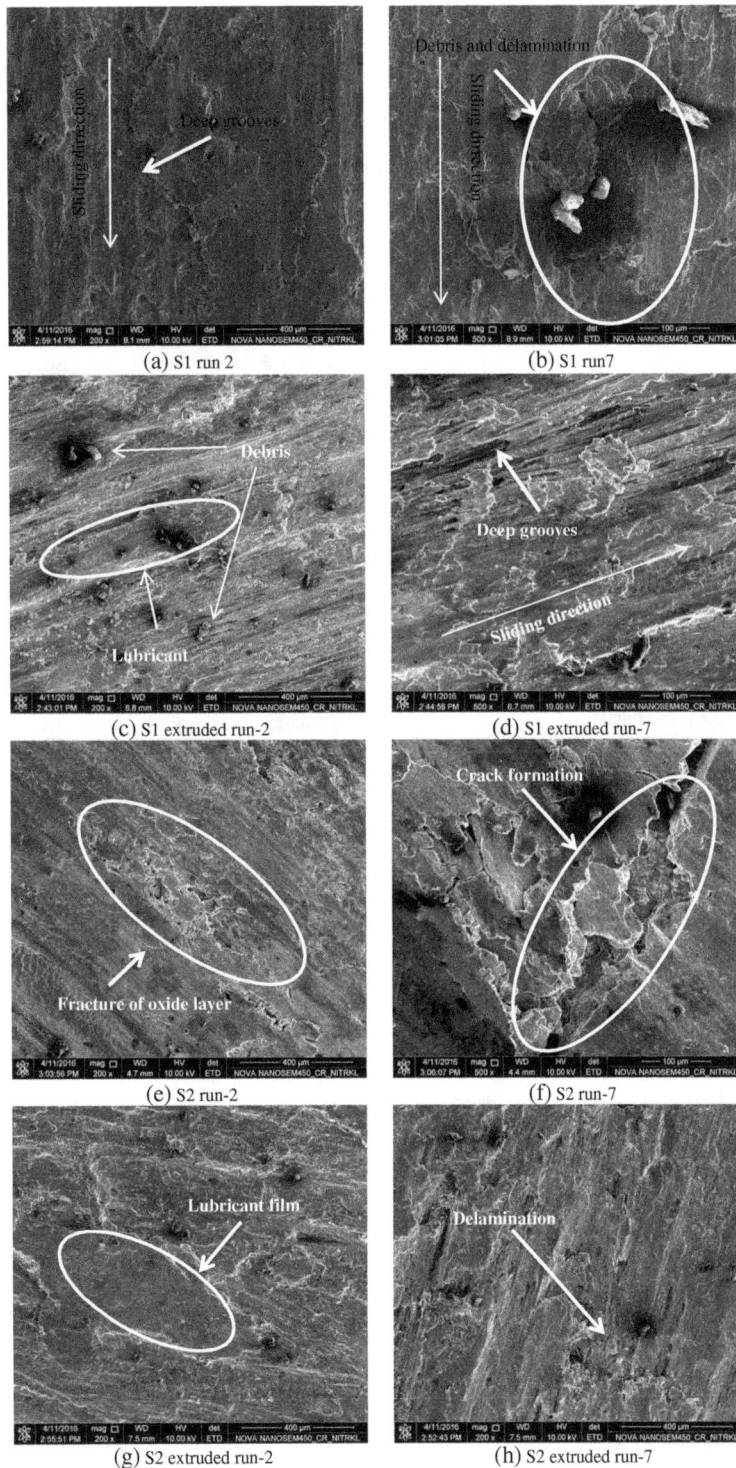

(a) S1 run 2

(b) S1 run7

(c) S1 extruded run-2

(d) S1 extruded run-7

(e) S2 run-2

(f) S2 run-7

(g) S2 extruded run-2

(h) S2 extruded run-7

Fig. 10 SEM micrographs of worn surfaces

Due to frequent repetitive sliding behaviour, subsurface crack has been induced due to the fatigue failure of the pin. These subsurface cracks grow with increasing travel distance, and eventually, shear deformation occurs to the surface. Moreover, at the adverse conditions, melting, thermal softening and adhesion take the predominant role to cause plastic deformation. In the case of AMCs, the mechanism of wear is less severe than the base metal alloys. Metal/graphite composite forms a lubricating layer on the tribosurface due to the shearing of graphite particles which prevents the metal to metal contact, causing the reduction of friction and wear.

Conclusions

The effect of extrusion on the improvement of the mechanical and tribological properties of the AMCs was investigated. The results of this investigations are summerised as follows:

(a) Mechanical properties of both types of AMCs are improved due to the improved bond strength after extruding it through mathematically contoured cosine profiled die.

(b) It was found very less amount of surface defects (few cracks at the corner zone) in the extruded product and supports improvement of flexural strength and tribological properties. Wear rate of the extruded specimen is lesser compared to that of the sintered specimen for each run.

(c) The addition of zinc causes liquid phase sintering because of its low melting temperature that leads higher bond strength and density and gives a comparative better performance for the composition. The addition of more amount of fine titanium may improve the properties manifold by compromising with thermal conductivity.

(d) Shearing of the well-distributed graphite particles at the tribosurface acts as a lubricant. So, the addition of graphite particle improves the wear resistance by compromising the little amount of hardness.

(e) At higher loading and sliding velocity condition, a mixed type of wear mechanism (oxidative, delamination, adhesive and abrasion) takes place. But oxidative and delamination are the predominating wear mechanism found on the surface.

Authors' contributions
All authors read and approved the final manuscript.

Competing interests
The authors declare that they have no competing interests.

References
Abdollahi, A., Alizadeh, A., & Baharvandi, H. R. (2014). Dry sliding tribological behavior and mechanical properties of Al2024–5 wt.%B4C nanocomposite produced by mechanical milling and hot extrusion. *Materials & Design, 55,* 471–481.

Anilkumar, H., Hebbar, H., & Ravishankar, K. (2011). Mechanical properties of fly ash reinforced aluminium alloy (Al6061) composites. *International Journal of Mechanical and Materials Engineering, 6*(1), 41–45.

Baradeswaran, A., & Perumal, A. E. (2014). Wear and mechanical characteristics of Al 7075/graphite composites. *Composites Part B: Engineering, 56,* 472–476.

El-Kady, O., & Fathy, A. (2014). Effect of SiC particle size on the physical and mechanical properties of extruded Al matrix nanocomposites. *Materials & Design, 54,* 348–353.

Gopalakrishnan, S., & Murugan, N. (2012). Production and wear characterisation of AA 6061 matrix titanium carbide particulate reinforced composite by enhanced stir casting method. *Composites Part B: Engineering, 43*(2), 302–308.

Jabbari Taleghani, M. A., Ruiz Navas, E. M., & Torralba, J. M. (2014). Microstructural and mechanical characterisation of 7075 aluminium alloy consolidated from a premixed powder by cold compaction and hot extrusion. *Materials & Design, 55,* 674–682.

Min, K. H., Kang, S. P., Kim, D.-G., & Kim, Y. D. (2005). Sintering characteristic of Al2O3-reinforced 2xxx series Al composite powders. *Journal of Alloys and Compounds, 400*(1–2), 150–153.

Ozdemir, I., & Toparli, M. (2003). An investigation of Al-SiCp composites under thermal cycling. *Journal of Composite Materials, 37*(20), 1839–1850.

Padmavathi, C., Upadhyaya, A., & Agrawal, D. (2011). Effect of microwave and conventional heating on sintering behavior and properties of Al–Mg–Si–Cu alloy. *Materials Chemistry and Physics, 130*(1), 449–457.

Prasad, V. B., Bhat, B., Mahajan, Y., & Ramakrishnan, P. (2001). *Effect of extrusion parameters on structure and properties of 2124 aluminum alloy matrix composites.*

Ramesh, C. S., Seshadri, S. K., & Iyer, K. J. L. (1992). A model for wear rates of composite coatings. *Wear, 156*(2), 205–209.

Ramesh, C., Keshavamurthy, R., & Naveen, G. (2011). Effect of extrusion ratio on wear behaviour of hot extruded Al6061-SiC p (Ni–P coated) composites. *Wear, 271*(9), 1868–1877.

Ravindran, P., Manisekar, K., Narayanasamy, P., Selvakumar, N., & Narayanasamy, R. (2012). Application of factorial techniques to study the wear of Al hybrid composites with graphite addition. *Materials & Design, 39,* 42–54.

Ravindran, P., Manisekar, K., Narayanasamy, R., & Narayanasamy, P. (2013). Tribological behaviour of powder metallurgy-processed aluminium hybrid composites with the addition of graphite solid lubricant. *Ceramics International, 39*(2), 1169–1182.

Schatt, W., Association, E.P.M., & Wieters, KP. (1997). Powder metallurgy: processing and materials. European Powder Metallurgy Association. KIT Scientific Publishing ist Mitglied der Arbeitsgemeinschaft der Universitätsverlage und der Association of European University Presses (AEUP).

Soltani, N., Jafari Nodooshan, H. R., Bahrami, A., Pech-Canul, M. I., Liu, W., & Wu, G. (2014). Effect of hot extrusion on wear properties of Al–15 wt.% Mg2Si in situ metal matrix composites. *Materials & Design, 53,* 774–781.

Suresha, S., & Sridhara, B. (2010). Wear characteristics of hybrid aluminium matrix composites reinforced with graphite and silicon carbide particulates. *Composites Science and Technology, 70*(11), 1652–1659.

Torralba, J. M., da Costa, C. E., & Velasco, F. (2003). P/M aluminum matrix composites: an overview. *Journal of Materials Processing Technology, 133*(1–2), 203–206.

Yigezu, B. S., Jha, P., & Mahapatra, M. (2013). The key attributes of synthesizing ceramic particulate reinforced Al-based matrix composites through stir casting process: a review. *Materials and Manufacturing Processes, 28*(9), 969–979.

Zhang, C., Zhao, G., Chen, H., Guan, Y., Li, H. (2012a). Optimization of an aluminum profile extrusion process based on Taguchi's method with S/N analysis. *The International Journal of Advanced Manufacturing Technology, 60*(5), 589–599. doi:10.1007/s00170-011-3622-x.

Zhang, C., Zhao, G., Chen, H., Guan, Y., Kou, F. (2012b). Numerical simulation and metal flow analysis of hot extrusion process for a complex hollow aluminum profile. *The International Journal of Advanced Manufacturing Technology, 60*(1), 101–110. doi:10.1007/s00170-011-3609-7.

5

Surface characterization of austenitic stainless steel 304L after different grinding operations

Nian Zhou[1,2*], Ru Lin Peng[3] and Rachel Pettersson[2,4]

Abstract

Background: The austenitic stainless steel 304L is widely used as a structural material for which the finished surface has significant effect on the service performance. A study of the grinding process with regard to the quality of the ground surfaces is therefore interesting from the point of view of both industrial application and scientific research.

Method: This work investigates the influence of grinding parameters including abrasive grit size, machine power, and grinding lubrication on the surface integrity of the austenitic stainless steel 304L. The induced normal grinding force, grinding surface temperature, metal removal rate, and surface property changes have been investigated and compared.

Results and Conclusion: Using grinding, lubrication significantly enhanced the metal removal rate. Surface defects (deep grooves, smearing, adhesive chips, and indentations), a highly deformed thin surface layer up to a few microns in thickness, and high surface tensile residual stresses parallel to the grinding direction have been observed as the main damage induced by the grinding operations. Surface finish and deformation were found to be improved by using smaller abrasive grits or by using lubrication during grinding. Increasing the machine power increased surface deformation while reducing surface defects. The results obtained can provide a reference for choosing appropriate grinding parameters when machining 304L; and can also help to understand the failure mechanism of ground austenitic stainless steel components during service.

Keywords: Austenitic stainless steel 304L, Grinding, Surface characterization, Microstructure, Residual stress

Background

It is well known that the geometrical, physical, and mechanical properties of the affected surface layer from machining have significant effects on the functional performance of machined components. Service failure related to fatigue or corrosion almost always starts from the surface or near the surface of the components. Compared with mechanical polishing, surface grinding of 304L has been found to induce a higher degree of damage leading to an obvious fatigue life reduction (Poulain et al. 2013); they stated that the surface finish should be taken into account when designing components against fatigue. Stress corrosion cracking in chloride environment has been observed in 304L stainless steel in both machined and ground conditions

without any externally applied loading (Acharyya et al. 2012); the high magnitude of tensile residual stresses and plastic deformation on the surface and sub-surface layers have been proposed as the main factor. A propensity for pitting to initiate at surface defect areas has been demonstrated by (Turnbull et al. 2011), and clear evidence of small embryonic cracks emerging at pits has been observed. Considering that the properties of the surface layers will be largely affected and could be controlled by the machining operation parameters, studies with various materials have been carried out to find the correlation under different manufacturing processes, such as Inconel 718 (Yao et al. 2013), Ti-6Al-4V (Guo et al. 2010), duplex stainless steel 2304 (Zhou, Peng Ling, et al. 2016).

Grinding is mainly regarded as a finishing operation, though it can also be used for bulk material removal. It is a complex cutting process with geometrically unspecified cutting edges (Dieter 1989); the grinding zone involves

* Correspondence: nzh@du.se
[1]Department of Material Science, Dalarna University, SE-79188 Falun, Sweden
[2]KTH, SE-10044 Stockholm, Sweden
Full list of author information is available at the end of the article

contact between an abrasive with a randomly structured topography and the workpiece material. The grinding process is largely influenced by the friction of the interface, the flow characteristics of the material, and the grinding speed (Kopac and Krajnik 2006). Deformation of the workpiece material is mainly introduced by the metal removal process and the rubbing contact from grinding operations (Turley and Doyle 1975). Grinding requires a high specific energy and can generate a high grinding zone temperature (Outwater and Shaw 1952). Surface properties of the components can be largely affected by the selected grinding parameters. Surface roughness, which is the foremost characteristic in surface property issues, have been reported to be significantly affected by the grinding depth and feed rate of Ti-6Al-4V (Guo et al. 2010). The extent of the deformed surface layer has been found to depend on the form of the grit as well as the rake angle for 70:30 brass (Turley and Doyle 1975). High tensile residual stresses have been observed in Inconel 718 ground surfaces by different grinding wheels (Yao et al. 2013). Compressive residual stresses have also been found in medium carbon steel when grinding using miniature monolayer electroplated CBN (cubic boron nitride) wheels (Vashista et al. 2010). The surface integrity of GH4169 was shown to be susceptible to the magnitude of cutting depth (Zeng et al. 2015). All these surface property changes due to the varied grinding parameters may largely affect the materials' performance during application. Thus, there is a need for controlled grinding processes to give a desirable surface finish to components.

However, stainless steels are commonly recognized as materials difficult to machine because of their high toughness, high work hardening rate, and low thermal conductivity. They have been characterized as 'gummy', i.e., rubbery or adhesive, during machining, showing a tendency to produce stringy chips, which degrades the surface finish and reduces tool life (Jang et al. 1996). Austenitic stainless steels are particularly challenging to machine because of their high degree of work hardening and their galling tendency (Boothroyd and Knight 2005). Grindability of four different stainless steels has been compared by (Jiang et al. 1996); high adhesion of alumina particles from alumina wheels was observed on the ground surfaces of PM 2205 and PM 2507 stainless steels. Different grinding operations have been reported to introduce different types and extent of damage in the surface and sub-surface layers of stainless steels, while some beneficial effects have also been reported. Surface burns (Zhang et al. 2012), microcracks, and micro-voids (Jiang et al. 1996) have been observed on ground stainless steel surfaces. Formation of a deformed zone near the surface with highly fragmented grain structure as well as martensitic transformation has been found for ground austenitic stainless steel 304L (Acharyya et al.

2012). Significant improvements in surface roughness and surface defects have been reported by the application of cryogenic cooling when grinding austenitic stainless steel 316 (Manimaran et al. 2014). High tensile residual stresses have shown to be introduced during grinding of austenitic stainless steel 304 (Turnbull et al. 2011), although a lower tensile residual stress was measured when grinding fluid was used (Fredj et al. 2006).

Increasing quality demands for grinding are required with respect to productivity, precision, and cost; thus, surface property issues due to grinding have become a serious cause of concern in manufacturing of stainless steel components. However, little research work has been published on the surface integrity of ground austenitic stainless steel prepared with a well-controlled grinding process. The present work aims to connect the industrial interest with scientific research by studying the grinding behavior and surface properties of austenitic stainless steel 304L. Grinding operations using different abrasive grit sizes, machine powers, and grinding lubrication, have been compared. Normal grinding force, grinding surface temperature, and metal removal rate were measured, surface properties including surface roughness and defects, cross-sectional microstructure, and surface residual stresses have been investigated. The results obtained in this study can be used as a reference to choose more effective grinding parameters for a required surface finishing when fabricating austenitic stainless steel components. The results can also help to understand the failure mechanism of austenitic stainless steels during service.

Methods
Material
The investigated material was 304L (UNS S30403) austenitic stainless steel with a chemical composition: 0.019 C, 0.32 Si, 1.55 Mn, 0.029 P, 0.001 S, 18.22 Cr, 8.11 Ni, 0.011 Nb, 0.31 Cu, 0.16 Co, 0.071 N, and balance Fe (all in wt%). During production, the as-received material had been solution annealed at 1100 °C followed by forced air cooling and water quenching. It was then pickled in an acid bath to remove oxide scale caused by annealing and restore the corrosion resistance. After that, the material was roll leveled to improve the flatness.

The main measured mechanical properties perpendicular to the rolling direction at room temperature are yield strength $R_{P0.2}$ 230 MPa, ultimate tensile strength R_m 642 MPa, elongation 54%, and hardness 170 HB. The material was delivered in the form of test coupons with dimensions $400 \times 150 \times 2$mm. An EBSD mapping showing the microstructure of the material is presented in Fig. 1. Around 1.7% of ferrite was present in the material, which is normal for 304L. The ferrite measurement was done according to ASTM E1245 after etching in

Fig. 1 Cross-section microstructure and grain orientation of 304L austenitic stainless steel in the as-received condition

40% NaOH solution, using 2.5 V for 3 s and was calculated over 10 fields at ×1000 magnification. The fields were chosen from one edge to the other, i.e., throughout the thickness. Less ferrite was seen close to the edge/surface, compared to the middle section/bulk.

Grinding operations

A Chevalier FSG-2A618 grinding machine was used for the grinding operations; the set-up is shown in Fig. 2. As shown in the figure, grinding belts (50 mm in width, 473 mm in length) with conventional aluminum oxide grits were mounted on the grinding wheel and test coupons ($400 \times 150 \times 2mm$) were fixed by screws on the edges to the working table during grinding. The grinding wheel (50 mm in width, 150 mm in diameter) used was from the Kemper Radix Go series, which is an expanding roller made of 20-mm-thick rubber. All grinding operations were conducted along the rolling direction of the material. A fixed grinding speed, $v_s = 23m/s$, and fixed feed rate, $v_w = 8m/min$, were used, based on the recommendation from the material supplier. During grinding, the grinding wheel rotated clockwise, and the working table was moved back and forth; thus, both up and down grinding were performed during the operation. The grinding parameters which were varied were

Fig. 2 Grinding set-up

the abrasive grit size, the machine power, and the grinding lubrication; these are also parameters that can be varied in an industrial grinding process to achieve different grinding results in addition to cutting speed and feed rate. The machine power means a certain percentage of the total motor power, which is 1 kW, is used to drive the grinding belt around. During grinding, a given machine power was used, and the grinding force was adjusted manually by a hand wheel to reach the given machine power. The grinding lubrication used was 3% of Mobilcut 321, which is a synthetic fluid with specific gravity of 1.10 at 20 °C and a pH value of 9.4. During grinding, the lubrication was poured onto the workpiece surface, and the path of the lubrication has been described in detail in a previous paper (Zhou, Peng Ling, et al. 2016).

Three groups of ground samples were prepared to investigate the influence of three grinding parameters, including (I) abrasive grit size, (II) machine power, and (III) grinding lubrication. The detailed grinding parameters and procedures given in Table 1 were selected based on recommendations by the material supplier. Grinding started with a coarse grit (60#) for 5 min to remove the original surface. Further grinding was carried out in steps with finer grit in each step until the desired surface finish had been reached (Outokumpu 2013). For each step, a new abrasive was used and grinding was performed long enough to remove the deformation induced from the previous step.

For the investigation of the influence of the abrasive grit size, 60# (165~405 μm), 180# (25~114 μm), and 400# (11~45 μm) grits were used as the final surface finish, respectively. In this case, 60% machine power was maintained and no grinding lubrication was used. To investigate the machine power effect, the same grit size (180#) was used as the final surface finish and no lubrication was employed. The machine power used was 30%, 60%, or 90%. The study of lubrication influence was performed by grinding with and without lubrication while keeping the final surface finish (180#) and machine power (60%) the same. In order to avoid too large a test matrix, and since grinding without lubricant is more interesting from the perspective of both production cost and environmental issues, the effect of grinding lubrication was restricted to only one condition.

Characterization methods

Since the grinding force was adjusted by the machine power, a piezo-electric transducer-based dynamometer (Kistler 7257B) was mounted under the working table to measure the normal force during grinding operations, as shown in Fig. 2. The Kistler 7257B was connected to a PC using a National Instruments data acquisition device. The measured force values were read manually. Due to

Table 1 Grinding parameters and procedures for each sample

Group no.	Comparison	Grit size	Machine power	Lubrication	Grinding procedures
I	Abrasive grit size	60#	60%	Without	60#(5 min) + 60#(2.5 min)
		180#			60#(5 min) + 180#(5 min)
		400#			60#(5 min) + 180#(5 min) + 280#(5 min) + 400#(5 min)
II	Machine power	180#	30%	Without	60#(5 min) + 180#(5 min)
			60%		
			90%		
III	Lubrication	180#	60%	Without	60#(5 min) + 180#(5 min)
				With	60#(lubrication 5 min) + 180#(lubrication 5 min)

vibrations in the grinding table, the planar forces were difficult to measure accurately. Thus, in the current paper, the measured force is the force acting perpendicularly on the contact zone of grinding and is termed the normal grinding force. A FLIR i5 infrared camera was used to measure the ground surface temperature during operation. The emissivity setting of the camera was 0.95; the measured spots were near the contact area between the workpiece material and the grinding wheel.

A square sample of 30 mm in edge length was taken from as-delivered plate and each ground plate to check the plate thickness, and thus investigate the metal removal by different grinding conditions. Using a Mitutoyo digital indicator (1 μm resolution) mounted on a Mitutoyo granite plate; thicknesses at the four corners of the sample were measured manually. For each corner, 6 repeat measurements were made, i.e. 24 thickness values, and the mean value was taken as the thickness of the plate.

A 3D optical topometer (Wyko NT9100) was used to measure the surface roughness, choosing five points (1.3 × 0.95 mm in area) for each ground sample, and then averaging. To investigate the surface topography and surface defects, a scanning electron microscopy (SEM, FEG-SEM Zeiss Ultra 55) was used in this study.

Electron channeling contrast imaging (ECCI) is a technique that can provide diffraction contrast images to analyze deformation, damage, strain field, or even individual defects in crystalline materials (Johansson et al. 2013). In this study, a Hitachi FEG-SEM SU-70 was used to study the cross-section microstructural development by different grinding operations.

The residual stresses in the surface layer parallel ($\sigma_\parallel$) and perpendicular ($\sigma_\perp$) to the grinding direction were determined by X-ray diffraction. Cr-K_a radiation was used, giving a diffraction peak at $2\theta \sim 128°$ for the {220} lattice planes of the austenitic phase. Peaks were measured at nine ψ-angles (ψ = ±55°, ±35°, ±25, ±15°, 0°). The Pseud-Voigt profile (Hauk 1997) was used to determine simultaneously peak positions and peak width in full width at half maximum (FWHM) for both the K_{a1}

and K_{a2} diffraction peaks, and the results presented in this study are from the K_{a1} diffraction peaks. Residual stresses were calculated based on the $\sin^2\psi$ method (Hauk 1997; Noyan and Cohen 1987) with an X-ray elastic constant of $6 \times 10^{-6} MPa^{-1}$. Controlled electrolytic polishing of an area of 12 mm in RD direction and 15 mm along the TD direction was used on one sample to remove the surface layer in order to measure the residual stresses in the sub-surface layer. No correction has been made for possible stress relaxation due to polishing.

Results and discussion
Grinding force and measured surface temperature
The effects of the grinding parameters on the grinding force and the measured surface temperature are compared in Table 2. The results show that increasing the machine power increases the required normal grinding force. The application of grinding lubrication was observed to reduce the normal grinding force from 100 to 40 N even though the same machine power (60%) was used. A similar effect has also been observed in previous work by the current authors when grinding duplex stainless steel 2304 (Zhou, Peng Ling, et al. 2016). Using lubrication during grinding operations can help to retain abrasive grit sharpness, reduce friction between abrasive and workpiece material, and contribute to a favorable mode of chip forming, thus reducing the normal force (Paul and Chattopadhyay 1996).

As illustrated in Fig. 3, the surface temperatures were measured close to the contact area between the workpiece material and the grinding wheel. Since it is the temperature in the grinding zone that actually affects the ground surface properties and the measured temperature values are influenced by the surface conditions as well as the settings of the infrared camera, the results given here are used only to indicate the tendency of the temperature change for different grinding parameters. Grinding with a higher machine power significantly increased the surface temperature. The increase of machine power directly increased the grinding force and the resulted

Table 2 Measured grinding force and surface temperature by different grinding conditions

Group no.	Comparison	Grinding parameters			Measured normal grinding force (±10 N)	Measured surface temperature
		Final surface finish	Machine power	Lubrication		
I	Abrasive grit size	60#	60%	Without	100 N	60 °C
		180#			100 N	68 °C
		400#			100 N	70 °C
II	Machine power	180#	30%	Without	60 N	50 °C
			60%		100 N	68 °C
			90%		150 N	85 °C
III	Lubrication	180#	60%	Without	100 N	68 °C
				With	40 N	35 °C

higher friction between the abrasive and workpiece material caused an obvious increase in grinding heat. Somewhat counterintuitively, it was seen that grinding with finer final surface (group I) resulted in slightly higher surface temperatures. The most plausible explanation for this is that a finer finish added more steps of grinding procedures and increased the total grinding time. Because heat generated by each grinding step was accumulated, a slight increase in the surface temperature was observed. However, the machine power has a much bigger influence on temperature. The cooling effect with lubrication is notable although the temperature measured in the lubrication condition may possibly be that of the lubricant rather than the metal surface; lubrication can effectively reduce friction between the abrasive grits and the workpiece, and help remove grinding heat (Yao et al. 2013). In general, the measured grinding surface temperature close to the grinding zone is relatively low in this study.

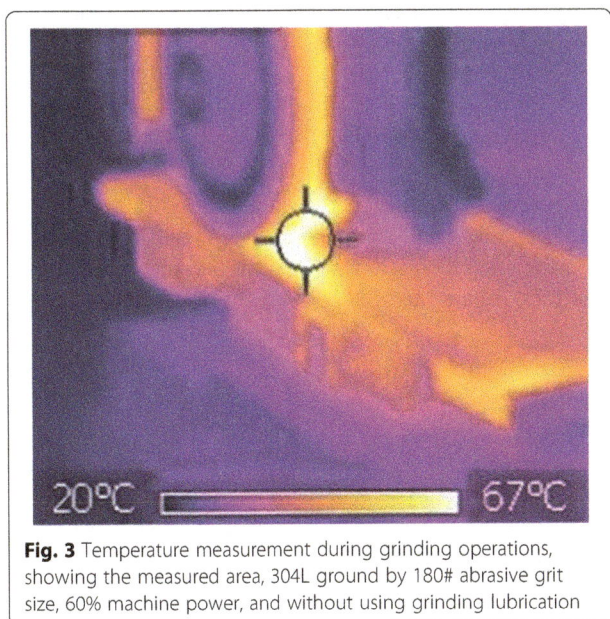

Fig. 3 Temperature measurement during grinding operations, showing the measured area, 304L ground by 180# abrasive grit size, 60% machine power, and without using grinding lubrication

Metal removal

Calculated metal removal results in different grinding conditions as well as standard errors are shown in Fig. 4. The metal removal is calculated according to the following equation:

$$\delta_{\text{metal removal}} = \delta_{\text{as delivered material}} - \delta_{\text{after grinding operation}}$$

$$(1)$$

where $\delta_{\text{metal removal}}$ is the thickness of metal removal due to different grinding operations

$\delta_{\text{as delivered material}}$ is the thickness of as delivered test coupon $\delta_{\text{after grinding operation}}$ is the thickness of test coupons after different grinding operations

Here, the thickness of the test coupons, both in the as-delivered and ground states, is the mean value of the 24 measured thickness values for each plate; the estimated standard errors of random intercept are calculated by the mixed effect model (Cheng 2014), they can be interpreted as production errors, which are only associated with plate thickness variation, the error from the reproducibility was excluded.

As shown from Fig. 4a, both 60# and 180# grit size abrasives have good metal removal ability. 28 μm of metal was ground away by using a 60# grit size abrasive; and 20 μm more in thickness was removed by adding one more step of the grinding process using the 180# grit size abrasive. However, the 280# and 400# grit size abrasives have relatively lower metal removal ability, they are used for the surface finish process; only 8 μm in thickness of metal was ground away by using a 280# grit size abrasive for 5 min plus another 5 min grinding by a 400# grit size abrasive. The machine power also has a large influence on the metal removal behavior. As illustrated in Fig. 4b, when using same grit size abrasives and same grinding procedures, a maximum metal removal rate was seen with the intermediate power (60%). A lower machine power (30%) gave 25 μm of metal removal compared to 48 μm at 60%, at 90% power the

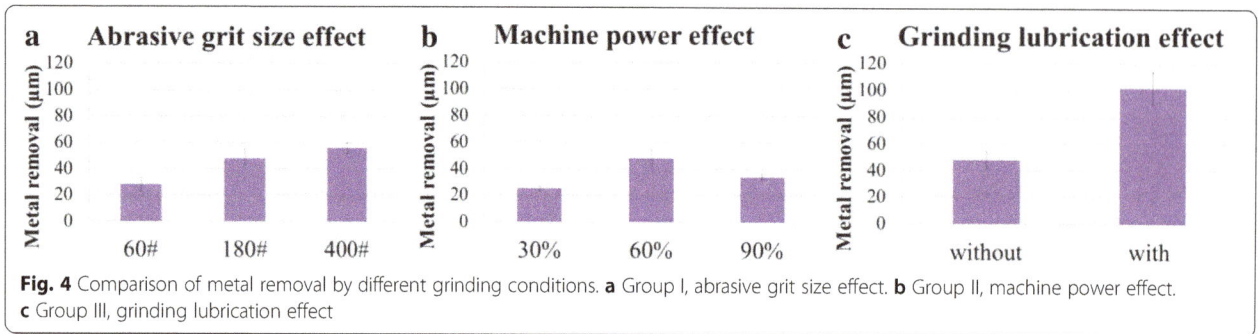

Fig. 4 Comparison of metal removal by different grinding conditions. **a** Group I, abrasive grit size effect. **b** Group II, machine power effect. **c** Group III, grinding lubrication effect

metal removal dropped to 34 µm. Higher machine power introduced higher grinding force. When grinding with a lower machine power (grinding force), the abrasive grits slide and rub over the material surface instead of effective cutting and ploughing. Using a higher machine power (grinding force) increased friction between the abrasive and the workpiece surface as well as wear of abrasive grits. Figure 4c shows the significant improvement of metal removal ability by using grinding lubrication; the removed thickness was more than doubled (from 48 µm increased to 102 µm) by using lubricant. This can be explained by the fact that grinding lubrication can help reducing friction at the contact surfaces, diffusing away heat, retaining sharpness of the abrasive grits, thus induces a favorable metal removal mode (Manimaran et al. 2014; Fredj et al. 2006).

Ground surface roughness and topography

In this study, surface roughness was measured using both R_a and R_z values. Figure 5 gives the results for different abrasive grit size (group I), error bars in the figure are standard deviations calculated from the five measurements for each sample. As shown from the figure, both R_a and R_z values are decreased by using a finer grit size abrasive as the final surface finish. For the coarse (60#) abrasive grit size, a R_a value of 1.81 µm and a R_z value of 18.4 µm were measured. With a finer (180#) grit size, both R_a and R_z values were reduced to around the half (R_a = 0.77 µm and R_z = 10.66 µm), while for the finest (400#) grit size, the R_a value decreased to 0.34 µm and R_z to 5.66 µm.

Figure 6 presents SEM images showing the surface topography and surface defects resulting from using different abrasive grit size (group I). Deep grooving, smearing, adhesive chips, and indentations, shown in Fig. 6a, are the main defects observed on the ground surfaces. Similar defects have also been observed when grinding duplex stainless steel 2304 in the authors' previous work (Zhou, Peng Ling, et al. 2016). The formation of such defects is related to different interactions between the grinding grits and workpiece surfaces. The non-uniform metal removal process, including chip forming and ploughing, introduced deep grooves on the ground surfaces. At the contact zone between the abrasive grit tops and the workpiece surface, material is pushed out and moved across the surface, leading to smearing areas (Totten et al. 2002) in addition to chip formation. The redeposition process (Turley and Doyle 1975) caused adhesive chips; the material was transferred to the grits by adhesion, and then was transferred back to the ground surface by friction welding. Rubbing contact between abrasive grits or cut down chips and workpiece surface caused formation of indentations. As shown from the figure, the surface finish was clearly improved by using finer grit size abrasives. Compared with using 60# grit size abrasive, it was clear that deep grooves, large smearing areas, and adhesive chips were reduced by using a 180# grit size abrasive. Surface defects were reduced even more by using the finest (400#) grit size. For all three samples, surface indentations have been observed.

The abrasive grit size has a major influence on the surface roughness and the surface finish, the improvement by using a smaller grit size abrasive is significant. During grinding, only a small top region of the abrasive grits is

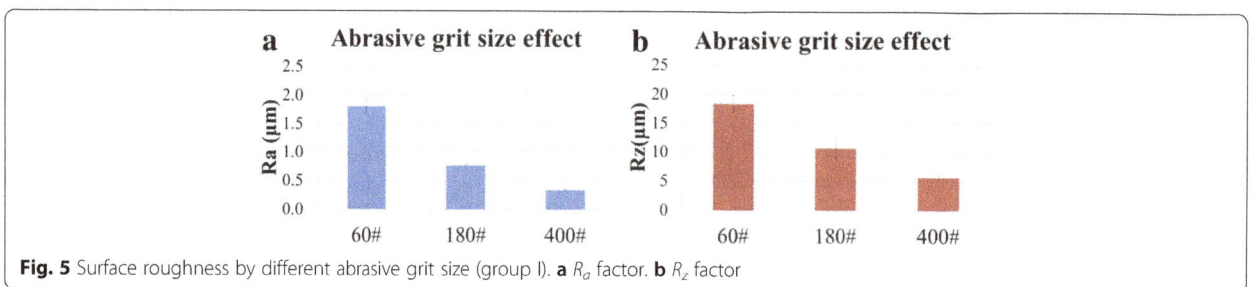

Fig. 5 Surface roughness by different abrasive grit size (group I). **a** R_a factor. **b** R_z factor

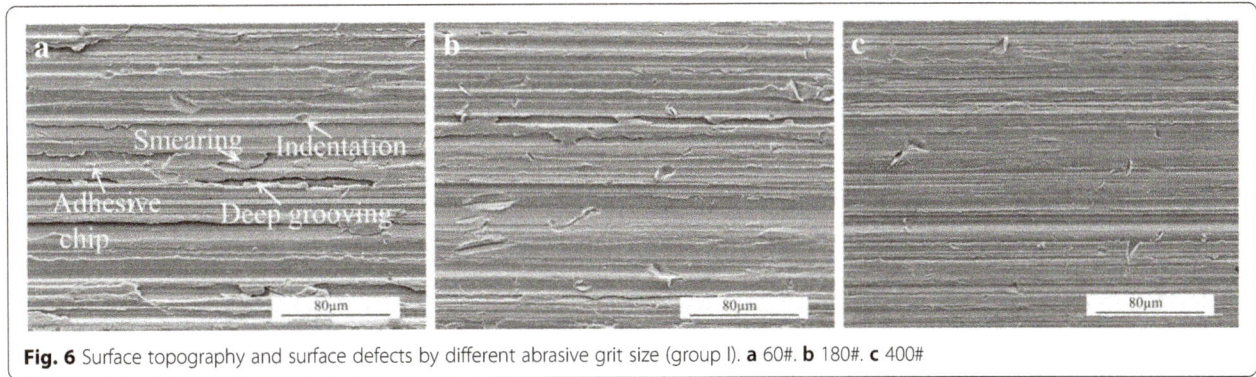

Fig. 6 Surface topography and surface defects by different abrasive grit size (group I). **a** 60#. **b** 180#. **c** 400#

efficient for metal removal, the remaining part is sliding and rubbing the workpiece material (Sin et al. 1979). The coarser grit size abrasives have bigger abrasive particles, causing larger areas of rubbing during grinding; meanwhile, the size and distribution of the abrasive grit grains are more uneven in coarser grit size abrasives; thus, more defects and a worse surface finish have been induced by using coarser grit size abrasives.

Figures 7 and 8 illustrate the effect of machine power (group II) on the ground surface roughness and surface topography, respectively. Compared with grinding by different grit size abrasives, the machine power has a much smaller influence on ground surface roughness. The R_a values shown in Fig. 7 varied very little although the machine power has been doubled and tripled while the R_z values show some variation. The effects on surface topography and surface defects are different; the SEM images (Fig. 8) show a clear improvement of surface finish by increasing the machine power. The ground surface appeared much smoother, with fewer surface defects when the machine power was increased to 90% (Fig. 8c), although many small smearing areas were still observed. Stainless steel 304L has high toughness and high ductility (Outokumpu 2013), and has been characterized as 'gummy', i.e., rubbery, or adhesive, during machining (Jang et al. 1996). A lower machine power means a lower (normal) grinding force, which led to more rubbing instead of the effective metal removal processes chip forming, cutting, or ploughing, as demonstrated by the results in Fig. 4. Together with the high

adhesion of the material on the grinding belt, large smearing areas as well as adhesive chips were generated. Meanwhile, a higher machine power means a higher downward force, therefore, instead of being cut, surface material was pressed down and slid along the grinding direction. As a result, a higher degree of strain hardening with the ground surface reduced the rubbery or adhesive behavior and improved surface finish after grinding, although a large amount of smearing areas remained. However, it should be noted that the effect of varying machine power also depends on the property of workpiece material. For example, in Zhou, et al.'s work (Zhou, Peng Ling, et al. 2016) on grinding of duplex stainless steel 2304, there is an optimum machine power for grinding, above and below which the surface finish becomes poorer.

Figures 9 and 10 present the influence of grinding with and without lubrication (group III) on surface roughness and surface topography, respectively. By using lubrication during grinding operations, large improvements of the surface roughness and surface finish are achieved, as seen in the figures. Both R_a and R_z values decreased to nearly a half, while the reduction of ground surface defects is very evident. This can be explained by decreased friction with good lubrication, which in turn retained abrasive sharpness and enhanced more uniform effective metal removal. Meanwhile, the reduction of surface temperature during grinding with lubrication also reduced the material's rubbery or adhesive behavior and it reduced redeposition.

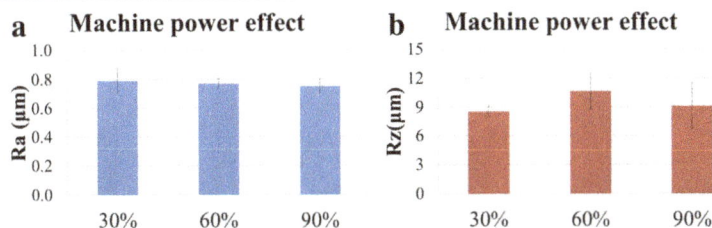

Fig. 7 Surface roughness by different machine power (group II). **a** R_a factor. **b** R_z factor

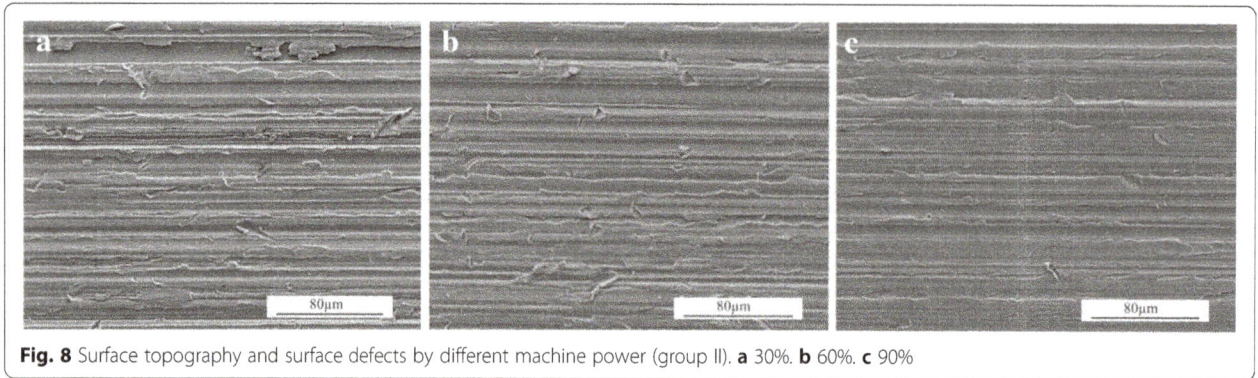

Fig. 8 Surface topography and surface defects by different machine power (group II). **a** 30%. **b** 60%. **c** 90%

Cross-section microstructure

Backscattered electron microscopy images revealing typical cross-section microstructures near the grinding surfaces when using different grit size abrasives (group I), different machine powers (group II), and wet/dry grinding (group III) are shown in Figs. 11, 12, and 13, respectively. Various magnifications have been used to investigate the microstructural development. For all the presented images, the grinding direction is perpendicular to the sample's cross-section. Similar features were observed for all the ground samples, which are illustrated in Fig. 11a1, a2. Depending on the grinding parameters used, smearing of different size and amount, adhesive chips, or cold welded chips with an irresolvable microstructure was observed along the ground surfaces. A heavily deformed surface layer extending up to a few microns from the ground surface was formed, followed by a much thicker sub-surface layer showing less plastic deformation. The heavily deformed surface layer comprised fragmented grains and dislocation sub-cells. Such a surface layer has also been observed in other ground materials, for example, in ground duplex stainless steel 2304 in work by the current authors (Zhou, Peng Ling, et al. 2016). The sub-surface region is characterized by slip bands and strain contrast (different grey shades) from plastic deformation. Next to the surface layer, densely populated and deformed slip bands of multiple orientations were observed, while further away slip bands became fewer and straighter as the degree of plastic deformation decreased. For all the investigated samples, the grinding induced deformation zone was much smaller than the abrasive grit size used.

Corresponding to the surface topography observations, a high amount of large smearing areas or adhesive chips (Fig. 11a1) was observed in the cross section from grinding by using 60# grit size abrasive. In addition, cold welded chips with formation of microcracks (Fig. 11a2) were also visible. Both the amount and size of such defects decreased when finer (180# and 400#) grit size abrasives were used. Images in Fig. 11 show that the abrasive grit size has major influence on both the degree of deformation and the deformation depth. Grinding with a 60# grit size abrasive induced a heavily deformed surface layer of 3–4 μm in thickness with a clear fragmented grain structure. The thickness of the heavily deformed layer was reduced to nearly the half when using 180#, and around one third by 400#. Meanwhile, coarse abrasive grit (60#) induced a deformed sub-surface layer of over 20 μm in thickness. The thickness of the deformation-affected layer reduced between 15 and 10 μm when using 180# and 400# abrasive grit sizes, respectively. Both the density of slip bands and the number of deformation slip systems was reduced by using finer grit size abrasives, which again indicated smaller deformation.

The low machine power (30%) induced a large amount of small smearing areas or adhesive chips along the grinding surface, which agrees with the results from the surface topography investigation, while very few cold welded chips resulted. As shown before, the normal grinding force was low. As a result, metal removal became less effective; abrasive grits slid over the material surface, introducing many smearing areas as well as

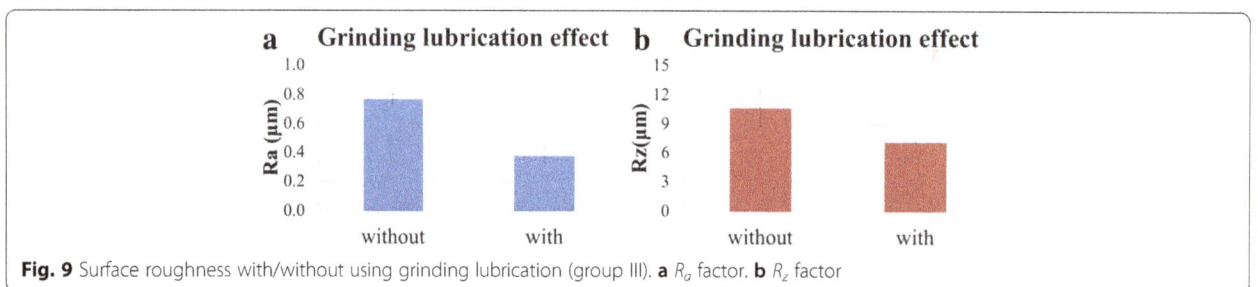

Fig. 9 Surface roughness with/without using grinding lubrication (group III). **a** R_a factor. **b** R_z factor

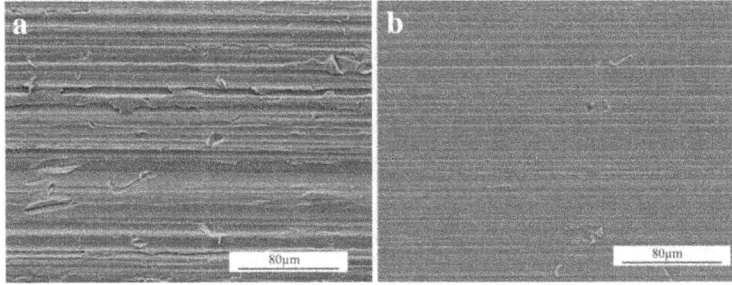

Fig. 10 Surface topography and surface defects (**a**) without and (**b**) with using grinding lubrication (group III)

adhesive chips on the ground surface. On the other hand, as the friction between the abrasive and workpiece material was low, the grinding temperature was low and thus few cold welding chips could form. In the case of grinding at 90% machine power, more grinding heat was generated due to increased grinding friction and promoted the formation of cold welding chips. As shown in Fig. 12, which presents typical cross-section microstructures for different machining powers, the thicknesses of the highly deformed surface layers are similar. In the sub-surface region, the formation of slip bands indicated that the deformation intensity and deformation depth are also similar. However, deformation was found to be very non-uniform when grinding using 30% machining power. As illustrated in Fig. 12a1, a2, which are from the same sample, some surface areas were highly deformed while in others the deformation was very small.

Using grinding lubrication largely reduced the formation of smearing, adhesive chips, and cold welded chips on the ground surface, which can be simply explained by the measured lower normal grinding force and lower temperature. From the electron channeling images presented in Fig. 13, the reduction of deformation is also obvious; although the thickness of the deformed surface layer is similar, the deformation in the sub-surface region is much lower. As shown in Fig. 13b, continuous deformation fringes were observed with a few slip bands formed only in a range of several microns. Meanwhile,

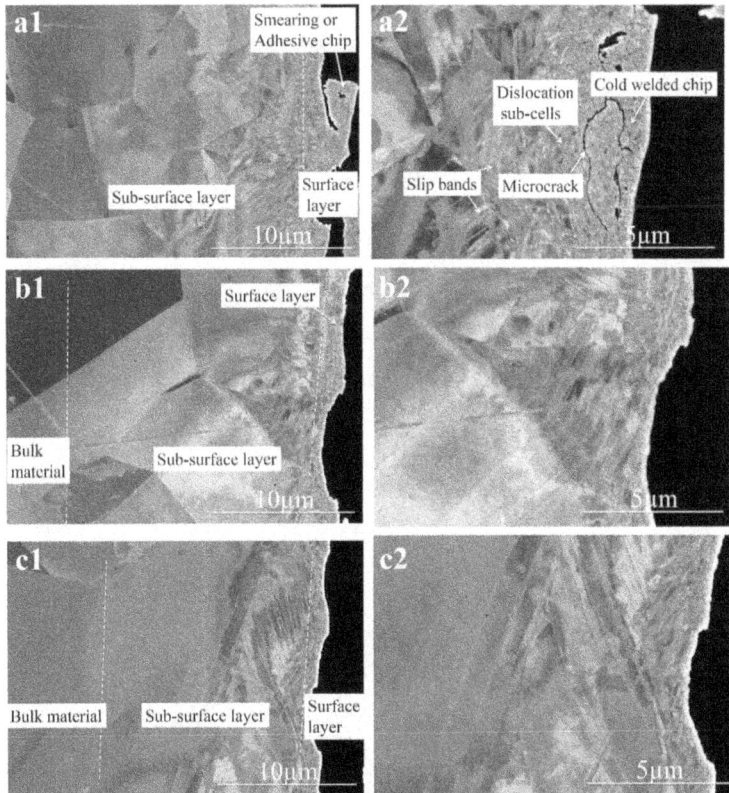

Fig. 11 Cross-section microstructure after using different abrasive grit size (group I). **a** 60#. **b** 180#. **c** 400#

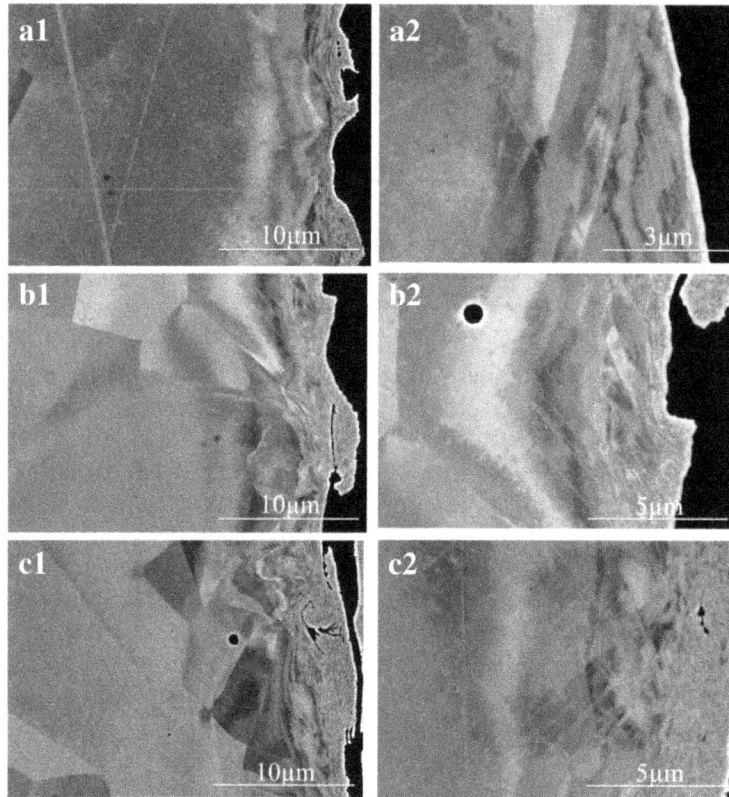

Fig. 12 Cross-section microstructure resulting from use of different machine power (group II). **a** 30%. **b** 60%. **c** 90%

deformation was found to be more uniform by using grinding lubrication.

Ground surface residual stresses

The full width at half maximum method (FWHM) reveals the broadening of a diffraction peak, which can be related to an increased density of crystal defects in a material (Chen et al. 2014). Here the observed difference in FWHM is considered to be related to different surface deformations from the grinding processes. In Fig. 14, diffraction peaks measured on the as-delivered surface and on the ground surface by 180# abrasive grit size,

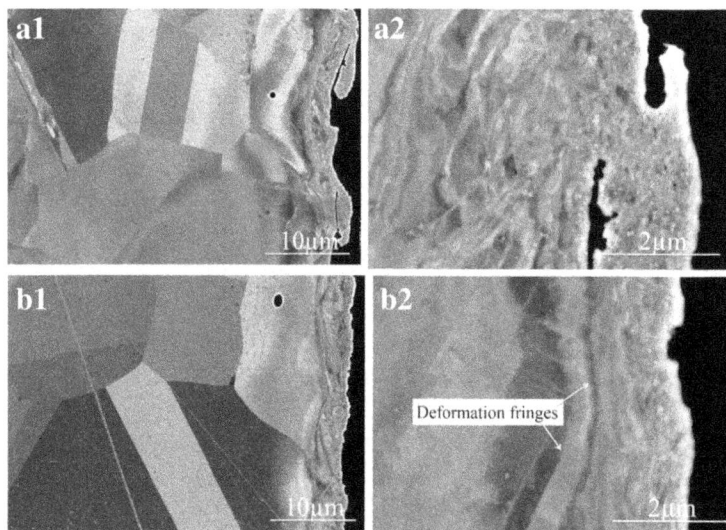

Fig. 13 Cross-section microstructure (**a**) without and (**b**) with grinding lubrication (group III)

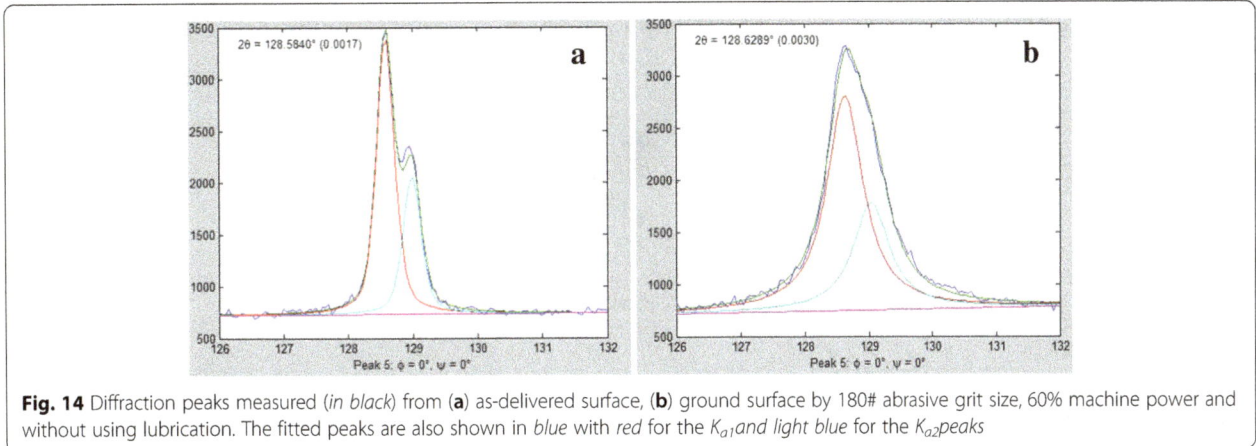

Fig. 14 Diffraction peaks measured (*in black*) from (**a**) as-delivered surface, (**b**) ground surface by 180# abrasive grit size, 60% machine power and without using lubrication. The fitted peaks are also shown in *blue* with *red* for the K_{a1} and *light blue* for the K_{a2} peaks

60% machine power, without using lubrication are illustrated respectively. The FHWM of the K_{a1} peak is 0.332° for the as-delivered surface and 0.68° for the ground surface. The FWHM results of all ground surfaces by different grinding parameters as well as the as-delivered material are presented and compared in Fig. 15. A lower FWHM by using finer grit size abrasives or grinding lubrication was observed (Fig. 15a and c), indicating smaller surface deformation induced by these grinding operations. On the other hand, as illustrated in Fig. 15b, the influence of the machine power on the surface deformation is relatively small while the influence from the abrasive grit size is more dominant. The FWHM results agree with the observed microstructural changes in the ground surface.

The surface residual stresses parallel ($\sigma_\parallel$) and perpendicular ($\sigma_\perp$) to the grinding directions by different grinding conditions are presented in Fig. 16. Uncertainties of the obtained residual stresses, given as error bars in the figures, are derived from standard deviations in diffraction peak fitting as well as for deviations of the measured diffraction peak positions from expected distributions with measurement directions according to the elastic theory for a quasi-isotropic material.

As shown from the figure, the grinding operations in this study generated tensile $\sigma_\parallel$ and compressive $\sigma_\perp$

stresses in the ground surfaces. The highest surface tensile residual stress, up to 374 ± 44 MPa, was induced by using the coarse (60#) grit size abrasive, as seen in Fig. 16a. The tensile $\sigma_\parallel$ was reduced nearly by one third, to 228 ± 25 MPa, by using a finer (180#) grit size abrasive, while the compressive $\sigma_\perp$ stress was similar. However, when using the even finer (400#) grit size abrasive, the surface tensile $\sigma_\parallel$ stress increases but the compressive $\sigma_\perp$ stress decreases. The trend of the machine power influence on surface residual stresses is very clear from Fig. 16b ,i.e., higher machining power generated a higher tensile $\sigma_\parallel$ stress but lower compressive $\sigma_\perp$ stress in the ground surfaces. The effect of using grinding lubrication is also obvious, as shown in Fig. 16c: using grinding lubrication reduced residual stresses, both $\sigma_\parallel$ and $\sigma_\perp$, in the surface layer.

For the sample ground with 60# grit size abrasive, the in-depth residual stress profile parallel ($\sigma_\parallel$) and perpendicular ($\sigma_\perp$) to the grinding direction as well as the FWHM profile were also measured. As shown in Fig. 17, the grinding induced tensile $\sigma_\parallel$ stress decreased dramatically from 374 to 44 MPa at the surface to compression within a depth of around 15 μm beneath the ground surface. The compressive $\sigma_\perp$ stress increased from 91.1 to 18 MPa at the surface to over 200 MPa in the subsurface region and then dropped gradually to zero in the

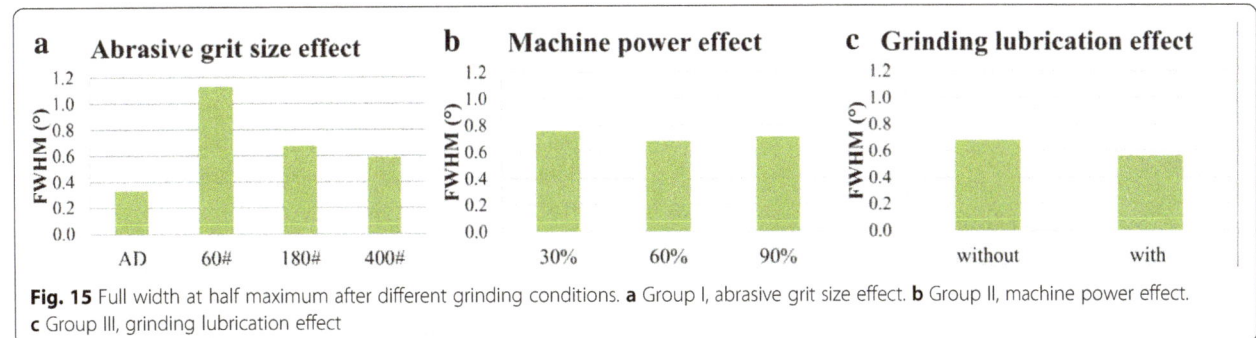

Fig. 15 Full width at half maximum after different grinding conditions. **a** Group I, abrasive grit size effect. **b** Group II, machine power effect. **c** Group III, grinding lubrication effect

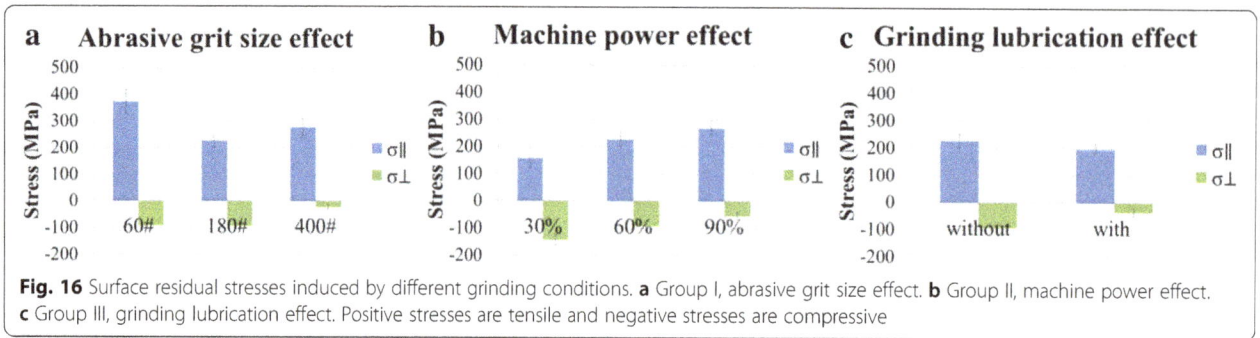

Fig. 16 Surface residual stresses induced by different grinding conditions. **a** Group I, abrasive grit size effect. **b** Group II, machine power effect. **c** Group III, grinding lubrication effect. Positive stresses are tensile and negative stresses are compressive

bulk material. The results indicated the trend that residual stresses both along and transverse to the grinding direction largely shifted towards compression under the ground surfaces. High tensile residual stresses exist only in a very thin surface layer along the grinding direction. Such large residual stress gradients in the near surface region are typical for machined workpieces (Zhou, Peng Ling, et al. 2016). Meanwhile, a decrease of FWHM with increasing depth was also observed and reveals plastic deformation gradient under the ground surface.

During grinding, mechanically and thermally induced residual stresses are predominant and they exist simultaneously (Davim 2010). The mechanical interactions between the abrasive and the workpiece material tend to cause anisotropic residual stresses in the ground surface. The surface layer of the workpiece material parallel to the grinding direction experiences compressive plastic deformation; while perpendicular to the grinding direction, tensile deformation dominates (Guo et al. 2010; Noyan and Cohen 1987). As a result, the interaction between the surface layer and the bulk material leaves the ground surface in a condition of tension along the grinding direction but compression in the transverse direction. This is also seen in the present work.

Heat was generated in the contact zone between the abrasive and the workpiece material during grinding

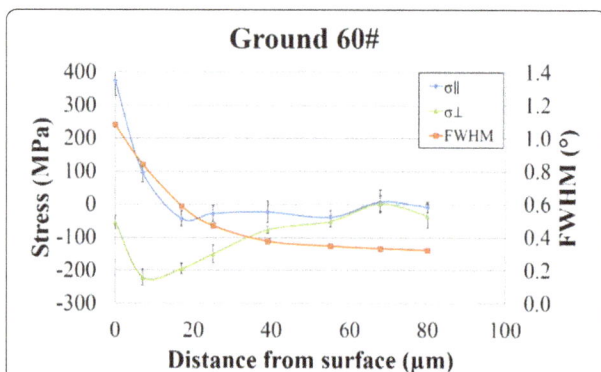

Fig. 17 In-depth residual stresses and full width at half maximum profiles after grinding by 60# as final surface finish (Zhou, Pettersson, et al. 2016)

operations, because of the low thermal conductivity of the stainless steel 304L, a temperature gradient was formed from the surface to the bulk material. This results in tensile stresses in both the directions during the cooling period after the grinding zone moved away (Fredj et al. 2006; Davim 2010). In this study, the measured highly anisotropic nature of the surface residual stress field, as well as the relatively low surface temperature in all of the ground samples, indicates that mechanically induced residual stresses dominate over thermally induced residual stresses.

When grinding with a 60# grit size abrasive, the high surface plastic deformation as well as grinding heat induced a high tensile ($\sigma_{\parallel}$) stress, which was reduced by using the 180# grit size abrasive because of the lower mechanical effect. However, compared with the 180#, the use of the 400# grit size abrasive generated more tensile residual stresses, both ($\sigma_{\parallel}$) and $\sigma_{\perp}$. This is probably attributable to the larger number of grinding steps when grinding with finer (280# and 400#) grit size abrasives, which introduces more accumulated heat on the surface, thus the increased thermal effect plus the decreased mechanical effect led to higher tensile residual stresses in the surface. The trend of the surface residual stresses with different machining powers is very obvious. A higher machining power increased the grinding temperature; but both the microstructure investigation and the FWHM results showed that the deformation of the surface and the sub-surface layer are similar at different used machining powers. Thus, an increasing thermal effect can be considered as the main factor that introduces higher tensile residual stresses, both parallel and perpendicular to the grinding directions when grinding at a higher machining power. Using grinding lubrication decreased both surface deformations and heat from the grinding operation; however, the decreased surface tensile ($\sigma_{\parallel}$) stress and compressive $\sigma_{\perp}$ stress results indicate that the reduction of mechanical effects was more significant than that of the thermal effects in this study.

Conclusions

The effect of different grinding parameters on the processing and the surface properties of the austenitic stainless steel 304L has been investigated. The following conclusions can be drawn:

- A higher machining power results in a larger normal grinding force; while using grinding lubrication reduces the normal force.
- A higher machining power increases the grinding temperature, while the grinding lubrication has an effective cooling effect. A slight increase in temperature with finer grit sizes was attributed to the cumulative effect of a longer total grinding time.
- A coarser grit size abrasive gives a higher metal removal ability, and lubrication significantly increases the metal removal rate. For the investigated 304L austenitic stainless steel, there is an optimum machining power for the effective removal of material.
- Using smaller grit size abrasives or grinding lubrication results in fewer surface defects. Although the influence of the machining power on the surface roughness was small, a higher machining power was seen to improve the surface finish of austenitic stainless steel 304L.
- Cross-section microstructure investigations indicate that a coarser grit size, a higher machining power and grinding without lubrication introduce higher plastic deformation in the ground surface. A low machine power causes a non-uniform deformation zone while grinding lubrication gave a more uniform deformation zone. Microcracks were observed along the ground surface for all samples; although the number and size vary with the grinding parameters. These microcracks may be liable to initiate crevice corrosion.
- The grinding operation induces anisotropic residual stresses in the surface layer, with a tensile stress parallel and a compressive stress perpendicular to the grinding direction. The residual stress level depends on the grinding conditions. Residual stresses both along and transverse to the grinding direction are largely shifted towards compression in the sub-surface layer. The high tensile residual stresses generated parallel to the grinding direction are likely to affect the material's susceptibility to stress corrosion cracking. The measured residual stress anisotropy and the observed relatively low grinding surface temperature indicate that mechanical effect dominated over thermal effects in this study. It is suggested that the grinding direction could be optimized to give compressive stresses in the direction of the highest applied load and thus reduce the risk for stress corrosion cracking.

Acknowledgements
The authors are grateful to Outokumpu Stainless Research Foundation, Region Dalarna, Region Gävleborg, Länsstyrelsen Gävleborg, Sandvikens kommun, Jernkontoret and Högskolan Dalarna for financial support. Professor Stefan Jonsson is acknowledged for his contribution to the initial plan of this work. The authors express our gratitude towards Mikael Schönning and Timo Pittulainen at Outokumpu Stainless AB for providing the test materials and material data. The assistance with part of the sample preparation work by Annethe Billenius is also appreciated.

Authors' contributions
All the authors contributed to the design of the work. NZ contributed to the major part of the experiments, evaluation and writing. RLP ran part of the XRD measurements and commented on the manuscript. RP revised the work critically. All authors read and approved the final manuscript.

Authors' information
Nian Zhou is a PhD student at the Material Science Department of Dalarna University and KTH (Royal Institute of Technology), Sweden. She works on the influence of grinding operations on surface integrity and chloride-induced stress corrosion cracking of stainless steels for her PhD project, and the project is sponsored by Outokumpu Stainless Research Foundation. She has published three papers for the PhD work and co-authored six papers for her master degree.
Ru Lin Peng is currently associate professor at Linköping University, Sweden. She received her PhD in Engineering Materials from Linköping University in 1992 and worked at the Studsvik Neutron Research Laboratory of Uppsala University before joining the Department of Management and Engineering of Linköping University as a faculty member in 1997. Peng's research concerns enhancing surface properties of engineering materials, and she has coauthored over 130 articles.
Rachel Pettersson is research manager at Jernkontoret (The Swedish Steel Producers' Association) and adjunct professor in Corrosion Science at KTH (Royal Institute of Technology). Formerly, she was department manager with Outokumpu Stainless and section manager at Swerea KIMAB. She has over 130 publications on pitting corrosion, stress corrosion cracking, high temperature corrosion, stainless steels.

Competing interests
The authors declare that they have no competing interests.

Author details
[1]Department of Material Science, Dalarna University, SE-79188 Falun, Sweden. [2]KTH, SE-10044 Stockholm, Sweden. [3]Department of Management and Engineering, Linköping University, SE-58183 Linköping, Sweden. [4]Jernkontoret, SE-11187 Stockholm, Sweden.

References
Acharyya, S., Khandelwal, A., Kain, V., Kumar, A., & Samajdar, I. (2012). Surface working of 304L stainless steel: impact on microstructure, electrochemical behavior and SCC resistance. Materials Characterization, 72, 68–76.
Boothroyd, G., & Knight, W. (2005). Fundamentals of metal machining and machine tools, 3 ed. CRC Press
Chen, Z., Peng, R., Avdovic, P., Moverare, J., Karlsson, F., Zhou, J., & Johansson, S. (2014). Analysis of thermal effect on residual stresses of broached Inconel 718. Advanced Materials Research, 996, 574–579.
Cheng, H. (2014). Analysis of panel data, 3 ed. Cambridge University Press
Davim, J. P. (2010). Surface intergrity in machining. Springer
Dieter, G. E. (1989). Mechanical Metallurgy (pp. 699-704). McGraw-Hill
Fredj, N. B., Sidhom, H., & Braham, C. (2006). Ground surface improvement of the austenitic stainless steel AISI 304 using cryogenic cooling. Surface & Coatings Technology, 200, 4846–4860.
Guo, G., Liu, Z., Cai, X., An, Q., & Chen, M. (2010). Investigation of surface integrity in conventional grinding of Ti-6Al-4V. Advanced Materials Research, 126–128, 899–904.
Hauk, V. (1997). Structural and Residual Stress Analysis by Nondestructive Methods. Elsevier Science

Jang, D., Watkins, T., Kozaczek, K., Hubbard, C., & Cavin, O. (1996). Surface residual stresses in machined austenitic stainless steel. *Wear, 194,* 168–173.

Jiang, L., Paro, J., Hänninen, H., Kauppinen, V., & Oraskari, R. (1996). Comparison of grindability of HIP austenitic 316L, duplex 2205 and Super duplex 2507 and as-cast 304 stainless steels using alumina wheels. *Journal of Materials Processing Technology, 62,* 1–9.

Johansson, S., Moverare, J., & Peng, R. (2013). Recent applications of scanning electron microscopy. *Practical Metallography, 50,* 810–820.

Kopac, J., & Krajnik, P. (2006). High-performance grinding - a review. *Journal of Materials Processing Technology, 175,* 278–284.

Manimaran, G., Pradeep Kumar, M., & Venkatasamy, R. (2014). Influence of cryogenic cooling on surface grinding of stainless steel 316. *Cryogenics, 59,* 76–83.

Noyan, I., & Cohen, J. (1987). *Residual Stress Measurement by Diffraction and Interpretation.* Springer

Outokumpu, (2013). Handbook of stainless steel.

Outwater, J., & Shaw, M. (1952). Surface temperature in grinding. *Transaction of the ASME, 74,* 73–86.

Paul, S., & Chattopadhyay, A. (1996). The effect of cryogenic cooling on grinding forces. *International Journal of Machine Tools and Manufacture, 36,* 63–72.

Poulain, T., Mendes, J., Henaff, G., & De Baglion, L. (2013). Influence of surface finish in fatigue design of nuclear power plant components. *Procedia Engineering, 6,* 233–239.

Sin, H., Saka, N., & Shu, N. (1979). Abrasive wear mechanisms and the grit size effect. *Wear, 55,* 163–190.

Totten, G., Howes, M., & Inoue, T. (2002). *Handbook of residual stress and deformation of steel.*

Turley, D., & Doyle, E. (1975). Factors affecting workpiece deformation during grinding. *Material Science and Engineering, 21,* 261–271.

Turnbull, A., Mingard, K., Lord, J., Roebuck, B., Tice, D., Mottershead, K., Fairweather, N., & Bradbury, A. (2011). Sensitivity of stress corrosion cracking of stainless steel to surface machining and grinding procedure. *Corrosion Science, 53,* 3398–3415.

Vashista, M., Kumar, S., Ghosh, A., & Paul, S. (2010). Surface integrity in grinding medium carbon steel with miniature electroplated monolayer cBN wheel. *Journal of Materials Engineering and Performance, 19,* 1248–1255.

Yao, C., Jin, Q., Huang, X., Wu, D., Ren, J., & Zhang, D. (2013). Research on surface integrity of grinding Inconel 718. *The International Journal of Advanced Manufacturing Technology, 65,* 1019–1030.

Zeng, Q., Liu, G., Liu, L., & Qin, Y. (2015). Investigation into grindability of a superalloy and effects of grinding parameters on its surface integrity. *Journal of Engineering Manufacture, 229*(2), 238–250.

Zhang, H., Chen, W., Fu, X., & Huang, L. (2012). Temperature measurement and burn mechanism of stainless steel 1Cr11Ni2W2MoV in grinding. *Materials Science Forum, 723,* 433–438.

Zhou, N., Peng Ling, R., & Pettersson, R. (2016). Surface integrity of 2304 duplex stainless steel after different grinding operations. *Journal of Materials Processing Technology, 229,* 294–304.

Zhou, N., Pettersson, R., Peng, R., & Schönning, M. (2016). Effect of Surface Grinding on Chloride Induced SCC of 304L. *Materials Science and Engineering A, 658,* 50–59.

6

Numerical and heat transfer analysis of shell and tube heat exchanger with circular and elliptical tubes

J. Bala Bhaskara Rao[1][*] and V. Ramachandra Raju[2]

Abstract

Background: Heat exchanger is a device in many industrial applications and energy conversion systems. Various heat exchangers are designed for different industrial processes and applications. Shell and tube heat exchanger (STHE) has its own importance in the process industries.

Methods: Experimental and numerical simulations are carried for a single shell and multiple pass heat exchangers with different tube geometries i.e. circular tubes to elliptical tubes. The experiment was carried out with hot fluid in tube side and cold fluid in shell side with circular tubes at 600 tube orientation and 25 % baffle cut. Heat transfer rates and pressure drops are calculated for various Reynolds numbers from 4000 to 20000. Fluent software is used for numerical investigations. Both circular and elliptical tube geometries with 450,600 and 900 orientations are used for the numerical studies. In addition to 25 % baffle cut, quarter baffle cut and mirror quarter baffle cut arrangements are used for comparison. The experimental values of heat transfer rates and pressure drops over shell side and tube side along the length of STHE are compared with those obtained from fluent software.

Results and Conclusion: It is found that the elliptical tube geometry with mirror quarter baffle cut at 450 tube orientation is 10 % higher than existing shell and tube heat exchanger and the pressure drop decrement in tube side shows up to 25 %.

Keywords: Shell and tube heat exchanger, Elliptical tubes, Heat transfer, Pressure drop

Background

Heat exchanger is a universal device in many industrial applications and energy conversion systems. Various heat exchangers are designed for different industrial processes and applications. In heat exchangers, shell and tube heat exchanger presents great sustainability to meet requirements and gives efficient thermal performance. Shell and tube heat exchanger (STHE) is widely used in petro-chemical industry, power generation, energy conservation, and manufacturing industry (Qian 2002). The baffle member plays an important role in STHE, and it supports tube bundle and also equally distribute the fluid in the shell side. When segmental baffles are used in STHE which have many disadvantages (Kern 1950a; Li & Kottke 1998a), the low heat transfer is achieved due to the flow stagnation, i.e., dead zones which are created at the corners between the baffle and the shell wall (Li & Kottke 1998b). It requires higher pumping power, and it creates a high pressure drop under the same heat load. The orientation of tubes will influence the annular surface area surrounded by the fluid. It is also influences the heat transfer rate. The new baffle cut arrangement achieved higher heat transfer rates and lower pressure drops (Master et al. 2006; Mukherjee 1992; Li & Kottke 1998c; Lei et al. 2008), so it is required to develop a new type STHE using different baffle cut arrangements to achieve a higher heat transfer rate. For the last few years, already described methods have been used to calculate heat transfer and pressure drop in the shell side of the STHE with different baffles (Peng et al. 2007). The different calculation procedures have been checked against experimental measurements on a small-scale heat exchanger. Kern method, Tinker method, and Delaware

* Correspondence: raobasijarajapu@gmail.com
[1]Department of Mechanical Engineering, SISTAM College, Kakinada, India
Full list of author information is available at the end of the article

method (Palen & Taborek 1969; Kern 1950b; Tinker 1951) gave the best results in comparison to other methods in literature. The shell side design under the inside flow phenomenon must be understood by experimental and numerical analysis. The shell and tube heat exchanger design was explained by Gay et al. (Bell 1963) who worked on heat transfer, while Halle et al. (Gay et al. 1976) and Pekdemir et al. investigated pressure drop (Halle et al. 1988; Pekdemir et al. 1994; Li & Kottke 1998d). Nowadays, the numerical methods have become an economical alternative for the research of STHE, and through a detailed flow pattern and a temperature field, it could be obtained with much less difficult (Seemawute & Eiamsa-ard 2010; Rhodes & Carlucci 1983; Huang et al. 2001; Stevanovic et al. 2001). The collective effect of all the above parameters on heat transfer is quite interesting to design STHE with the optimistic approach. Baffle cuts are placed to increase the flow rate in the shell side and also reduce the vibrations of the shell and tube heat exchanger. A STHE with 12 copper tubes and a stainless steel shell is used for the proposed system. Numerical analysis is conducted with elliptical tubes which are replaced by circular tubes. The modifications in the tube and shell overall pressure losses in the shell from the entrance to the exit points of the fluid are determined. The pressure drops over the tube and shell sides are altered with tube orientations to give maximum heat transfer efficiency. The baffle cuts are provided at 25 % with respect to the diameter and quarter baffle cut with respect to the cross sectional area. Mirror quarter baffle cut also is considered for effective heat transfer. It is observed that heat transfer rate increases with the increase of the surface area.

Methods

Experimental setup

The shell and tube heat exchanger with circular tubes and 60° tube orientation for baffle using 25 % cut was taken as an experimental device. The dimensions of the STHE are shown in Fig. 1. The shell is made of stainless steel, and the tube is made of copper. The circulated hot fluid was used in the tubes and the cold fluid in the shell. The proposed model gives a flexibility to conduct several experiments; the scale of the model can be implemented for industrial applications.

The water is pumped from the water tank, and it is divided into two streams. One stream is a cold fluid which flows into the shell side of the STHE. The other stream is a hot fluid, which is heated by an electrical heater with a temperature control device, and it passes through the tube side of the STHE. The

flow rate is calculated by flow meters, and the flow was regulated by using valves. Thermocouples are used to measure the temperature of the inlet and outlet for the hot and cold fluids. The flow path of the hot and cold fluids and the attachments of thermometers, flow meters, pumps, and valves are as shown in Fig. 2.

Shapes of tubes and baffles

Elliptical tubes with different orientations The copper tubes in the shell side are arranged with elliptical geometry. The elliptical tubes are prepared by equal volumes of elliptical and circular geometry. The ratio of the major axis to the minor axis of the elliptical tube is 2:1; 45°, 60°, and 90° orientation is maintained for the tubes as shown in Fig. 3a–c, respectively.

Different shape of the baffles The flow of the cold fluid is controlled by the shape of the baffle. The degree of turbulence created and energy utilized in the turbulence are governed by the baffles. Different baffle cuts, i.e., 25 % baffle cut, quarter baffle cut, and mirror quarter baffle cut are used for this analysis. The baffles are not only meant for structural support but also increase the external surface area of the tubes which resulted in improvement in the heat transfer.

Numerical analysis and validation of the work

For the numerical analysis, the actual model is represented as a virtual computer model using CATIA software package and analysis is performed with the help of a finite volume method as a computational fluid dynamics (CFD) tool. The inlet and outlet boundary conditions for the analysis are carried out as follows.

Geometric modeling and fluid properties for the analysis

Based on the above experimental model, a virtual model is prepared for CFD analysis and whatever the geometric parameters are for the actual model, the same dimensions are considered for virtual also. Hence, the geometry scale between the actual model and the virtual model is 1:1. As a result, one can minimize the deviation of CFD analysis values and the practical values obtained. The heat exchanger is designed with water working fluid for both hot and cold conditions. The properties of water are directly implemented for the analysis. For the shell, stainless steel is considered, and for the tube, copper material is considered as per the actual model. The diameter and thickness of the tube of the heat exchanger consider the Reynolds number calculated for different mass flow rates of hot water and cold water in the heat exchanger. The Reynolds numbers are greater than 4000

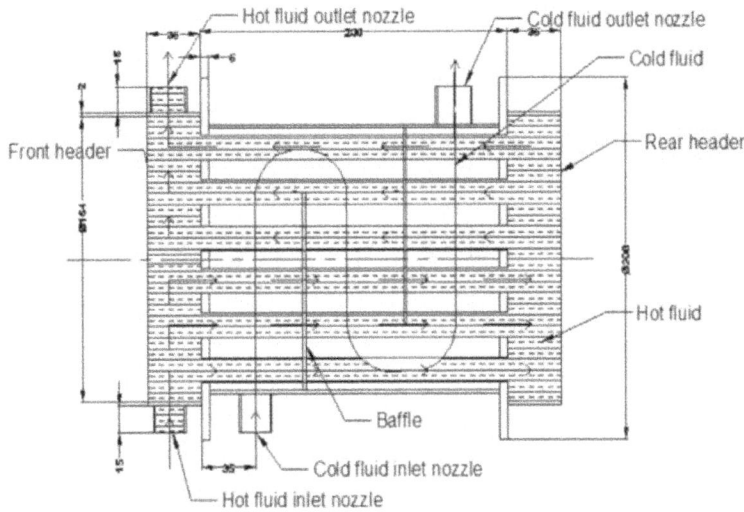

Fig. 1 Shell and tube heat exchanger for the experimentation

for the different inlet mass flow rates as mentioned in the boundary conditions. Hence, the flow in the pipe as well as the shell is considered as turbulent. The maximum temperature of the hot fluid is 348 K at atmospheric pressure, and it is in the liquid phase and water is an incompressible fluid; hence, Mach number is considered in the incompressible region. The model is prepared as a three-dimensional geometric model for the analysis.

Inlet and outlet boundary conditions for the analysis of shell and tube heat exchanger

Inlet and outlet conditions for hot fluid The hot water is entered at a temperature of 348 K, and different mass flow rates such as 0.15785, 0.3827,

0.55763, and 0.71782 kg/s are given at the inlet of the heat exchanger tube side nozzle. From the tube, the flow is considered to atmosphere pressure only. Based on the given inlet and outlet conditions, the inbuilt program of Fluent software calculated the remaining parameters.

Inlet and outlet conditions for cold fluid The cold fluid enters at a temperature of 298 K, and different mass flow rates such as 0.34589, 0.8403, 1.2245, and 1.5762 kg/s are given at the inlet of the heat exchanger shell side nozzle. The flow of the cold water is guided by the baffles provided over the tubes; the water is entered at 298 K and atmospheric pressure and it is exited from the shell outer nozzle into atmospheric pressure. Hence, the pressure boundary is defined at the outlet of the shell.

Other boundary conditions and grid generation No slip condition for the tube shell inner and outer surfaces is given. The tube shell and inlet outlet nozzles with uniform cross sections are defined for the analysis. The direction of the flow was defined normal to the boundary. Hydraulic diameter and turbulent intensity were specified at the inlet nozzle of both hot and cold fluids. The flow is assumed as an incompressible turbulent flow; the gradient of temperature is required at all the points of the heat exchanger. Hence, the separate grid is generated for both tube and shell side fluids, and they are separated by the boundaries.

Grid generation
Due to the complicated structure of STHE, the computational zone was meshed with the structured and unstructured tetrahedral grid. The grid system is

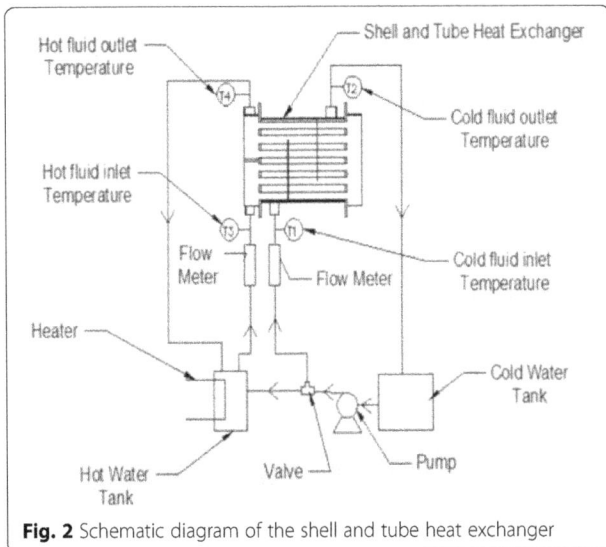

Fig. 2 Schematic diagram of the shell and tube heat exchanger

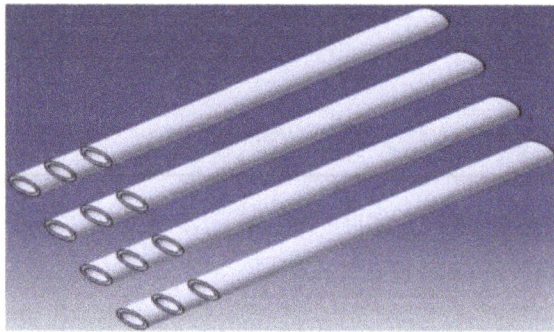

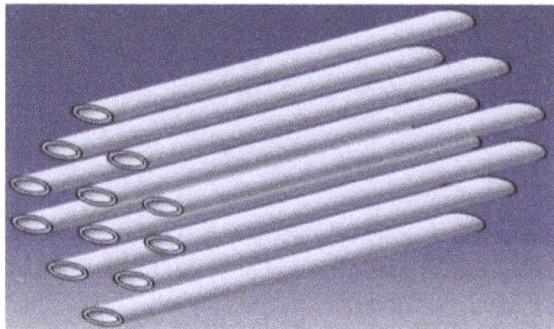

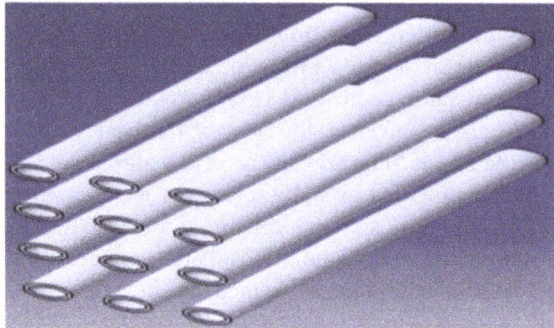

Fig. 3 a Elliptical tubes with 45° tube orientation. **b** Elliptical tubes with 60° tube orientation. **c** Elliptical tubes with 90° tube orientation

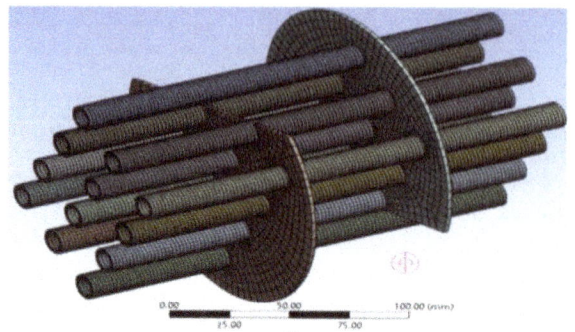

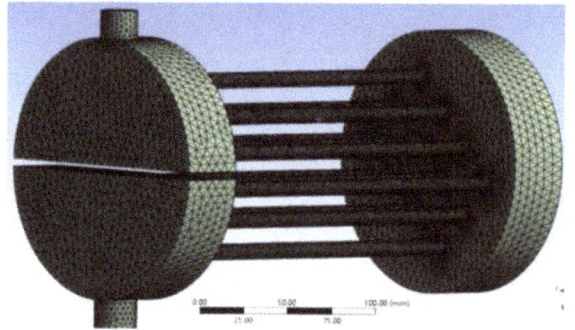

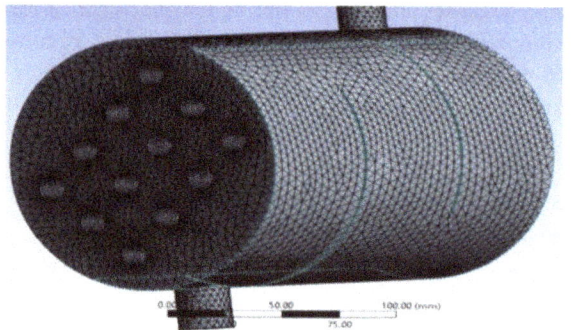

Fig. 4 a Meshed models for tubes and baffles at grid density 1,409,610. **b** Meshed model for tube side fluid at grid density 1,409,610. **c** Meshed model for shell side fluid at grid density 1,409,610

generated by ANSYS Workbench. During the verification of the grid system, the standard K-ε model is chosen. The Near wall region is meshed much finer (5.829e–002 mm) than shell wall regions in size to obtain optimized results as shown in Fig. 4a–c. For the high accuracy of numerical solutions, the grid independent test was performed for the model. Three different grid densities (1,409,610; 1,319,030; and 1,221,673) are generated for the STHE (Reynolds number = 15,640 and mass flow rate = 0.55763 kg/s). It is found that the deviation between the results of the different grid systems was less than 3 %. The final grid system of 1,409,610 was adopted for the remaining computational models.

Numerical solution chosen for the analysis

The CFD solver FLUENT was employed to solve the governing equations. The SIMPLE algorithm was used to resolve the coupling between pressure and velocity field. The second-order upwind scheme was

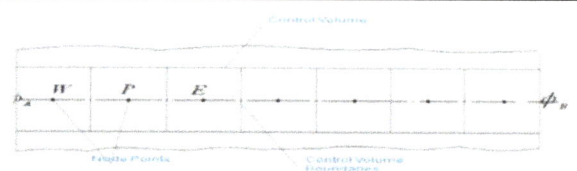

Fig. 5 Nodal points in the flow domain

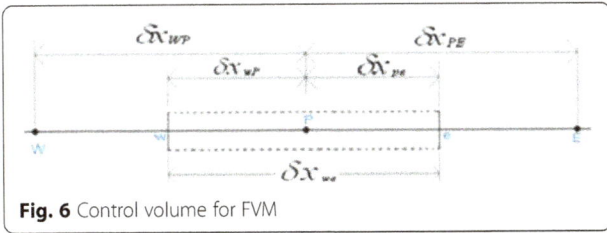

Fig. 6 Control volume for FVM

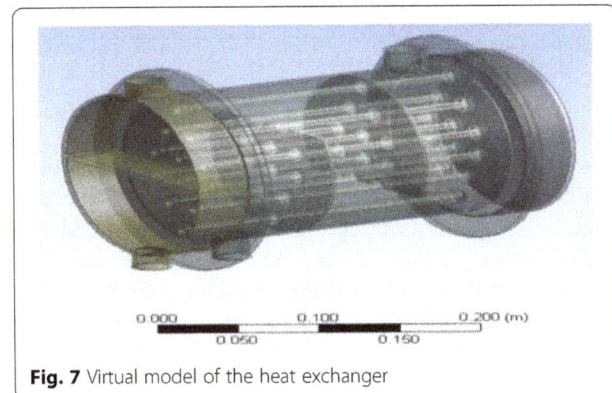

Fig. 8 Shell side vector diagram of heat transfer

used for discretization pressure, momentum, energy, turbulent kinetic energy, and turbulent dissipation rates. The analysis is related to the fluid flow; therefore, the finite volume method is chosen. There are three basic steps involved in the finite volume technique:

1. Integration of fluid governing differential equations over each control volume of computational domain.
2. Discretization of an integrated equation into an algebraic equation/form which will be converted into a solution algorithm.
3. The control volume that will be solved in this discretized form equation will be used to write a solution algorithm for every iterative process until it satisfies the convergence criteria and stability.

In finite volume method (FVM), the domain can be divided into number of control volumes and place number of nodal points between points as shown in Fig. 5.

In discretization of an integrated equation into an algebraic equation, the nodal point Pis is shown in Fig. 6.

The general equation for the fluid, when the fluid is incompressible, viscous, turbulent is

$$\int_{cv} \frac{\partial}{\partial t}(\rho\phi)dv + \int_{cv} \mathrm{div}(\rho\phi u)dv = \int_{cv} \mathrm{div}(\Gamma\,\mathrm{grad}\ \phi)dv$$

$$+ \int_{cv} S(\phi)dv - 1$$

where ϕ = property of fluid

ρ = density of fluid

Γ = diffusion coefficient

S = source term

The general transport equation for a variable Ø in finite volume method is

(Rate of increase of Ø of fluid element)

+ (Net rate of fluid flow of Ø out of fluid element)

= (Rate of increase of Ø due to diffusion)

+ (Rate of increase of Ø due to source)

where $\qquad a_w = \dfrac{\Gamma_w\ A_w}{\delta x_{\mathrm{WP}}}, a_E = \dfrac{\Gamma_e\ A_e}{\delta x_{\mathrm{PE}}} \qquad a_P = (a_W + a_E - S_P)$

$a_P\ \phi_P = a_W\ \phi_W + a_E\ \phi_E + S_u - 2$

The resulting system of the linear algebraic equations is solved to obtain the distribution of the property Ø at the nodal point. The algebraic equations are solved by using direct methods (Cramer's rule, matrix inversion, and Gauss elimination) and indirect or iterative methods (Jacobi and Gauss-Seidel). The flow is incompressible turbulent; hence, the K-ε model is considered for the analysis. The K-ε model is the simplest turbulence model for which only initial boundary conditions need to be supplied and used for 3D

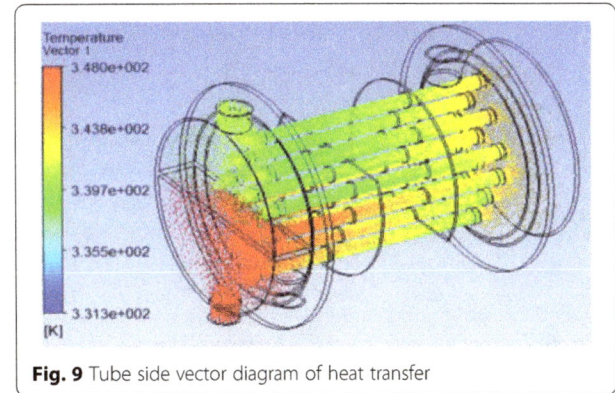

Fig. 7 Virtual model of the heat exchanger

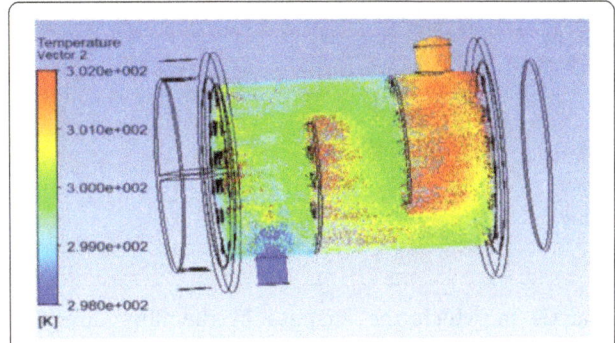

Fig. 9 Tube side vector diagram of heat transfer

Table 1 Comparison of the numerical analysis with the experimental results

Reynolds number	Inlet temp (K)		Heat transfer (watts)		Percent of error
	Cold fluid	Hot fluid	Value by expt	Value by numbers	
4418	298	348	3419.85	3691.38	7.93
10,733	298	348	6799.85	7170.84	5.45
15,640	298	348	9018.42	9485.87	5.18
20,132	298	348	10164.25	10965.60	7.8

analysis in which the changes in the flow direction are always so slow that the turbulence can adjust itself to local conditions. As per the grid density, the number of algebraic equations are generated for unite volume and the algebraic equations are solved numerically.

Validation of the analysis

The above boundary conditions, inlet and outlet conditions, and fluid properties are considered for the analysis, and the results are compared to the real model. The virtual model of shell and tube heat exchanger is used for numerical analysis as shown in Fig. 7. Figures 8 and 9 show the variations of heat transfer in the shell side and the tube side of the shell and tube heat exchanger in vector form. The inlet temperature of the hot and cold fluids at different flow rates which are utilized to calculate the corresponding outlet temperatures using calculation method is given in data appreciation. The heat transfer performance and pressure drop are evaluated, and these experiment values are tabulated in Table 1. The percentage of error is as shown in Fig. 10. The percentage of error is less than 8 %; hence, the considered boundary conditions are valid for the analysis of virtual model with modifications. Consequently, experimental results are closed to numerical results and the final concept is formulated towards CFD analysis for various configurations.

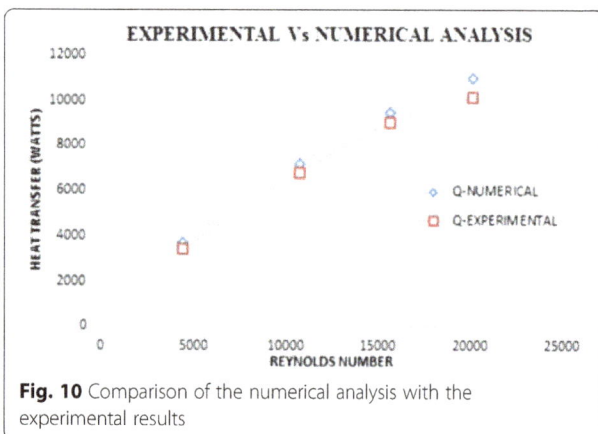

Fig. 10 Comparison of the numerical analysis with the experimental results

Data appreciation

The heat transfer rate and ultimately pressure drop were calculated from the initial data on the flow rates and the inlet and outlet temperatures of both shell and tube side of STHE. The heat transfer rate is obtained by the following equation.

$$Q = AU\Delta T_{m} \tag{1}$$

where Q = average heat flux between the cold and hot fluid in watts

$$Q = (Q_{C} + Q_{h})/2$$
$$Q_{C} = m_{c}c_{pc}(T_{co} - T_{ci})$$
$$Q_{h} = m_{h}c_{ph}(T_{hi} - T_{ho})$$

m_{c} and m_{h} are the cold and hot fluid mass flow rates in kg/s, respectively

c_{pc} and c_{ph} are the specific heats under constant pressure of cold and hot fluids in kJ/kg K, respectively

A = surface area based on the tube outside diameter in m^{2}

$$A = (\Pi \, Do \, L) \, N$$

where N = number of tubes

Do = outlet tube diameter in meters

L = effective length of the tube in meters

ΔT_{m} = logarithm mean temperature difference for hot and cold fluid in kelvins

The heat exchanger is one shell and multiple tube passes $\Delta T_{m} = \dfrac{(t_{2}-t_{1})\,(\sqrt{1+R2})}{ln\frac{[(2-p(1+R-\sqrt{1+R2})]}{(2-p(1+R+\sqrt{1+R2})]}}$

$$R = \frac{T_{1}-T_{2}}{t_{2}-t_{1}} \quad P = \frac{t_{2}-t_{1}}{T_{1}-t_{1}}$$

where t_{1} and t_{2} are the shell side inlet and outlet temperatures, respectively, and T_{1} and T_{2} are the tube side inlet and outlet temperatures in kelvins, respectively.

From the above calculations, the overall heat transfer coefficient can be obtained by Eq. (1).

Heat transfer efficiency on the tube side can be obtained by

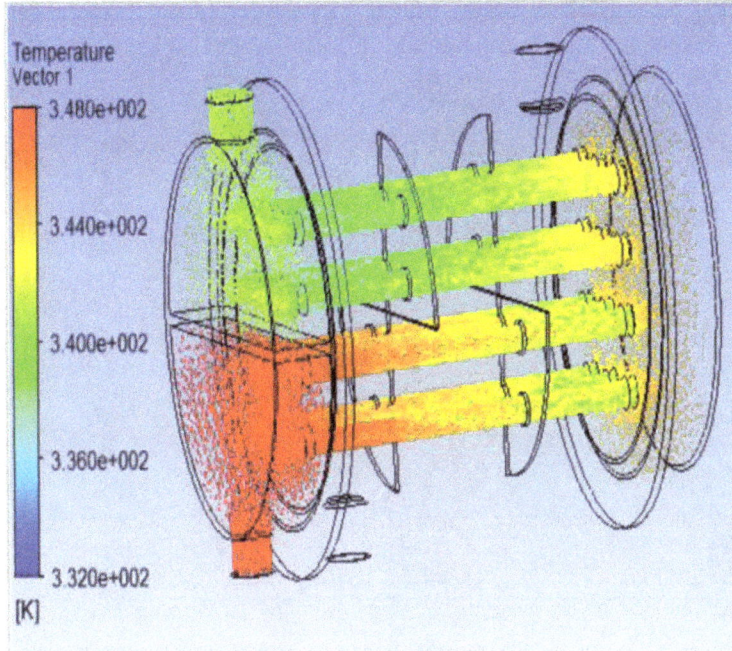

Fig. 11 Velocity vector of tube side heat transfer at mirror quarter baffle cut with circular tubes

$$N_U = 0.027 \, \text{Re}^{0.8} \, \text{Pr}^{0.333} \left(\frac{\mu_f}{\mu_w} \right)^{0.14}$$

For water $\frac{\mu_f}{\mu_w} = 1$

The properties of water and flow parameters at 348 K for the tube side are used, and accordingly, Re and Pr are calculated. The tube side film coefficient can be calculated by using

$$N_U = \frac{\text{hidi}}{k}$$

where N_U = Nusselt number for the tube side
di = inner diameter of the tube in meters

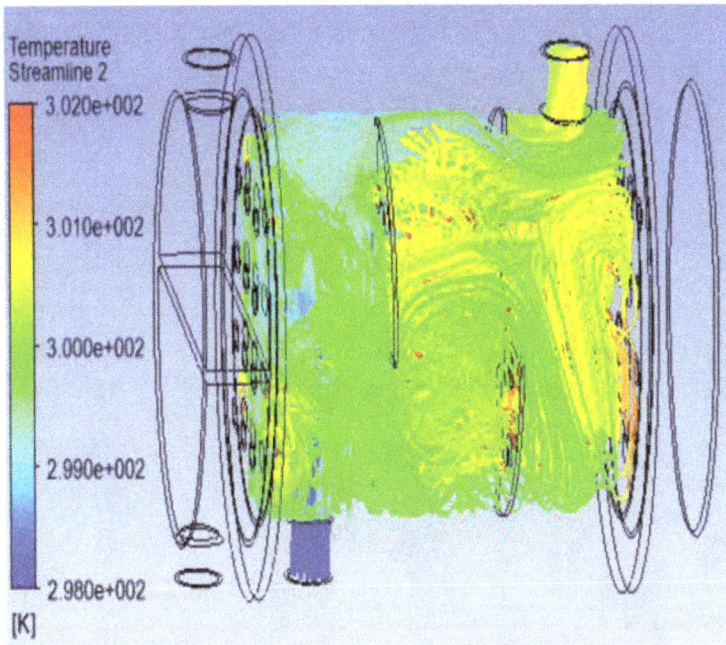

Fig. 12 Stream lines of shell side heat transfer at mirror quarter baffle cut with circular tubes

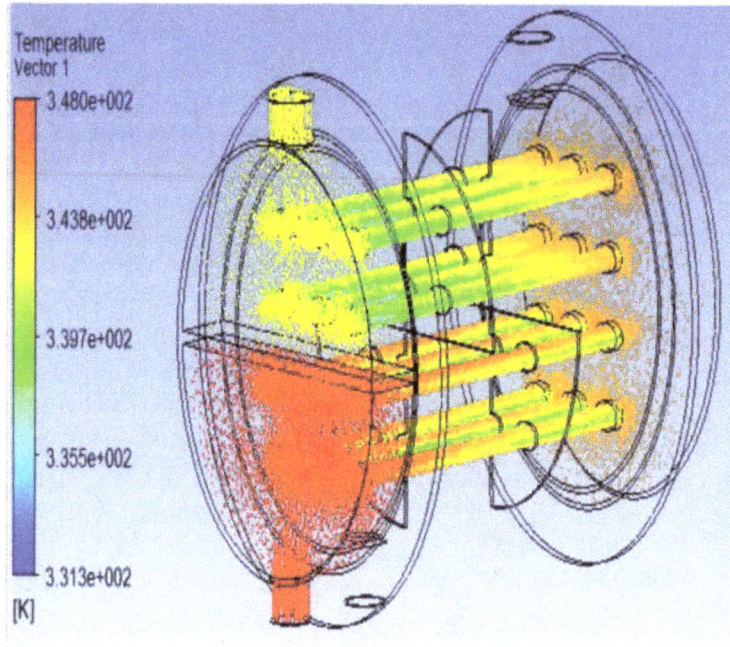

Fig. 13 Velocity vector of tube side heat transfer at mirror quarter baffle cut with elliptical tubes

K = thermal conductivity of hot fluid in W/m K

The overall heat transfer coefficient for the shell and tube heat exchanger can be found by using

$$U = \frac{1}{\frac{d0}{di h_i} + \frac{d0}{2k1} \log \frac{d0}{di} + \frac{1}{h_o}}$$

where h_i and h_o are heat transfer coefficients for the tube side and the shell side in W/m^2 K, respectively.

k_1 = thermal conductivity of the copper tube in W/m K

The shell side film coefficient (h_o) can be obtained from the above equation.

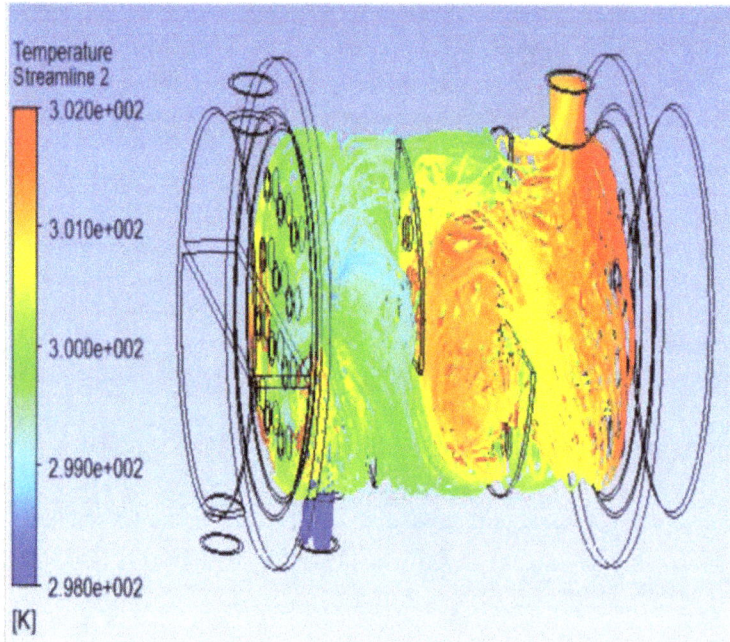

Fig. 14 Stream lines of shell side heat transfer at mirror quarter baffle cut with elliptical tubes

The pressure drop on the shell side of shell and tube heat exchanger is calculated by the equation

$$f = \frac{\Delta p0}{\frac{1}{2}\rho u2} \frac{d0}{l}$$

where f = friction factor $e^{(0.576 - 0.19 \ln Re)}$

(From *Heat Exchangers: Selection, Rating and Thermal Design*, by Sadik Kakac and Hongtan Liu).

Results and discussion

The flow parameters are varied by introducing different tube geometries, different tube orientations, and different baffle cuts. Subsequently, more heat transfer rate is created in the flow by introducing more surface area, and a mathematical model is reformed for more numerical analysis. As per this type of analysis, observation is found that more heat transfer efficiency for dropping the pressures along the length of the shell and tube of the STHE.

Circular tube and elliptical tubes
Heat transfer analysis
Heat transfer with different baffle cuts In this experiment, three types of baffle cuts are used for the analysis. The effects of heat transfer in the tube side and the shell side are investigated by using 25 % baffle cut, quarter baffle cut, and mirror quarter baffle cut. The cross section of the tubes which contain geometry such as circular, square, triangle, and elliptical will influence the thermal analysis due to variation of the surface area. Generally, a higher surface area gives higher heat transfer performance and vice versa. In the analysis, the circular tubes have a low surface area than the elliptical tubes. The vector contours in the mathematical model shown in Figs. 11, 12, 13, and 14 for circular tubes without strip (CWOS) and elliptical tube without strip

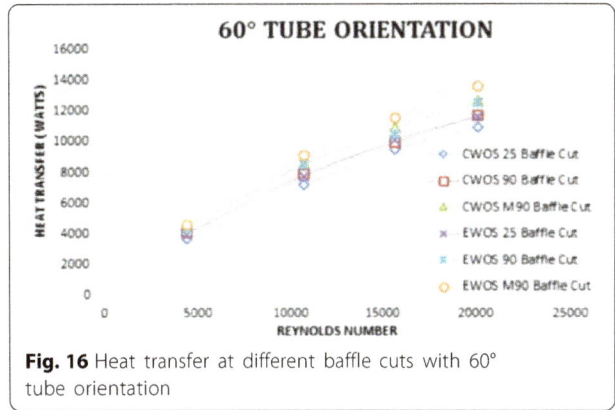

Fig. 16 Heat transfer at different baffle cuts with 60° tube orientation

(EWOS) indicate heat transfer distribution in the tube side and the shell side of the shell and tube heat exchanger. Visualizations are made on the elliptical model in comparison to the circular model mathematical analysis. Temperature distribution for vector diagram indicates the heat transfer rate increase if tube geometry changes from circular to elliptical. The temperature of the hot water in the tubes at the inlet is higher than all other regions. In the case of the shell, the temperature of the cold water at the outlet is higher. From all graphs (Figs. 15, 16, 17, 18, 19, and 20), it shows that the heat transfer rate increases with an increase of Reynolds number. The baffle cut changes the direction of the flow of the cold fluid and the elliptical tube geometry increases the heat transfer rate due to the increase of the surface area. The heat transfer rate increases from 25 % baffle cut to quarter baffle cut and also quarter baffle cut to mirror quarter baffle cut.

In the case of the quarter baffle cut, the direction of the flow of the cold fluid changes in the *X-Y* plane and also in the *Z* direction. As a result, the total length of the flow increased and the heat transfer rate also proportionately increased. In the circular tube, the heat transfer rate increases up to 11 % when the 25 % baffle cut is replaced by quarter

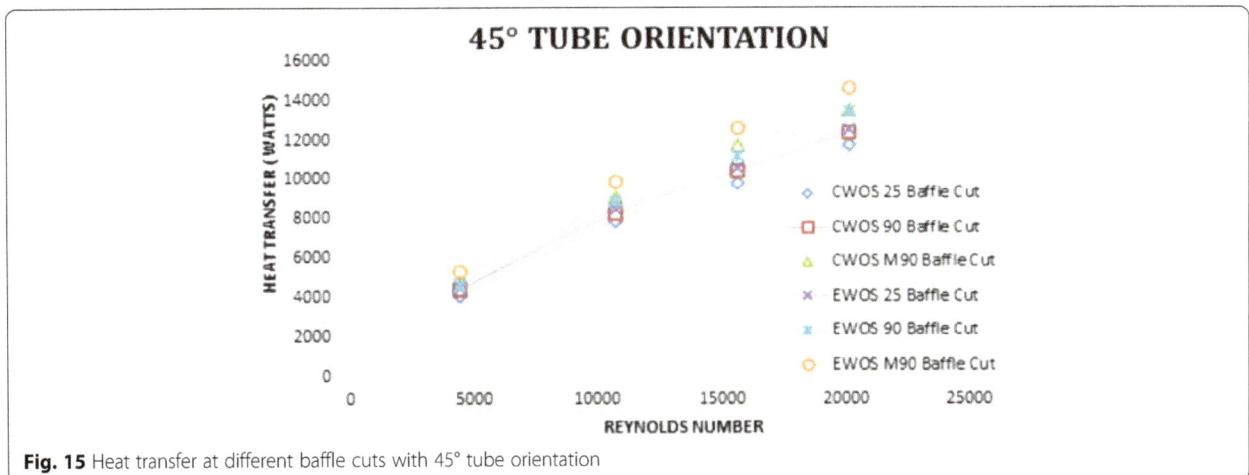

Fig. 15 Heat transfer at different baffle cuts with 45° tube orientation

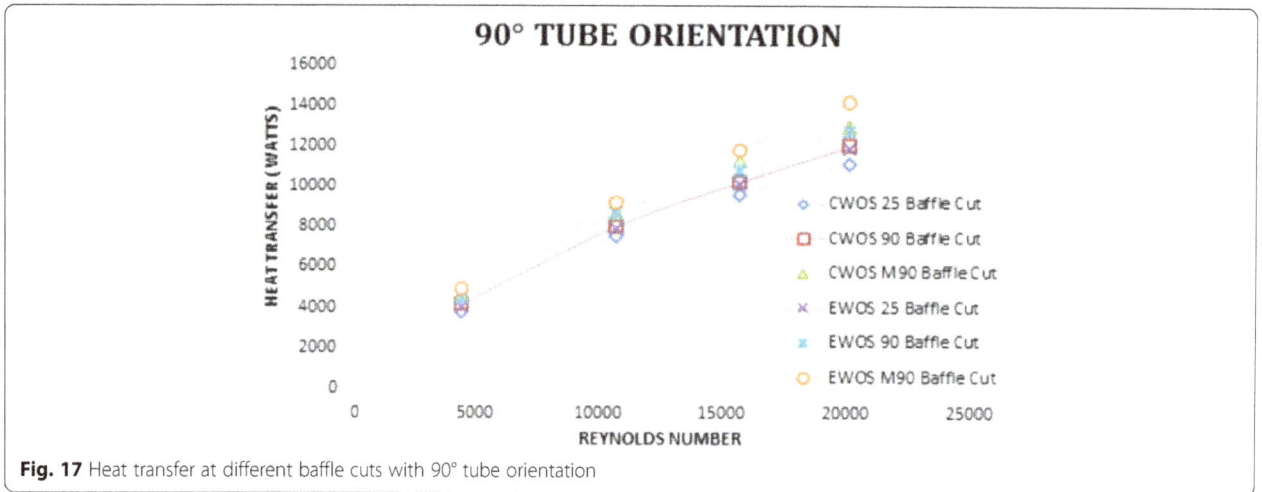

Fig. 17 Heat transfer at different baffle cuts with 90° tube orientation

baffle cut (90 baffle cut) and it is up to 10 % when the elliptical tubes are used from Figs. 15, 16, and 17. This is due to the increase flow length in the Z direction. In the case of the mirror quarter baffle cut (M90 baffle cut), the flow of the cold fluid splits in two streams and follows the helical path due to the baffle shape. Separation of the cold fluid and increase in flow length leads to a better heat transfer rate. Figures 15, 16. and 17 indicate that the heat transfer rate increases up to 14 % when quarter baffle cut is replaced by mirror quarter baffle cut in the circular tube, and it is up to 16 % if elliptical tubes are used.

Heat transfer with different tube orientation The tube orientation influences the diagonal, vertical, and horizontal space between the tubes. At particular arrangement, the space is uniform and it may help for a better heat transfer rate. The temperature difference between the hot fluid and the cold fluid also influences the heat transfer rate. Hence, for the tubes, different orientations 45°, 60°, and 90° are considered to find heat transfer variation along the tube side as well as the shell side. From all the graphs in Figs. 18, 19, and 20, values impose tube orientation from 60° to 90° heat transfer rate increases and the same against 90° to 45°. The heat transfer rate increases up to 6 % when the tube orientation changes from 60° to 90° in the circular tubes, and it is up to 8 % in the elliptical tubes. If the tube orientation changes from 90° to 45°, the heat transfer rate increases up to 6 % in the circular tubes and 9 % in the elliptical tubes.

Comparison of heat transfer on the shell side and the tube side of the shell and tube heat exchanger If comparison is made between the circular tubes and the elliptical tubes, the heat transfer rate always increases in the elliptical tube geometry for all Reynolds numbers. Among all three baffle cuts, mirror quarter baffle cut gives

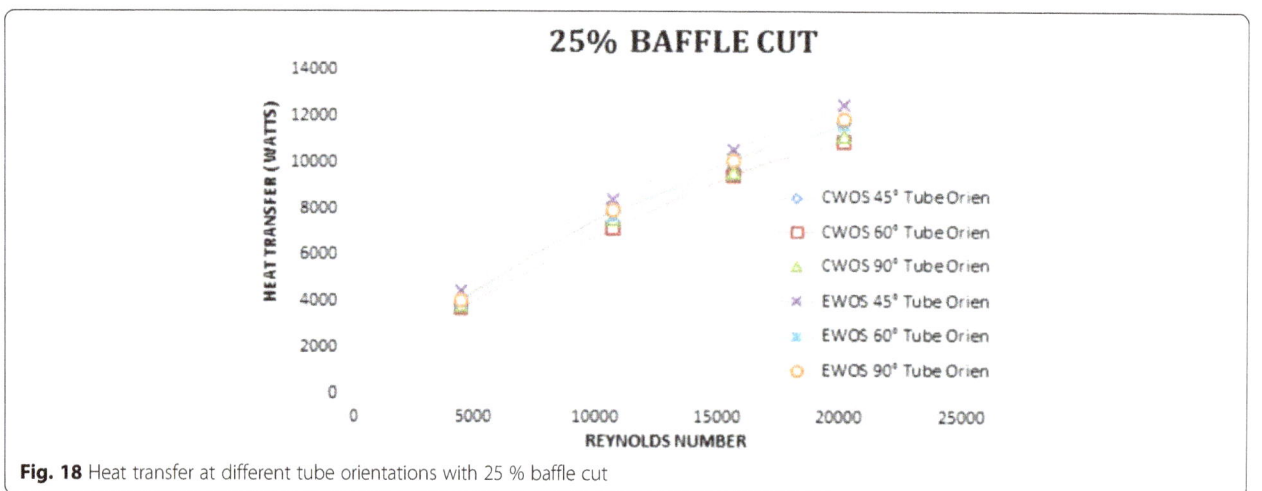

Fig. 18 Heat transfer at different tube orientations with 25 % baffle cut

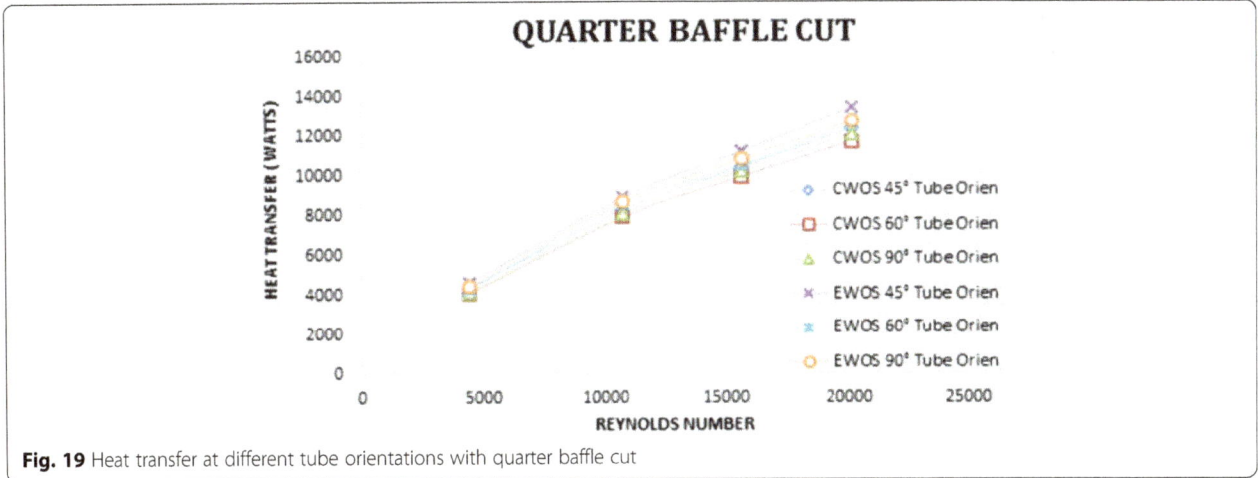

Fig. 19 Heat transfer at different tube orientations with quarter baffle cut

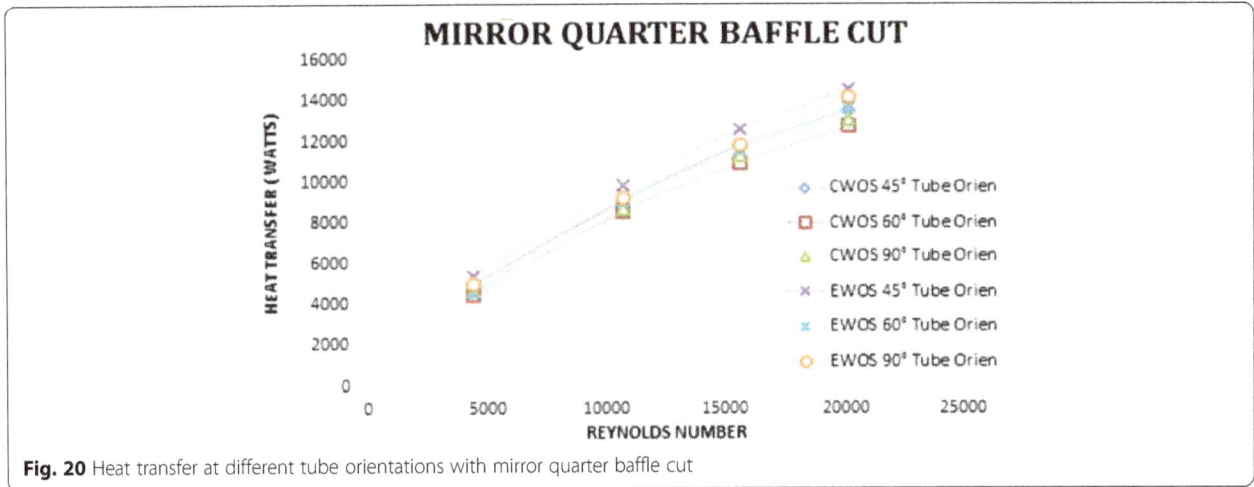

Fig. 20 Heat transfer at different tube orientations with mirror quarter baffle cut

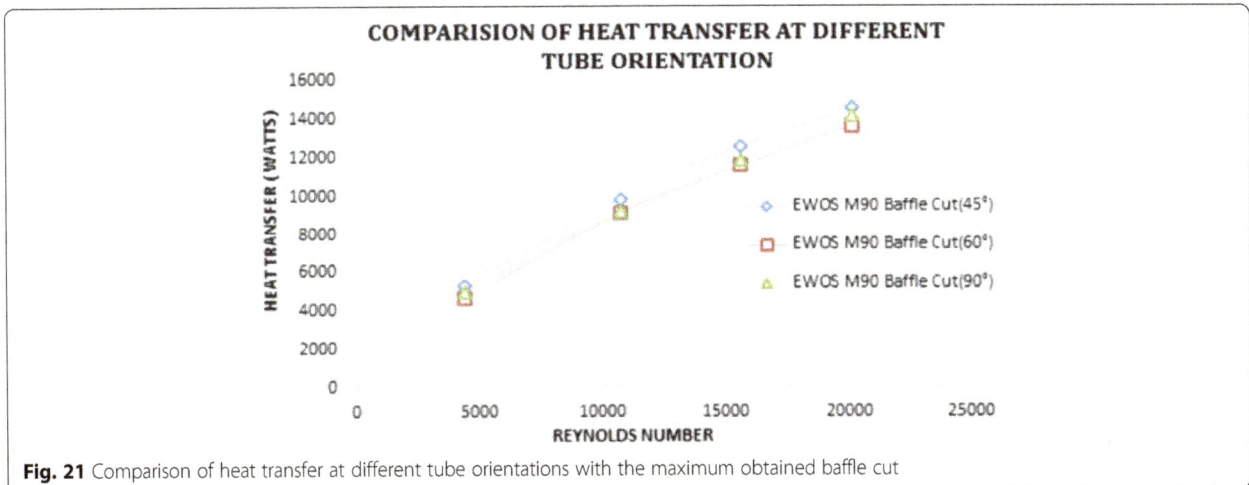

Fig. 21 Comparison of heat transfer at different tube orientations with the maximum obtained baffle cut

better heat transfer rates as it shows in Figs. 21 and 22. As variation is made between the circular tubes and the elliptical tubes, the heat transfer rate increases up to 10 % for all three baffle cuts. While considering tube orientation, the elliptical tubes and different baffle cuts of 25 %, quarter and mirror quarter are maintained. Among all the tube orientations, 45° tube orientation gives the maximum heat transfer rate as it is shown in Fig. 21. Observations are found that the heat transfer rate increases up to 8–9.5 % of different orientations with the circular and elliptical tubes, respectively, and then the overall heat transfer improvement with mirror baffle cut 45° tube orientation and elliptical tube is 10 % higher than the existing shell and tube heat exchanger.

Pressure drop analysis on shell side

Fluid flow pressure, volume, and temperature are the three inter dependent parameters. The heat transfer is high when the temperature difference in hot and cold fluids is more and pressure drop is also high. But in the case of the heat exchanger, sudden high pressure drops are not to be encouraged. To avoid a high pressure drop in the heat exchanger, a uniform transfer of heat energy through the heat exchanger is required. With this view, the pressure drop is calculated with different arrangements for the better performance of the heat exchanger.

Pressure drop with different baffle cuts The baffle cuts influence the direction of the flow of the cold fluid, and the baffles give structural support to the tubes. The vector contours in the mathematical model shown in Figs. 23 and 24, for the circular tubes and the elliptical tubes, indicate pressure drop variation in the tube side of the shell and tube heat exchanger. The pressure drop increases with an increase of Reynolds number. The pressure drop decreases from 25 % baffle cut to quarter baffle cut and also from quarter baffle cut to mirror quarter baffle cut. The total length of the flow increases in quarter baffle cut as compared to 25 % baffle cut, and it leads to a decrease of the pressure drop in the shell side of the shell and tube heat exchanger. The pressure drop reduced to 13 % for quarter baffle cut as compared to 25 % baffle cut in the circular tubes and to 15 % in the elliptical tubes as shown in Figs. 25, 26, and 27. In the case of the mirror baffle cut, the fluid in the shell splits into two streams, increases the flow path, and reduces the pressure drop in the shell side. The pressure drop reduced to 20 % from quarter baffle cut to mirror quarter baffle cut in the circular tubes and it is to 25 % in the elliptical tubes.

Pressure drop with different tube orientations The tube orientation for space distribution in the shell side influences pressure drop by different zones between the tubes. Tube orientations are created by putting 45°, 60°, and 90°. Clearance space flow zone is more for 45°, 90°, and 60° and declined gradually. These flow zones gives less pressure drop as the free space creates less turbulence. Observations are made for the circular tube and the elliptical tubes at different orientations such as from 60° to 90° and 90° to 45°. In the tube orientation for the circular tube at 60° to 90°, the pressure drop decreases to 30 % while the same at 90° to 45°, the pressure drop decreases by 20 % as shown in Figs. 28, 29, and 30. Similar formulations are made for the elliptical tubes at 60° to 90° tube orientation, the pressure drop decreases to 30 % while the same at 90° to 45° tube orientation, the pressure drop decreases by 25 % as shown in Figs. 28, 29, and 30.

Comparison of shell side pressure drop on the shell and tube heat exchanger If comparison is made in circular tubes and elliptical tubes, pressure drop decreases in the elliptical tube geometry for all Reynolds numbers. Among all three baffle cuts, mirror quarter baffle cut gives lesser pressure drop as shown in Figs. 31 and 32. While considering tube orientation, the elliptical tube and different baffle cuts of 25 %, quarter, and mirror quarter are maintained. Among all tube orientations, 45° tube orientation gives minimum pressure drop as shown in Fig. 31. When observations are found for elliptical tubes and circular tubes, then the pressure drop decreases to 20 % for different baffle cuts and it is to 18 % for different orientations. The overall pressure drop variation with mirror baffle cut 45° tube orientation and elliptical tube is 35 % lower than the existing shell and tube heat exchanger.

Pressure drop analysis on tube side

Pressure drop with different tube orientation The tube orientation influences pressure drop on the tube side by the diagonal, vertical, and horizontal space in the tubes. Tube orientations are created by putting 45°, 60°, and 90°. The pressure drop increases with increase of Reynolds number. Observations are made for the circular tubes and the elliptical tubes at different orientations such as from 60° to 90° and 90° to 45°. In the tube orientation for the circular tubes at 60° to 90°, the pressure drop decreases to 15 % while the same at 90° to 45°, the pressure drop decreases by 11 % as shown in Figs. 33, 34, and 35. Similar formulations are made for elliptical tubes at 60° to 90° tube orientation, the pressure drop decreases to 11 %

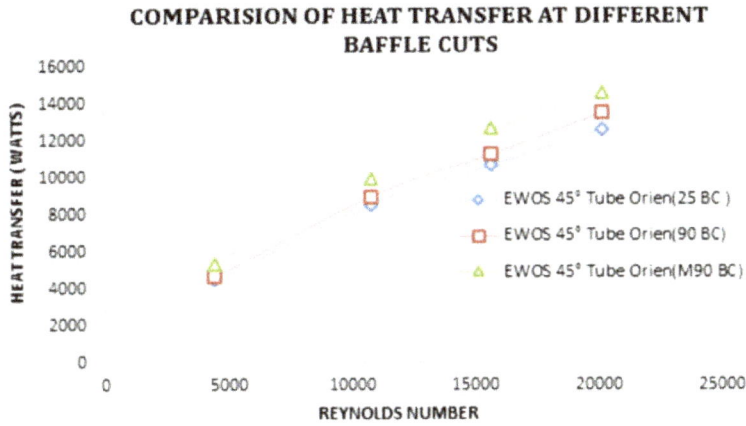

Fig. 22 Comparison of heat transfer at different baffle cuts with the maximum obtained tube orientation

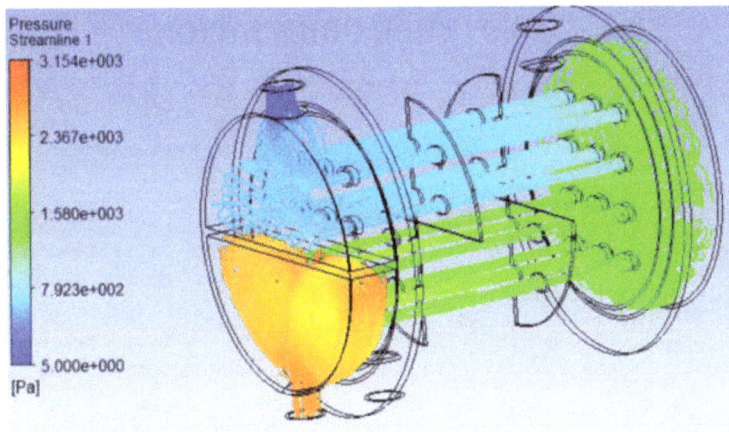

Fig. 23 Stream lines of tube side pressure drop at mirror quarter baffle cut with circular tubes

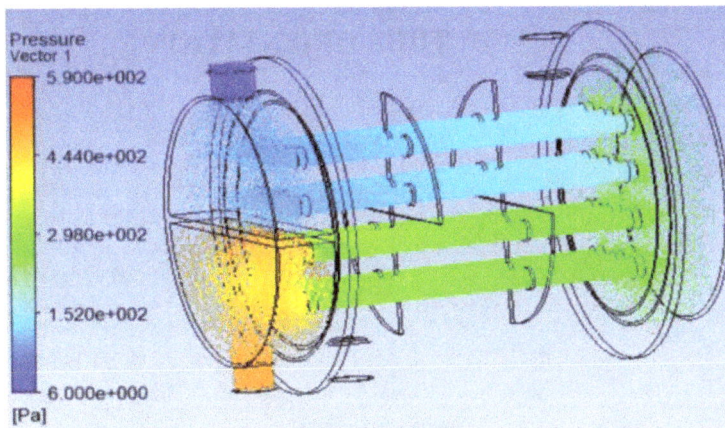

Fig. 24 Velocity vector of tube side pressure drop at mirror quarter baffle cut with elliptical tubes

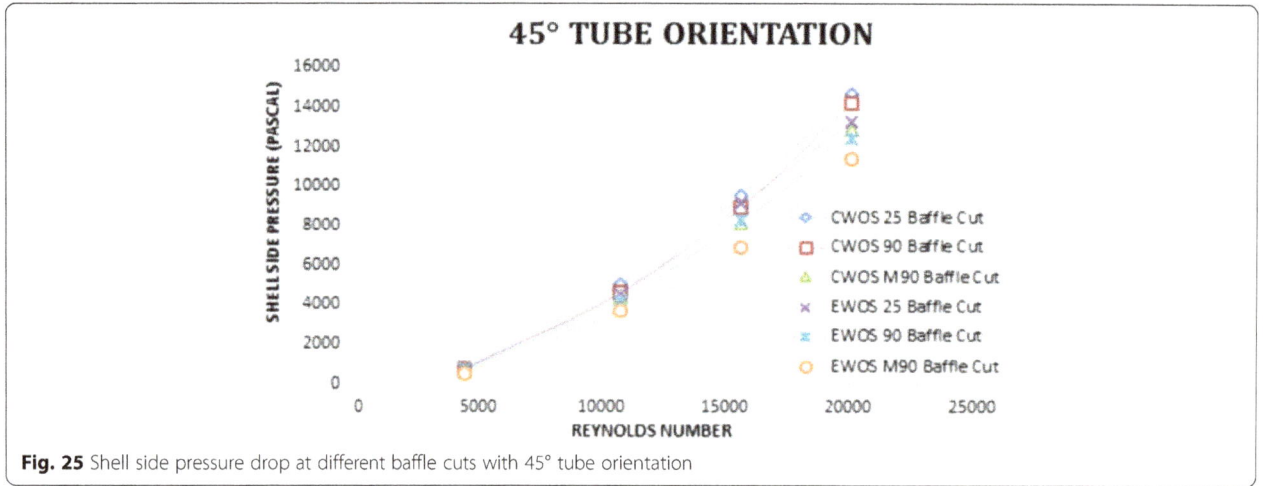

Fig. 25 Shell side pressure drop at different baffle cuts with 45° tube orientation

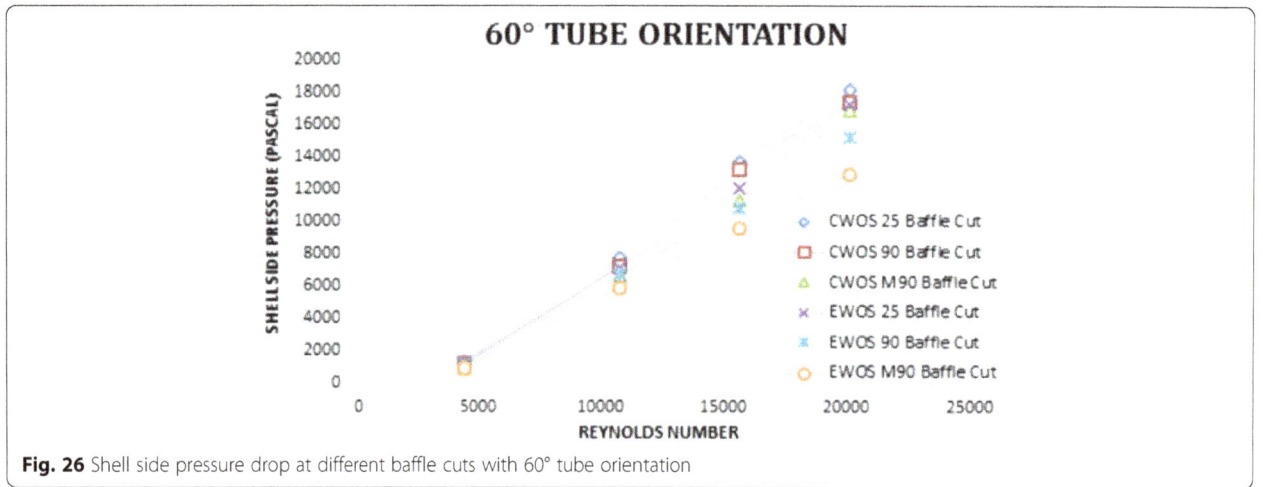

Fig. 26 Shell side pressure drop at different baffle cuts with 60° tube orientation

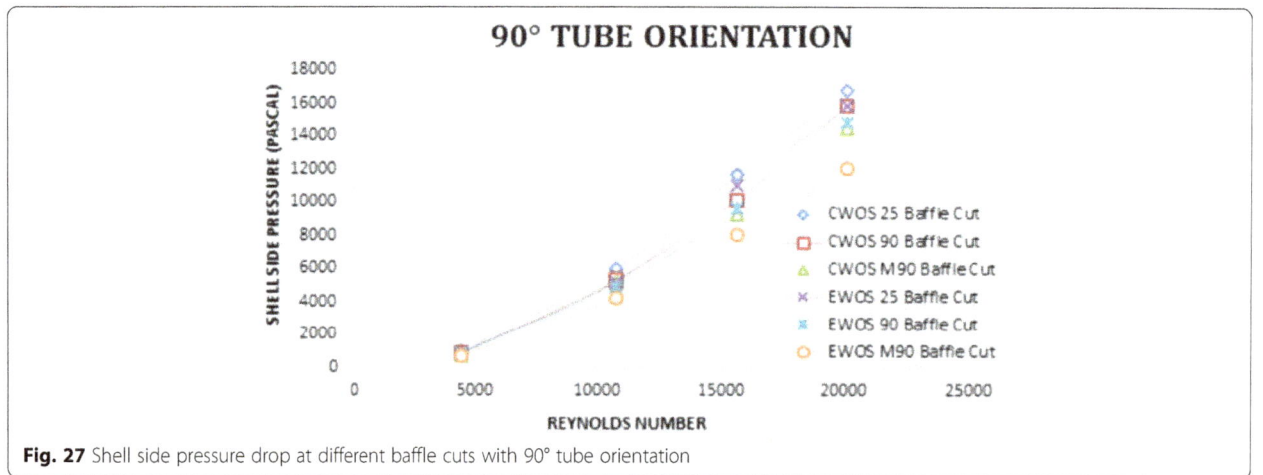

Fig. 27 Shell side pressure drop at different baffle cuts with 90° tube orientation

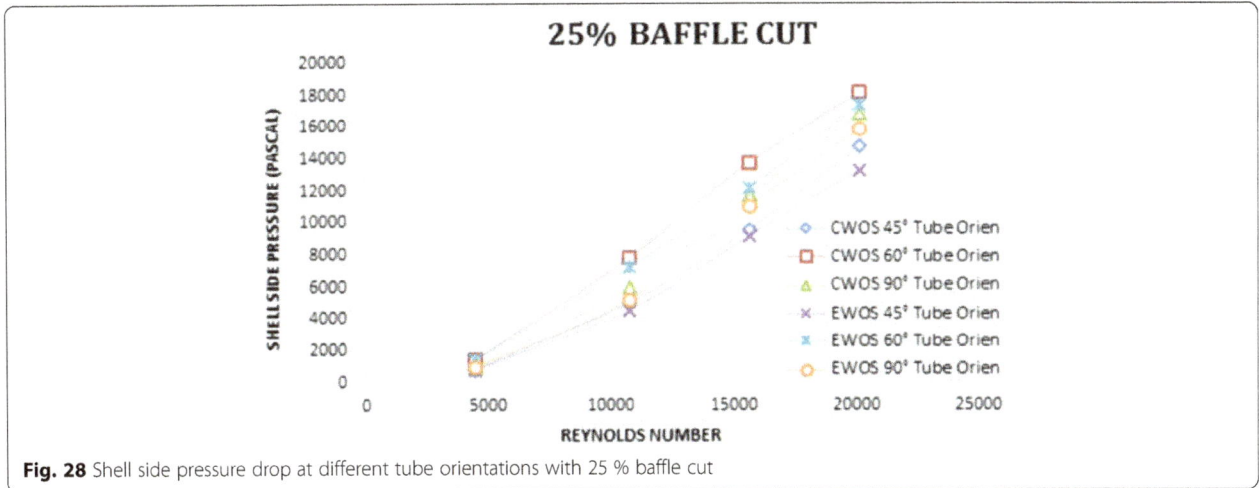

Fig. 28 Shell side pressure drop at different tube orientations with 25 % baffle cut

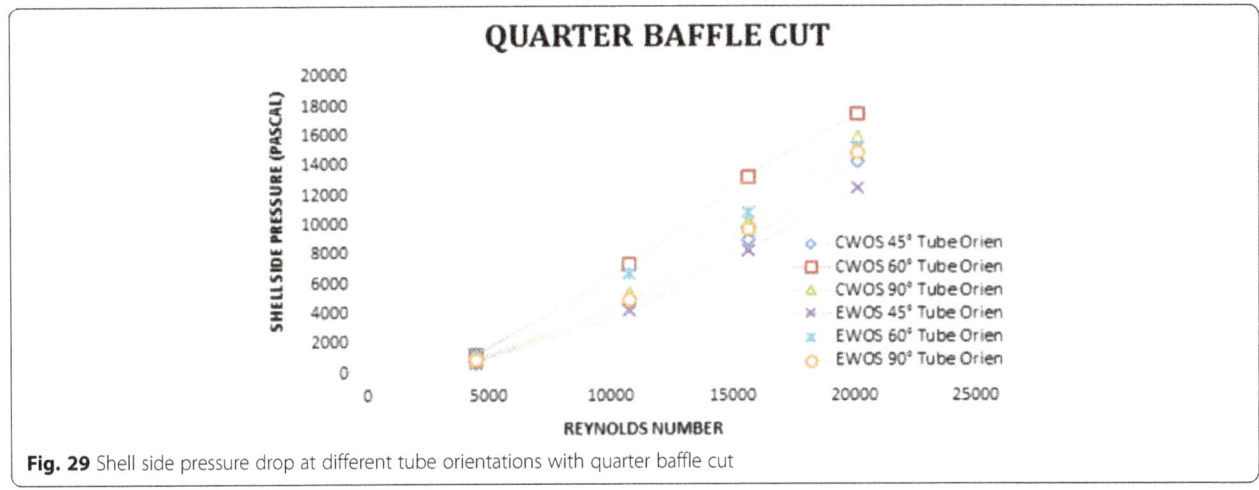

Fig. 29 Shell side pressure drop at different tube orientations with quarter baffle cut

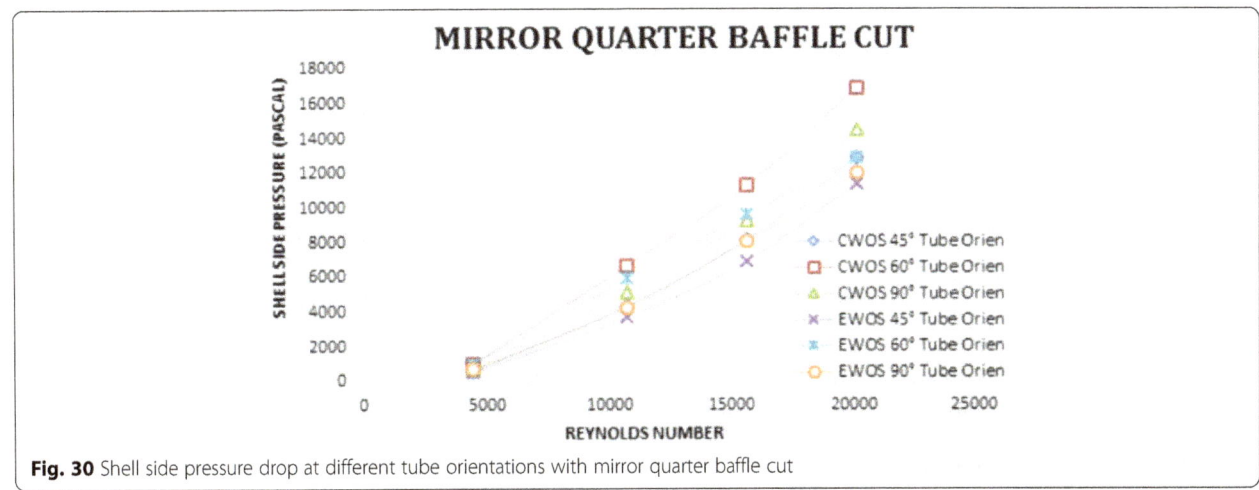

Fig. 30 Shell side pressure drop at different tube orientations with mirror quarter baffle cut

COMPARISION OF SHELL SIDE PRESSURE AT DIFFERENT TUBE ORIENTATION

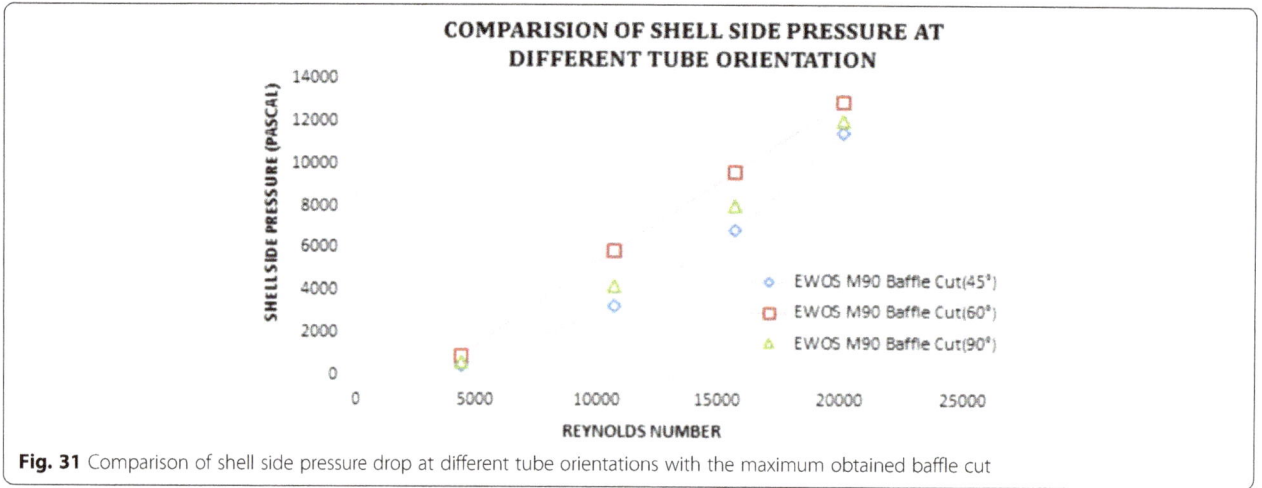

Fig. 31 Comparison of shell side pressure drop at different tube orientations with the maximum obtained baffle cut

COMPARISION OF SHELL SIDE PRESSURE AT DIFFERENT BAFFLE CUTS

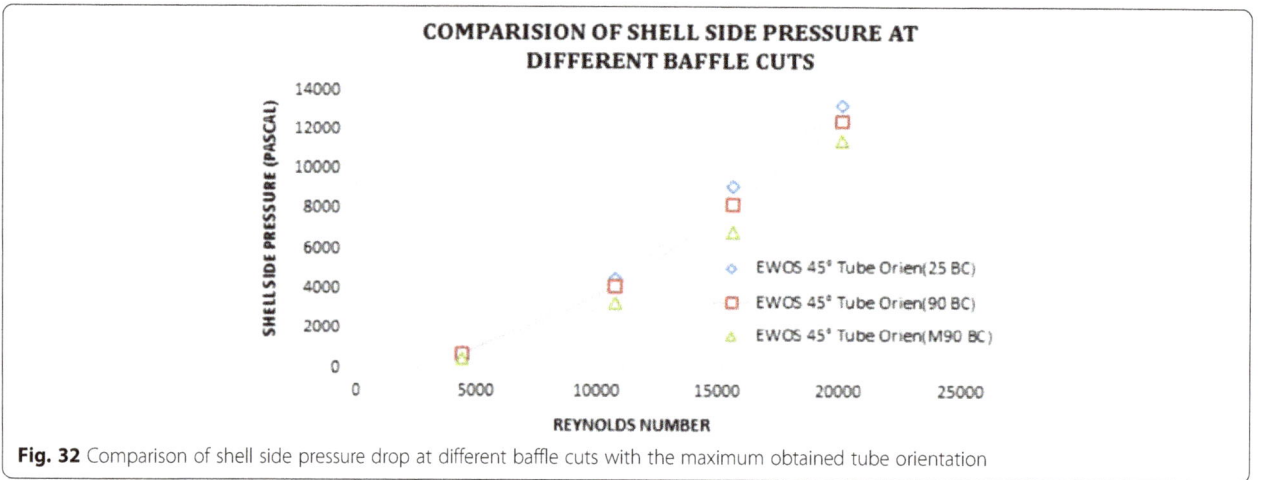

Fig. 32 Comparison of shell side pressure drop at different baffle cuts with the maximum obtained tube orientation

25% BAFFLE CUT

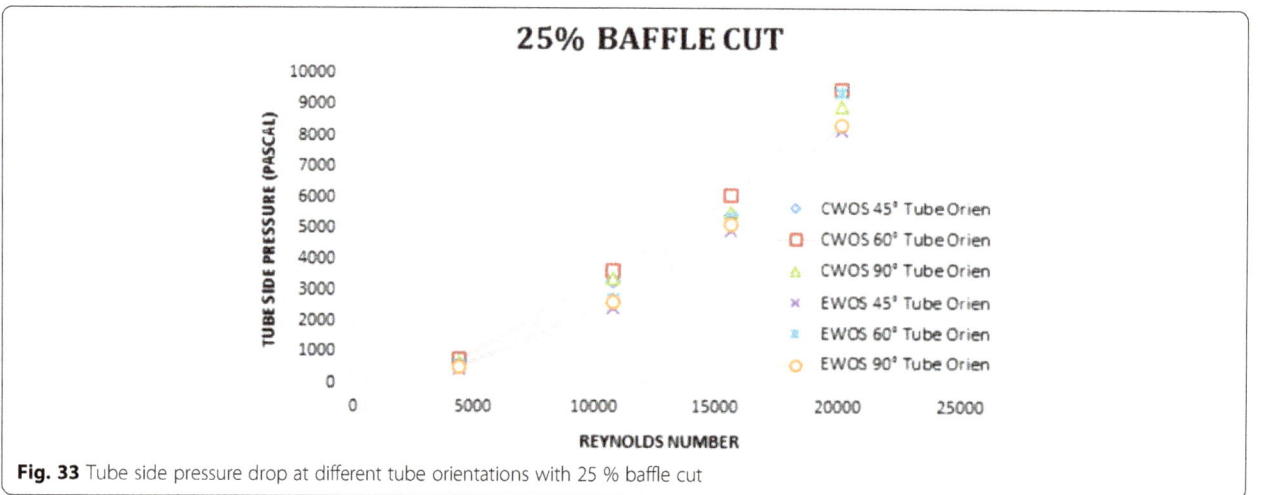

Fig. 33 Tube side pressure drop at different tube orientations with 25 % baffle cut

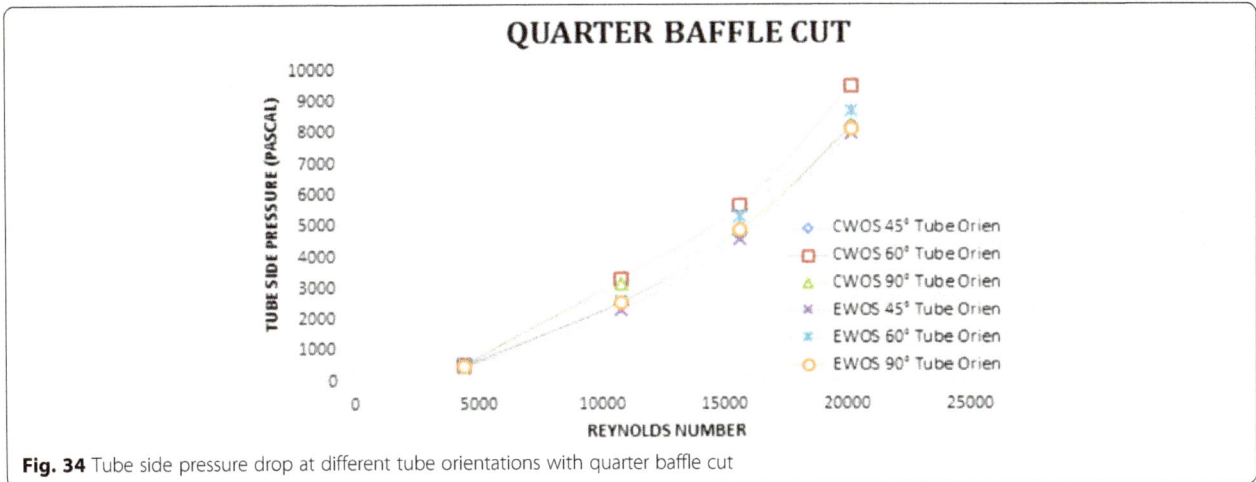

Fig. 34 Tube side pressure drop at different tube orientations with quarter baffle cut

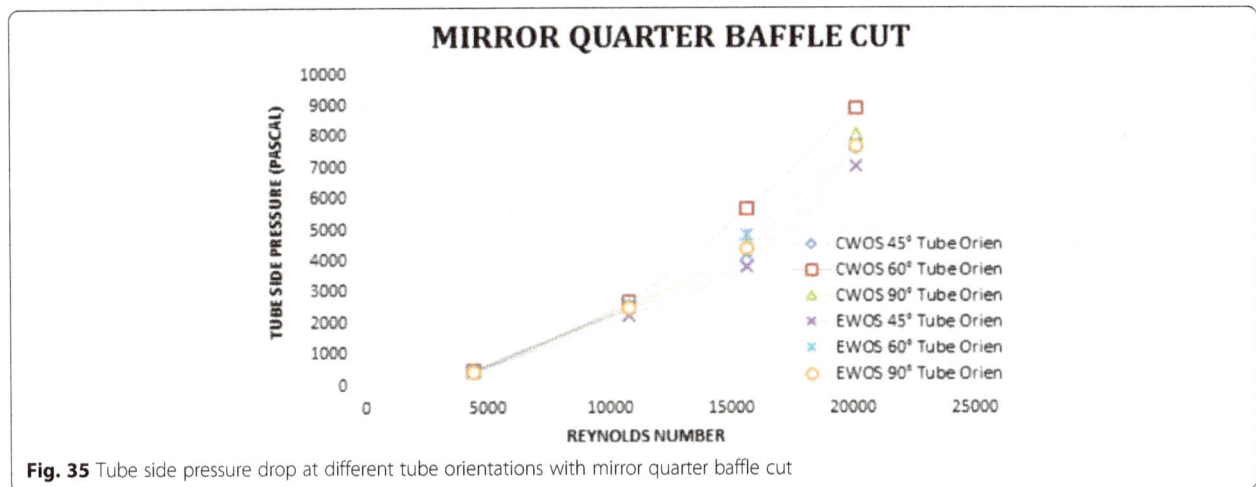

Fig. 35 Tube side pressure drop at different tube orientations with mirror quarter baffle cut

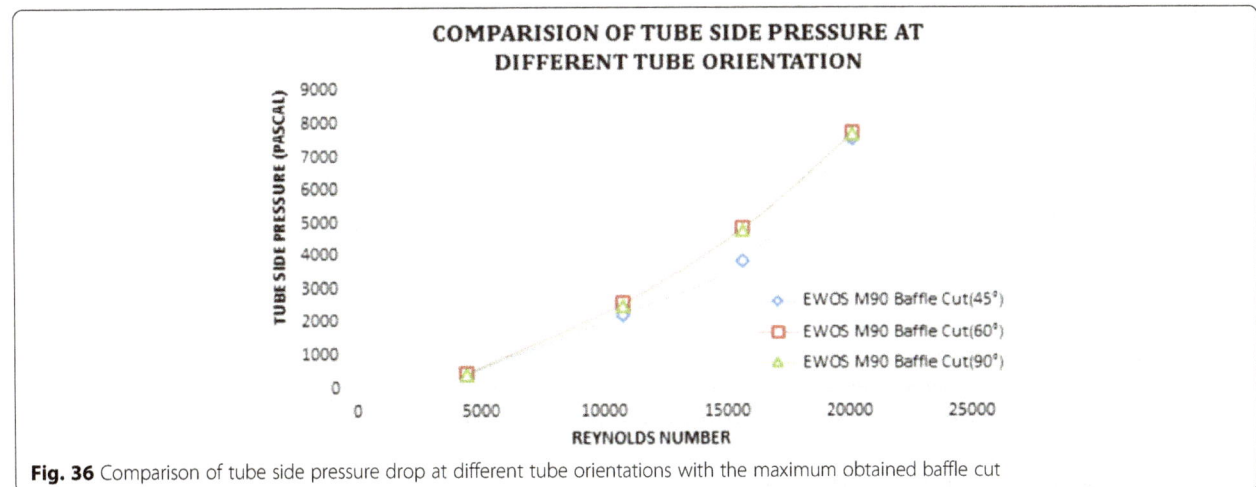

Fig. 36 Comparison of tube side pressure drop at different tube orientations with the maximum obtained baffle cut

while the same at 90° to 45° tube orientation, the pressure drop decreases by 9 % as shown in Figs. 33, 34, and 35.

Comparison of tube side pressure drop on shell and tube heat exchanger In comparison made in the circular tubes and the elliptical tubes, pressure drop decreases in the elliptical tube geometry for all Reynolds numbers. While considering tube orientation of the elliptical tubes, different baffle cuts of 25 %, quarter, and mirror quarter are maintained. Among all tube orientations, 45° tube orientation gives minimum pressure drop as shown in Fig. 36. Observations are found for the circular tubes and the elliptical tubes, then the pressure drop decreases to 24 % for different orientations. The overall pressure drop variation with mirror baffle cut, 45° tube orientation, and elliptical tubes is 25 % higher than existing shell and tube heat exchanger.

Conclusions

In the present study, CFD analysis is conducted on circular models and elliptical models of the STHE with various baffle cuts at different tube orientations. The tube geometry, tube orientation, and baffle cuts are considered to enhance the heat transfer rate by minimizing the pressure drop. Based on the simulation results, the following conclusions are derived.

a. The overall heat transfer rate in the Reynolds number range between 4000 to 20,000 increases with mirror baffle cut, 45° tube orientation, and the elliptical tube geometry is 10 % higher than the existing shell and tube heat exchanger.

b. Pressure drop reduction in the shell side is more influenced by baffle cuts. The pressure drop is minimum in the case of 45° tube orientation and mirror quarter baffle cut with elliptical tubes due to a uniform heat transfer through the heat exchanger.

c. If the tube geometry changes from circular to elliptical, the pressure drop over the tube side will be decreased. The pressure drop decrement for the circular tube to the elliptical tube in tube with mirror baffle cut, 45° tube orientation shows up to be 25 %.

Funding
This research received no specific grant from any funding agency from the public, commercial, or not-for-profit sectors.

Competing interests
The authors declare that they have no competing interests.

Author details
[1]Department of Mechanical Engineering, SISTAM College, Kakinada, India.
[2]Department of Mechanical Engineering, JNTU Kakinada, Kakinada, India.

References
Bell, KJ. (1963) Final report of the cooperative research programme on shell and tube heat exchangers, University of Delaware, Engineering Experimental Station, Bulletin No. 5.
Gay, B, Mackely, NV, & Jenkins, JD. (1976). Shell-side heat transfer in baffled cylindrical shell-and-tube exchangers-an electro chemical mass transfer modelling technique. *International Journal of Heat and Mass Transfer, 19*, 995–1002.
Halle, H, Chenoweth, JM, & Wabsganss, MW. (1988). Shell side water flow pressure drop distribution measurements in and industrial sized test heat exchanger. *Journal of Heat and Mass Transfer, 110*, 60–67.
Huang, X, Lu, Z, & Wang. (2001). Numerical modelling of the shell side flow in the shell and tube heat exchanger. In W. Liu (Ed.), *Proc. Int. Conf. on energy Conversion and application volume I, Wuhan, China* (pp. 301–304).
Kern, DQ. (1950a). *Process heat transfer* (pp. 127–171). New York: Mc-Graw-Hill.
Kern, DQ. (1950b). *Process heat transfer*. New York: McGraw-Hill.
Lei, YG, He, YL, Pan, C., et al. (2008). Design and optimization of heat exchangers with helical baffles. *Chemical Engineering Science, 63*(17), 4386–4395.
Li, HD, & Kottke, V. (1998a). Effect of leakage on pressure drop and global heat transfer in shell and tube heat exchanger for staggered tube arrangement. *International Journal of Heat and Mass Transfer, 41*(2), 425–433.
Li, HD, & Kottke, V. (1998b). Visualization and determination of local heat transfer coefficient in shell and tube heat exchangers for staggered tube arrangement by mass transfer measurements. *Experimental Thermal and Fluid Science, 17*(3), 210–216.
Li, H, & Kottke, V. (1998c). Effect of baffle spacing on pressure drop and local heat transfer in shell and tube heat exchangers for staggered tube arrangement. *International Journal of Heat and Mass Transfer, 41*(10), 1303–1311.
Li, HD, & Kottke, V. (1998d). Visualization and determination of local heat transfer coefficient in shell and tube heat exchangers for staggered tube arrangement by mass transfer measurements. *Experimental Thermal and Fluid Science, 17*, 210–216.
Master, BI, Chunangad, KS, Boxma, AJ, Kral, D, & Stehlik, P. (2006). Most frequently used heat exchangers from pioneering research to world wide applications. *Heat Transfer Engineering, 27*(6), 4–11.
Mukherjee, R. (1992). Use double-segmental baffles in the shell-and-tube heat exchangers. *Chemical Engineering Progress, 88*, 47–52.
Palen, JW, & Taborek, J. (1969). Solutions of shell side flow pressure drop and heat transfer by stream analysis method. *Chemical Engineering Progress Symposium Series, 65*, 53–63.
Pekdemir, T, Davies, TW, Haseler, LE, & Diaper, AD. (1994). Pressure drop measurements on the shell side of a cylindrical shell and tube heat exchanger. *Heat Transfer Engineering, 15*, 42–56.
Peng, B, Wang, QW, Zhang, C., et al. (2007). And experimental study of shell and tube heat exchanger with continuous helical baffles. *Journal of Heat Transfer, 129*, 1425–1431.
Qian, SW. (2002). *Handbook for heat exchanger design*. Beijing: Chemical Industry Press (in Chinese).
Rhodes, DB, & Carlucci, LN. (1983). Predicted and measured velocity distributions in a model heat exchanger. In *Int. Conf. on numerical methods in engineering Canadian nuclear society//American nuclear society* (pp. 935–948).
Seemawute, P, & Eiamsa-ard, S. (2010). Thermo hydraulics of turbulent flow threw a round tube by a peripheral-cut twisted tape with an alternate axes. *International Communications in Heat and Mass Transfer, 37*, 652–659.
Stevanovic, Z, Ilic, G, Radojkovic, N, Vukic, M, Stefanovic, V, & Vuckovic, G. (2001). Design of shell and tube heat exchanger by using CFD technique—part one: thermo hydraulic calculations. *Mechanical Engineering, 1*(8), 1091–1105.
Tinker, T. (1951). *Shell side characteristics of shell and tube heat exchangers, parts I, II, and III* (pp. 97–116). London: General Discussion on Heat Transfer Inst. Mech. Eng.

7

Effect of heating time of adsorber-collector on the performance of a solar adsorption refrigerator

Noureddine Cherrad[1,2]*, Adel Benchabane[1], Lakhdar Sedira[3] and Amar Rouag[1]

Abstract

This paper presents a numerical study of heat transfer inside the adsorber-collector of a solar adsorption refrigerator using the activated carbon AC35-methanol pair. The objective is to estimate the amount of the heat loss through the adsorber-collector, during the solar heating phase, and to determine the effect of heating time on the thermal efficiency of the system. The numerical results showed that the heating time is the most important factor affecting the amount of energy loss. It has shown that the shorter heating time corresponds to the higher efficiency of the adsorber-collector. In addition, a new optimal coefficient of performance, COP_{optm}, is proposed to determine the number of adsorbers to be added to a machine. This latter is considered for consuming an energy equivalent to that received by the adsorber-collector. These additional adsorbers use a heat transfer fluid, coming from the adsorber-collector, instead of direct heating by solar radiation. An application example is presented using experimental results obtained from the literature. It has shown that the number of the additional adsorbers can reach three adsorbers.

Keywords: Adsorber thermal efficiency, Heat loss, Heat recovery, Optimal coefficient of performance, Solar adsorption refrigerator

Background

The fossil energy is the foundation of the global economy; the consumption, the transport, and the delivery of this energy source are harmful to the human being, the animals, and the nature. We know that the combustion of oil rejects a thousand polluting substances in the atmosphere; one of the well-known direct impact is global warming.

The search for alternative energy sources brought together environmentalists and those who for other reasons do not want to be dependent on oil suppliers. Certainly, other sources of energy may be an alternative to oil, such as solar energy.

Solar power in itself does not present a problem. The problem of its use comes mostly from the investment cost which is relatively expensive compared to fossil energy because the profitability of certain types of solar

* Correspondence: chernoured@gmail.com
[1]Laboratoire de Génie Energétique et Matériaux, LGEM, Université de Biskra, B.P. 145 R.P., Biskra 07000, Algeria
[2]Département de Génie Mécanique, Université de Ouargla, Faculté des Sciences Appliquées, Ouargla 30 000, Algeria
Full list of author information is available at the end of the article

systems is still weak. The improvements are not intended only for the materials of the solar installation but also for the performance of the system, which includes all energy losses caused by its components.

The ordinary refrigerating machines, even those that are operated by electricity, are indirect consumer devices of petroleum energy, so a system that must be replaced in the future by an equipment belonging to solar power system technology. In the refrigeration machines powered by the electricity, the compressor is the driving force of the thermodynamic cycle of the system, while in refrigeration machines powered by the sun, the phenomenon of desorption by solar heating can be an alternative to the compression phase. This phenomenon takes place in the adsorber-collector of the solar machine; the more the desorption is important, the more the cold production is high.

The estimation of coefficient of heat loss per unit of time and temperature within the adsorber-collector is already developed by Klein 1975 and discussed by Duffie and Beckman 2013. Jing and Exell 1994 used the coefficient of loss developed by Klein 1975 for simulation and sensitivity analysis of an intermittent solar-powered charcoal/methanol refrigerator. Any-anwu et al. 2001 have made a study of an adsorption refrigerator using the charcoal/methanol pair, where the numerical model takes into account loss coefficient. The results of coefficient of performance (COP) are given according to the physical and dimensional characteristics of adsorber-collector. The loss coefficient was also considered by Leite and Daguenet 2000. It was a numerical simulation for instant transfers of heat and mass in each element of the refrigerating adsorption machine during the typical average day of each month. Another use of the coefficient of Klein was in the work of Chekirou et al. 2014, who introduced a modeling of heat and mass transfer in the tubular adsorbent of a refrigerating adsorption machine.

The objective of our study is to analyze the effect of heat loss within the adsorber-collector on the performance of solar adsorption refrigeration machines using AC35-methanol as working pair and to determine the criteria affecting the increase of this amount of energy.

The amount of heat loss through the adsorber-collector was calculated using the coefficient of loss over the solar heating time and an interval of temperatures between the temperature of adsorption (temperature at start of heating) and temperature of generation (temperature at end of heating). The study is done using a numerical program (Fortran) and validated with experimental results obtained from the literature.

Methods
Basic components
The main components of solar adsorption refrigerator are shown in the Fig. 1: the adsorber-collector containing the adsorbent-adsorbate pair (activated carbon AC35-methanol), the air-cooled condenser, and the evaporator within the cold chamber (Rouag et al. 2016).

A valve is set at inlet and outlet of adsorber-collector, to control the flow of the refrigerant from the adsorbent towards the condenser during the desorption phase, and from the evaporator towards the adsorbent during the adsorption phase.

Intermittent cycle
Four thermodynamic processes form the intermittent ideal cycle of solar-powered adsorption refrigerator are shown in Fig. 2.

First, the adsorber-collector is heated by solar energy, which increases the pressure of the adsorbent-adsorbate pair up to the condensing pressure P_c necessary for the condenser (processes A–B). In state B of cycle, the valve at the inlet of condenser is open, and the continuing of the heating at constant pressure induces the desorption of refrigerant (adsorbate) from the adsorbent towards the air-cooled condenser to be condensed at condensing temperature T_c (processes B–C).

The desorbed mass during processes B–C is the difference between the maximal mass of adsorbate in state B and the minimal mass of adsorbate in state C.

The maximum temperature (generating temperature T_g) reached by the adsorbent will be in state C. At this value of temperature, the valve is closed and solar irradiance

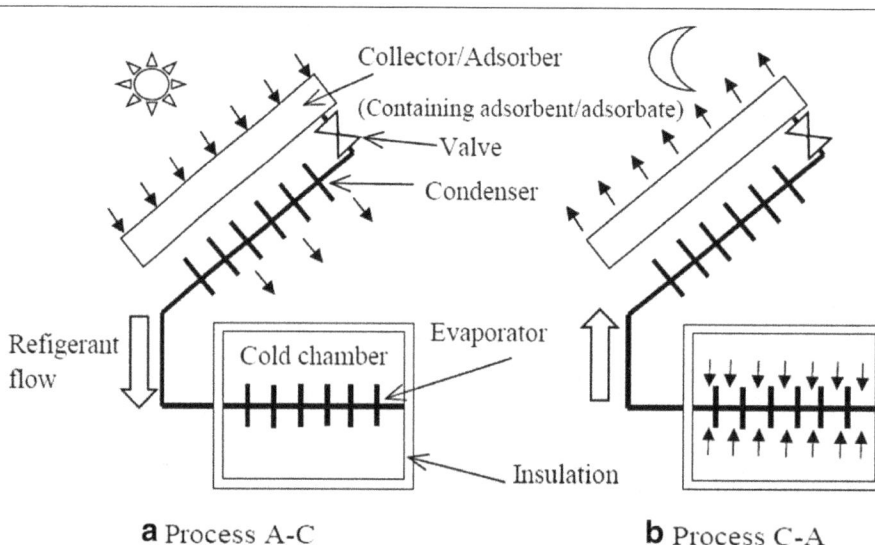

Fig. 1 Schematic representation of the solar adsorption refrigerator. **a** Process A-C. **b** Process C-A

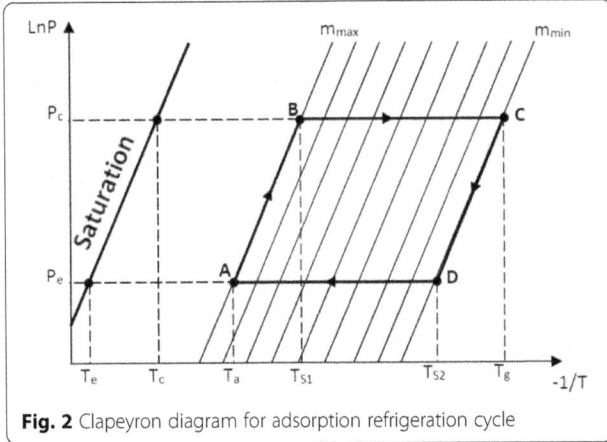

Fig. 2 Clapeyron diagram for adsorption refrigeration cycle

starts to decrease, which induces the drop of temperature and pressure of adsorber-collector (processes C–D).

When the pressure reaches the value P_e, which reigns in the evaporator, the valve is open, and the continuing of the cooling of adsorbent at constant pressure induces the evaporation of refrigerant at an evaporating temperature T_e from the evaporator towards the adsorbent by the adsorption phenomena (processes D–A). By this way, the evaporated refrigerant extracts heat and generates cold production within the cold chamber.

The adsorbed mass during processes D–A is the difference between the maximal mass of adsorbate in state A and the minimal mass of adsorbate in state D.

This cycle is said to be intermittent because the cold production starts only at sunset.

Adsorbed mass

Dubinin and Astakov proposed a state equation for mass adsorbed by microporous medium in equilibrium with polymodal distribution of pore size (Leite and Daguenet 2000):

$$X(T,P) = W_0 \rho_l(T) exp\left\{ -D\left[T.ln\left(\frac{P_s(T)}{P} \right) \right]^n \right\} \quad (1)$$

Where W_0 is the maximum volume adsorbed, ρ_l is the density of the adsorbate in the liquid state, D is the coefficient of affinity, T is the temperature of the adsorber-collector, P_s is the saturation pressure of the adsorbate, P is the equilibrium pressure of adsorbent-adsorbate pair, and n is the parameter of adjustment of the D–A equation.

The adsorbate mass concentration maximal X_{max} and minimal X_{min} are calculated by Eq. (1), respectively, at the adsorption temperature T_a and the generating temperature T_g as the following:

$$X(T_a, P_e) = X(T_{s1}, P_c) = X_{max} \quad (2)$$

$$X(T_g, P_c) = X(T_{s2}, P_e) = X_{min} \quad (3)$$

Where T_{s1} and T_{s2} are respectively the temperature at start of desorption and temperature at start of adsorption.

During the adsorption, the evaporating pressure is $P_e = P_s$ (T_e), and for the desorption, the condensing pressure is $P_c = P_s$ (T_c). T_e and T_c are respectively the evaporating temperature and the condensing temperature.

Thermal efficiency of the adsorber-collector

The total energy received by the adsorber-collector Q_{ge} can be divided in five parts:

$$Q_{ge} = Q_1 + Q_2 + Q_3 + Q_{des} + Q_L \quad (4)$$

The heats Q_1, Q_2, and Q_3 are respectively energy supplied to heat the adsorbent, energy supplied to heat the tubes in metal, and energy supplied to heat the adsorbate mass. They can be calculated as the following (Chekirou et al. 2011):

$$Q_1 = m_d Cp_d(T_g - T_a) \quad (5)$$

m_d is the adsorbent mass and Cp_d is the specific heat of adsorbent.

$$Q_2 = m_t Cp_t(T_g - T_a) \quad (6)$$

m_t and Cp_t are respectively the mass and specific heat of metallic tubes containing the adsorbent.

$$Q_3 = m_d \left(X_{max} \int_{T_a}^{T_{S1}} Cp_l(T)dT + \int_{T_{S1}}^{T_g} X(T)Cp_l(T)dT \right) \quad (7)$$

Cp_l is the specific heat of adsorbate in liquid state.

The heat necessary for desorption process Q_{des} is given by (Chekirou 2008):

$$Q_{des} = m_d nD \int_{T_{S1}}^{T_g} X(T)T^n \left(ln\frac{P_s(T)}{P_c} \right)^{n-1} \frac{q_{st}^2(T)}{rT^2} dT \quad (8)$$

Isosteric heat q_{st} is calculated as the function of the pressure and temperature:

$$q_{st} = L(T_c) + rTln\left(\frac{P_s(T)}{P_c} \right)$$
$$+ \left[\frac{\alpha rT}{nD} \right] \left[Tln\left(\frac{P_s(T)}{P_c} \right) \right]^{(1-n)} \quad (9)$$

L is the latent heat of the adsorbate, r is the particular gas constant of the adsorbate, and α is the thermal expansion coefficient of the adsorbate.

According to Anyanwu et al. 2001, the heat loss from the lateral surfaces of adsorber-collector is assumed to be negligible.

The global loss coefficient is given by Duffie and Beckman 2013 as follows:

$$U_L = U_t + U_b \qquad (10)$$

U_t is the loss coefficient from the top of the adsorber-collector, and U_b is the loss coefficient from the bottom of the adsorber-collector.

where:

$$U_t(T_p, T_{am}) = \left[\frac{n_g}{\frac{c_0}{T_p} \left(\frac{T_p - T_{am}}{n_{g+f}} \right)^e} + \frac{1}{h_w} \right]^{-1}$$

$$+ \left[\frac{\sigma (T_p + T_{am}) (T_p^2 + T_{am}^2)}{(E_p + 0.00591 n_g h_w)^{-1} + \frac{2n_g + f - 1 + 0.133 E_p}{E_g} - n_g} \right] \qquad (11)$$

with:

$$e = 0.43 \left(1 - \frac{100}{T_p} \right)$$

$$f = (1 + 0.089 h_w - 0.1166 h_w E_p)(1 + 0.07866 n_g)$$

$$c_0 = 520 (1 - 0.000051 \Omega^2)$$

$0° < \Omega < 70°$, and for $70° < \Omega < 90°$ we take $\Omega = 70°$

$$h_w = 2.8 + 3.0 W_V$$

T_p is the wall temperature of the top of adsorber-collector, T_{am} is the ambient temperature, n_g is the number of glass cover of adsorber-collector, E_p is the emissivity of the top wall of the adsorber-collector, E_g is the emissivity of the glass cover of the adsorber-collector, Ω is the adsorber-collector inclination, and W_v is the wind velocity.

Chekirou et al. 2014 took U_b as the constant.

To estimate the amount of total heat loss towards the air ambient during processes A–C (Fig. 2), the following equation is used:

$$Q_L = S.(U_t + U_b).(Tp_{max} - Tp_{min}).t_{max} \qquad (12)$$

Using Eq. (12), we cover the full time of heating process t_{max} on a range of wall adsorber-collector temperatures from Tp_{min} (temperature of adsorber-collector wall at the start of heating) to Tp_{max} (temperature of adsorber-collector wall at the end of heating).

The temperatures Tp_{max} and Tp_{min} given in Eq. (12) can be substituted respectively by the temperatures T_g and T_a, so:

$$Q_L = S.(U_t + U_b).(T_g - T_a).t_{max} \qquad (13)$$

And for Eq. (11), the temperature ambient T_{am} is considered to be equal to T_a.

It is known that the refrigerator adsorber-collector needs a useful energy Q_u and the remainder is a lost energy Q_L; therefore,

$$Q_u = Q_{ge} - Q_L \qquad (14)$$

While the recovery heat Q_r from the adsorber-collector is a part or all the lost energy Q_L, so $Q_r \le Q_L$, it varies from 0 to Q_L, then we can write:

$$Q_r = Q_L \left(1 - \frac{t_h}{t_{max}} \right) \qquad (15)$$

Equation (15) means that $Q_r = 0$ for a heating time $t_h = t_{max}$ and $Q_r = Q_L$ for heating time $t_h = 0$.

On the other hand, the thermal efficiency of the solar-powered adsorber-collector can be expressed as follows:

$$\eta = \frac{Q_u}{Q_{ge}} = \frac{Q_{ge} - Q_L}{Q_{ge}} = 1 - \frac{Q_L}{Q_{ge}} \qquad (16)$$

Coefficient of performance

The coefficient of performance of an adsorptive cooling system is equal to the heat extracted in the evaporator per the heat received in the adsorber-collector (Douss and Meunier 1988):

$$COP = \frac{Q_{ev}}{Q_{ge}} \qquad (17)$$

Where:

$$Q_{ev} = m_d (X_{max} - X_{min}) \left[L(T_e) - \int_{T_e}^{T_c} Cp_l(T) dT \right] \qquad (18)$$

The first term of Eq. (18) represents the heat absorbed by the evaporation of the refrigerant (adsorbate) at the evaporation temperature T_e. The second term represents the sensible heat necessary to bring the condensate of its condensing temperature T_c to evaporating temperature T_e (Chekirou et al. 2011).

Results and discussion

Tables 1 and 2 give all the relevant data of numerical computation.

The pressure of saturation P_s, the latent heat L, density ρ_l, and the mass specific heat of adsorbate (methanol) in the liquid state Cp_l are estimated by interpolation of the data given in Table 3.

For each reference of validation, i.e., Douss and Meunier 1988 and Lemmini and Errougani 2007, the mass m_d of the adsorbent (activated carbon AC35) has been taken the same as that considered by the authors. Lemmini and

Table 1 Data used in the present numerical study

Symbol	Parameter	Value
Cp_d	Specific heat of adsorbent	920 J kg^{-1} K^{-1}
Cp_t	Specific heat of tubes (copper) containing the adsorbent	380 J kg^{-1} K^{-1}
E_g	Emissivity of glass cover	0.88[a]
E_P	Emissivity of the top wall of adsorber-collector	0.1[a]
n_g	Number of glass cover	1[a]
r	Particular gas constant	259.34 J kg^{-1} K^{-1}
S	Adsorber-collector surface	0.73 m^2
t_{max}	Maximum time of heating process A–C (Fig. 2)	12 h
U_b	Loss coefficient from the bottom of the adsorber-collector	0.9 W K^{-1a}
W_v	Wind velocity	1 m s^{-1}
a	Thermal expansion coefficient of adsorbate	0.00126 K^{-1b}
Ω	Adsorber-collector inclination	34 °C[a]
σ	Stefan-Boltzmann constant	5.670367E-8 W m^{-2} K^{-4}

[a]Data obtained from the work of Chekirou et al. 2014
[b]Data obtained from the work of Goodwin 1987

Table 3 Thermophysical properties of methanol along the saturation line (Bejan and Kraus 2003)

T [K]	P_s [MPa]	ρ_l [kg m^{-3}]	Cp_l [kJ kg^{-1} K^{-1}]	L [kJ kg^{-1}]
200	6.1×10^{-6}	880.28	2.2141	1289.99
250	0.00081	831.52	2.3121	1234.79
300	0.0187	784.51	2.5461	1166.17
325	0.0603	760.74	2.7223	1124.564
350	0.1617	735.84	2.9362	1075.936
375	0.3748	708.86	3.1891	1017.33
400	0.7737	678.59	3.4912	944.57
425	1.4561	643.38	3.8713	852.59
450	2.5433	600.49	4.4067	739.19
475	4.1688	544.35	5.3805	610.48
500	6.5250	451.53	9.9683	391.09

Errougani 2007 have given the value of the adsorbent mass, which is equal to 14.5 kg, and for the machine studied by Douss and Meunier 1988, it is estimated by 12.5 kg.

The same for the mass m_t of metallic tubes containing the adsorbent, which are estimated to be equal to 45.68 and 26.024 kg for Douss and Meunier 1988 and Lemmini and Errougani 2007 respectively.

The maximum time t_{max} of heating processes A–C (Fig. 2) is considered for all day, so it is an average of 12 h.

Validation of numerical model
First, we have validated our numerical model by the results given by Douss and Meunier 1988. The authors have studied an experimental unit of an adsorption refrigeration machine that operates by electrical heating instead of the solar heating; therefore, heat loss Q_L cannot be estimated by Eq. (13), which is destined to solar

Table 2 Values of Dubinin-Astakhov equation parameters

Symbol	Parameter	Value
D	Coefficient of affinity	5.02E-07[a]
n	Parameter of adjustment of D-A equation	2.15[b]
W_0	Maximum adsorbed volume	0.000425 m^3 (of adsorbate) kg^{-1} (of adsorbent)[b]

[a]Data obtained from the work of Douss and Meunier 1988
[b]Data obtained from the work of Douss & Meunier 1988; Pons & Grenier 1986 and Critoph 1988

adsorption adsorber-collectors. Douss and Meunier 1988 have given this amount of heat; it is equal to 365 kJ. For the temperatures' data, we used the data of working fluid (WF) by Douss and Meunier 1988.

The compared results of *COP* are shown in Figs. 3, 4, and 5.

The average relative errors between the results compared in the Figs. 3, 4, and 5 are respectively 4.77, 5.22, and 9.96%.

The validation with the work of Douss and Meunier 1988 was made for the four calculated heats, i.e., Q_1, Q_2, Q_3, and Q_{des}, while Q_L is given by the author. Our numerical model takes into account the prediction of this quantity (Q_L), that is why we are going to carry out a second validation with Lemmini and Errougani 2007 to validate all the total energy Q_{ge}, ensuring by this way the validation of Eq. (13).

Lemmini and Errougani 2007 studied an experimental prototype of a solar adsorption refrigeration machine, where the recorded results are given in Table 4.

We have calculated the cooling production quantities Q_{ev} obtained by Lemmini and Errougani 2007 for eleven tested days by using Eq. (17), where *COP* and Q_{ge} are known. After that, we have predicted the operating temperatures of the studied machine.

The advantage of the developed numerical program is the prediction of operating temperatures T_a, T_c, and T_g for given cold production Q_{ev} and evaporating temperature T_e. The calculated temperatures will be used to estimate the amount of heat Q_{ge}.

The numerical results of *COP* variation using Eq. (17) compared with those obtained during the tested days by Lemmini and Errougani 2007 are shown in Table 4 and Fig. 6. The comparison shows an average relative error of 5.03%.

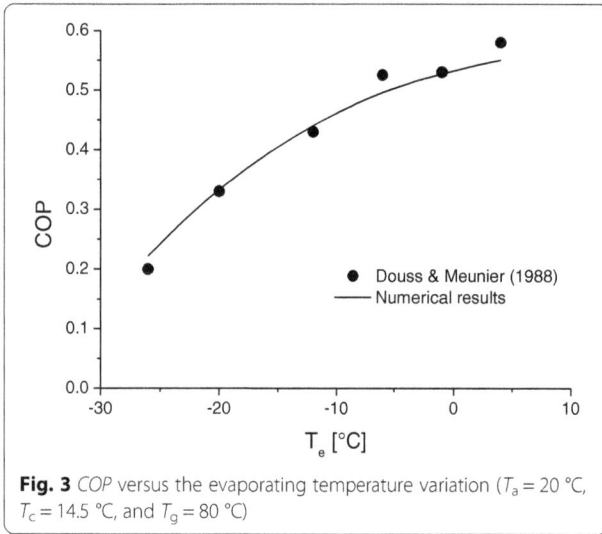

Fig. 3 COP versus the evaporating temperature variation ($T_a = 20$ °C, $T_c = 14.5$ °C, and $T_g = 80$ °C)

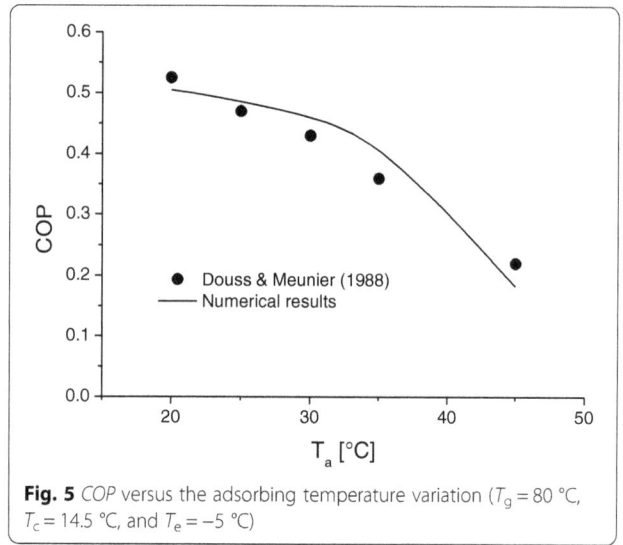

Fig. 5 COP versus the adsorbing temperature variation ($T_g = 80$ °C, $T_c = 14.5$ °C, and $T_e = -5$ °C)

Thermal efficiency and performance coefficient

The obtained results for the amounts of heats estimated, the thermal efficiency of the adsorber-collector, and the coefficient of performance of solar adsorption refrigerator in this subsection are based on the data considered for validation with the results of Lemmini and Errougani 2007.

As shown in Fig. 7, for one adsorber-collector, 25% of total received heat during t_{max} is useful and the rest must be recovered. The thermal efficiency takes two ends, maximum and minimum, which are 100% for th = 0 and 25% for $t_h = t_{max}$. This can be deduced from Eq. (16).

The center of this curve takes a parabolic form, where it must coincide with the recovered heat quantity $Q_r(t_h)/Q_{ge}(t_{max}) = 50\%$ and the quantity of heat received $Q_{ge}(t_h)/Q_{ge}(t_{max}) = 50\%$ at $t_h/t_{max} = 0.33$.

At this time where $Q_u(t_h)/Q_{ge}(t_{max}) = Q_L(t_h)/Q_{ge}(t_{max}) = 25\%$ and using Eq. (16), the value of thermal efficiency η is

50%. It is the moment where the three quantities must take the same value which equals to 50% and corresponds to $t_h/t_{max} = 0.33$.

The shorter heating times correspond to higher efficiencies of the adsorber-collector. It can allow to recover up to 75% of total heat received by the adsorber-collector during t_{max} with zero heat loss. This means that improving design of the adsorber-collector is required to ensure the useful amount of solar energy necessary for system functioning during a very short time. So, it is recommended to use a heat transfer fluid instead of direct heating by solar radiation. Thus, we can isolate the upper side of the adsorber-collector that represents the large hole of the heat, and we command the heating time by the temperature and the flow rate of the heat transfer fluid. The fluid will flow in

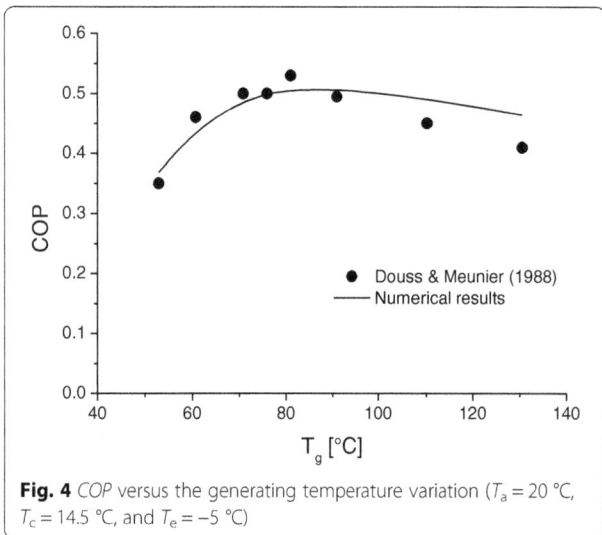

Fig. 4 COP versus the generating temperature variation ($T_a = 20$ °C, $T_c = 14.5$ °C, and $T_e = -5$ °C)

Table 4 Numerical results of COP variation compared with those calculated during the tested days by Lemmini and Errougani 2007

Day	COP[a]	COP[b]	Relative error [%]
02 April	9.00	9.65	7.22
03 April	8.00	7.86	1.75
04 April	11.00	10.10	8.18
05 April	9.00	8.49	5.67
07 April	6.90	7.54	9.28
10 April	6.00	5.73	4.50
11 April	10.20	10.69	4.80
12 April	8.60	8.55	0.58
13 April	7.90	7.73	2.15
14 April	6.00	6.24	4.00
15 April	9.00	9.65	7.22

[a]Data obtained from the work of Lemmini and Errougani 2007
[b]Numerical results obtained in the present work

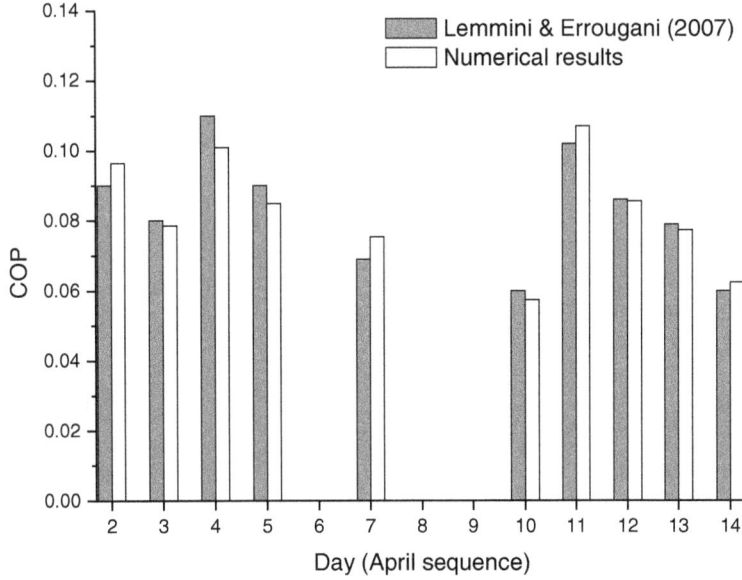

Fig. 6 Comparison of the numerical *COPs* of the present study with those obtained by Lemmini and Errougani 2007

the tubes of a heat exchanger set inside the adsorber-collector. Even if the prior heating of heat transfer fluid by solar energy takes time, we can use a system of energy storage in the form of heat. Thus, we ensure the heating of the adsorber-collector even at night, and the cycle of the machine becomes continuous instead of intermittent cycle with a very good performance.

With a recovery energy system, the number of additional adsorbers can be added to the machine with a total received energy Q_{ge} by only one additional adsorber is Num_{re}, so we can write:

One additional adsorber needs $\rightarrow Q_u$ (*)
Number of additional adsorbers (Num_{re}) to be added need $\rightarrow Q_r$ (**)

From (*), (**), Eqs. (14) and (15):

$$Num_{re} = \frac{Q_r}{Q_u} = \frac{Q_L\left(1-\frac{t_h}{t_{max}}\right)}{Q_{ge} - Q_L} \qquad (19)$$

As shown in Fig. 8, for a heating time t_h less than 4 h (t_{max} is assumed to be equal to 12 h), we can recover a

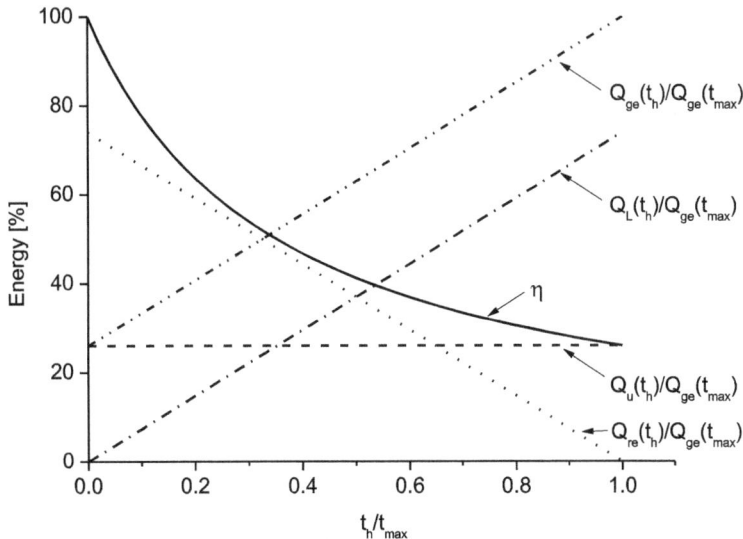

Fig. 7 Energy ratios and efficiency of adsorber-collector as function of the ratio of heating time and maximum heating time ($T_a = T_c = 25$ °C, $T_e = 0$ °C, and $T_g = 90$ °C)

useful energy for functioning of two to three additional adsorbers.

The recovery of energy loss Q_L leads us to propose a new optimal *COP*, it may be given by:

$$\text{COP}_{\text{optm}} = (\text{Num}_{\text{re}} + 1).\text{COP} \qquad (20)$$

By Eq. (20), we can see that a high optimal coefficient of performance COP_{optm} corresponds to a heating of shorter time t_h. If $t_h = t_{\text{max}}$, no energy recovery (by Eq. (19) $Qr = 0 \Rightarrow Num_{\text{re}} = 0$) and *COP* optm would take the value of the normal *COP*.

As shown in Fig. 9, with an energy recovery system from one adsorber-collector, the COP_{optm} value can reach up to four times the *COP* in the normal case.

Conclusions

In the present paper, a numerical model was developed to predict the different amounts of heat received by the adsorber-collector of a solar-powered adsorption refrigeration machine using activated carbon AC35/methanol as a working pair.

Four amounts of heat are useful for the operation of system, while the fifth amount is an energy loss. The estimate of this lost energy showed that the time of heating of the adsorber-collector is a very important factor affecting its thermal efficiency.

The shorter heating time allows to cut off the continuation of heat loss into the external environment and gives a good performance.

If we use a system heated by solar energy, the reduction of the heating time is somewhat difficult because it is related to the radiation intensity, which is as function of the time. So, it is useful to use a heat transfer fluid

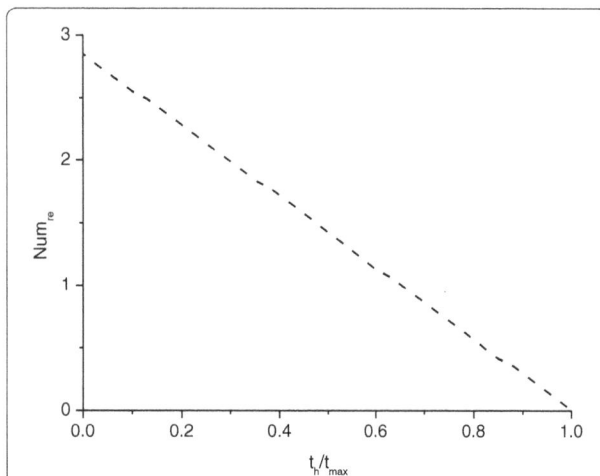

Fig. 9 Ratio of the new optimal COP_{optm} and the *COP* as a function of ratio of heating time and maximum heating time ($T_a = T_c = 25\ °C$, $T_e = 0\ °C$, and $T_g = 90\ °C$)

heated by the sun with storage system using the water for example instead the direct heating by solar radiation.

Furthermore, the performed study allows to define a new optimal coefficient of performance (COP_{optm}) more generalized. It takes the value of normal coefficient of performance (*COP*) for zero recovery of energy loss, and greater value for more energy recovery, keeping the same total energy necessary for the functioning of the system.

The usefulness of this optimal coefficient of performance is to determine the number of additional adsorbers that can be added to the machine for an equivalent consumption to energy received by the adsorber-collector.

The cold production increases with the number of additional adsorbers. Even though the days when solar radiation is low, the decrease in cold production will be better than that without an energy recovery system.

Nomenclature

COP Coefficient of performance
COP_{optm} Optimal coefficient of performance
Cp_d Specific heat of adsorbent, J kg^{-1} K^{-1}
Cp_l Specific heat of adsorbate in liquid state, J kg^{-1} K^{-1}
Cp_t Specific heat of tubes in metal containing the adsorbent, J kg^{-1} K^{-1}
D Coefficient of affinity
E_g Emissivity of glass cover of adsorber-collector
E_P Emissivity of top wall of adsorber-collector
L Latent heat of adsorbate, J kg^{-1}
m Adsorbate mass, kg
m_d Adsorbent mass, kg
m_t Mass of metallic tubes containing the adsorbent, kg
n Parameter of adjustment of D–A equation
n_g Number of glass cover of adsorber-collector

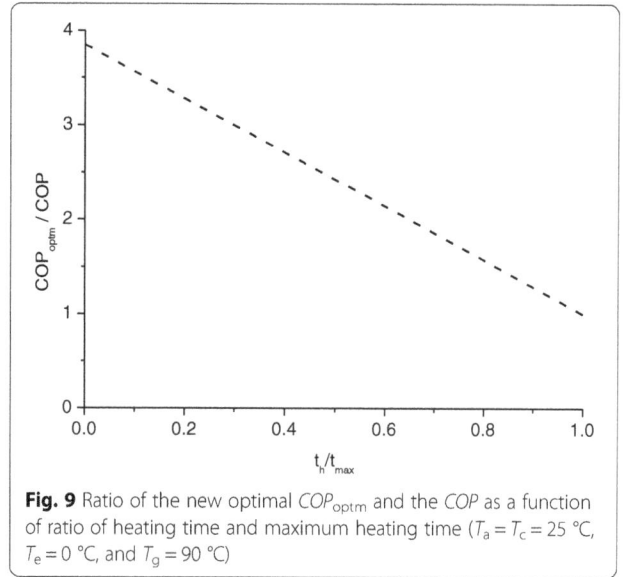

Fig. 8 Number of additional adsorbers and ratio of the recovery energy and the useful energy, as a function of the ratio of heating time and maximum heating time ($T_a = T_c = 25\ °C$, $T_e = 0\ °C$, and $T_g = 90\ °C$)

Num_{re} Number of additional adsorbers

r Particular gas constant of adsorbate, J kg^{-1} K^{-1}

P Equilibrium pressure of adsorbent-adsorbate pair, Pa

P_c Condensing pressure of adsorbate, Pa

P_e Evaporating pressure of adsorbate, Pa

P_s Saturation pressure of adsorbate, Pa

Q_{des} Heat necessary for desorption process, J

Q_{ev} Heat load in evaporator (cooling production), J

Q_{ge} Total energy received by adsorber-collector, J

q_{st} Isoteric heat, J kg^{-1}

Q_1 Energy supplied to heat the adsorbent, J

Q_2 Energy supplied to heat the tubes in metal containing the adsorbent, J

Q_3 Energy supplied to heat the adsorbate mass, J

Q_L Heat loss by adsorber-collector during the time of heating (processes A–C of Fig. 2), J

Q_r Recovered heat from adsorber-collector, J

Q_u Useful energy for functioning of system, J

S Adsorber-collector surface, m^2

T Temperature, K

T_a Adsorption temperature (temperature at end of adsorption), K

T_{am} Ambient temperature, K

T_c Condensing temperature of adsorbate, K

T_e Evaporating temperature of adsorbate, K

T_g Generating temperature (temperature at end of desorption), K

T_p Wall temperature of the top of adsorber-collector, K

Tp_{max} Maximal temperature of heated walls of adsorber-collector (including upper and bottom parts), K

Tp_{min} Minimal temperature of heated walls of adsorber-collector (including upper and bottom parts), K

t_h Heating time of adsorbent, s

t_{max} Maximum time of heating processes A–C (Fig. 2), s

U_b Loss coefficient from the bottom of the adsorber-collector, W K^{-1}

U_L Global loss coefficient from the adsorber-collector surfaces, W K^{-1}

U_t Loss coefficient from the top of the adsorber-collector, W K^{-1}

T_{s1} Temperature at start of desorption, K

T_{s2} Temperature at start of adsorption, K

W_v Wind velocity, m s^{-1}

W_0 Maximum adsorbed volume, m^3 (of adsorbate) kg^{-1} (of adsorbent)

X_{max} Maximal adsorbate mass concentration in 1 kg of adsorbent, kg kg^{-1}

X_{min} Maximal adsorbate mass concentration in 1 kg of adsorbent, kg kg^{-1}

Greek symbols

α Thermal expansion coefficient of adsorbate, K^{-1}

ΔT_{am} Variation of ambient temperature, K

ΔT_p Variation of upper wall temperature of adsorber-collector, K

η Thermal efficiency of adsorber-collector

Ω Adsorber-collector inclination, °

ρ_l Density of the adsorbate in the liquid state, kg m^{-3}

σ Stefan-Boltzmann constant, W m^{-2} K^{-4}

Authors' contributions
All authors read and approved the final manuscript.

Competing interests
The authors declare that they have no competing interests.

Author details
[1]Laboratoire de Génie Energétique et Matériaux, LGEM, Université de Biskra, B.P. 145 R.P., Biskra 07000, Algeria. [2]Département de Génie Mécanique, Université de Ouargla, Faculté des Sciences Appliquées, Ouargla 30 000, Algeria. [3]Laboratoire de Génie Mécanique, LGM, Université de Biskra, B.P. 145 R.P., Biskra 07000, Algeria.

References
Anyanwu, E. E., Oteh, U. U., & Ogueke, N. V. (2001). Simulation of a solid adsorption solar refrigerator using activated carbon/methanol adsorbent/refrigerant pair. *Energy Conversion and Management, 42*, 899–915.

Bejan, A., & Kraus, A. D. (2003). *Heat transfer handbook*. New York: Wiley.

Chekirou, W. (2008). *Etude et analyse d'une machine frigorifique solaire à adsorption*. Alegria: University of Constantine. Doctorat Thesis.

Chekirou, W., Boussehain, R., Feidt, M., Karaali, A., & Boukheit, N. (2011). Numerical results on operating parameters influence for a heat recovery adsorption machine. *Energy Procedia, 6*, 202–216.

Chekirou, W., Chikouche, A., Boukheit, N., Karaali, A., & Phalippou, S. (2014). Dynamic modelling and simulation of the tubular adsorber of a solid adsorption machine powered by solar energy. *International Journal of Refrigeration, 39*, 137–151.

Critoph, E. (1988). Performance limitations of adsorption cycles for solar cooling. *Solar Energy, 41*(1), 21–31.

Douss, N., & Meunier, F. (1988). Effect of operating temperatures on the coefficient of performance of active carbon-methanol systems. *Heat Recovery Systems & CHP, 8*(5), 383–392.

Duffie, J. A., Beckman, W. (2013). *Solar engineering of thermal Processes* (4th ed.). New York: Wiley.

Goodwin, R. D. (1987). Methanol thermodynamic properties from 176 to 673 K at pressures to 700 bar. *Journal of Physical Chemistry, 16*(4), 799–892.

Jing, H., & Exell, R. H. B. (1994). Simulation and sensitivity analysis of an intermittent solar-powered charcoal/methanol refrigerator. *Renewable Energy, 4*(1), 133–149.

Klein, S. A. (1975). Calculation of flat-plate collector loss coefficients. *Solar Energy, 17*, 79–80.

Leite, A. P. F., & Daguenet, M. (2000). Performance of a new solid adsorption ice maker with solar energy regeneration. *Energy Conversion and Management, 41*, 1625–1647.

Lemmini, F., & Errougani, A. (2007). Experimentation of a solar adsorption refrigerator in Morocco. *Renewable Energy, 32*, 2629–2641.

Pons, M., & Grenier, P. H. (1986). A phenomenological adsorption equilibrium law extracted from experimental and theoretical considerations applied to the activated carbon + methanol pair. *Carbon, 24*(5), 615–625.

Rouag, A., Benchabane, A., Labed, A., & Boultif, N. (2016). Thermal design of air cooled condenser of a solar adsorption refrigerator. *Journal of Applied Engineering Science & Technology, 2*(1), 23–29.

8

Structure-properties relationship in TRIP type bainitic ferrite steel austempered at different temperatures

Hoda Nasr El-Din[1], Ezzat A. Showaib[2], Nader Zaafarani[2] and Hoda Refaiy[1*]

Abstract

Background: Attractive properties of TRIP-type bainitic ferrite (TBF) steel ascribe to its unique microstructure of lath structure bainitic ferrite matrix and interlath retained austenite films. This work is concerned with obtaining ultra high-strength hot forged TBF steel with high elongation and excellent strength-elongation balance.

Methods: The effect of austempering temperature on the microstructure along with its retained austenite characteristics and tensile properties of a hot forged TBF steel was studied. A detailed investigation correlating the steel structure and its tensile properties was carried out.

Results: Tensile strength ranging from 1058 to 1552 MPa was achieved when the hot forged steel was austempered at (325 - 475 °C).

Conclusions: Ultra high tensile strength of 1058 MPa, large total elongation of 29% and excellent strength-elongation balance of 30 GPa % were attained when the steel was austempered at 425 °C. The large total elongation of this steel is mainly due to the uniform fine lath structure matrix and the pronounced TRIP effect of a large amount of retained austenite films which prevents a rapid decrease of strain hardening rate at low strain and leads to a relatively high strain hardening at high strain level. Rapid transformation of blocky retained austenite at low strain in the hot forged TBF steel austempered at higher temperatures results in a rapid increase of initial strain hardening. In addition, the coarse microstructure that contains large blocks of retained austenite / martensite and the insufficient numbers of bainitic ferrite lathes and retained austenite films deteriorate the total elongation and the strength-elongation balance of the TBF austempered at 475 °C.

Keywords: Ultrahigh-strength steels, TRIP-aided steels, Austempering heat treatments, Microstructure, Retained austenite characteristics, Tensile properties

Background

The reduction in weight of vehicle body can improve the fuel efficiency and environmental control. Therefore, there is an international attention to develop advanced high-strength steel (AHSS). The TRIP-aided multiphase (TMP) steel as a class of AHSS exhibits an excellent combination of strength and stretch-formability (Sugimoto et al. 1995), good deep drawability (Matsumura et al. 1992) and high fatigue strength (Sugimoto et al. 1997). The microstructure of this steel is mainly composed of bainitic

ferrite (bf) and carbon-enriched retained austenite (γ_r) embedded in a matrix of polygonal ferrite (Sugimoto et al. 1992).

The ideal energy absorption behaviour of TMP steels, which can be attributed to the transformation of metastable retained austenite into martensite under stress and strain (TRIP effect) (Bleck 2002; Sugimoto et al. 2006), and the high work-hardening response improve the crashworthiness of a vehicle through good distribution of strain during crash deformation (Bleck 2002). The superior formability of TMP steel (Sugimoto et al. 2006) can also be attributed to its high work-hardening properties. TMP steel has been applied to some impact members (Ojima et al. 1998) due to its high ability of energy

* Correspondence: eng_huda_refaiy@yahoo.com
[1]Plastic Deformation Department, Central Metallurgical R& D institute (CMRDI), Cairo, Egypt
Full list of author information is available at the end of the article

absorption under dynamic load. In spite of the excellent mechanical and technological properties of TMP steel, it cannot be applied to the automotive underbody press parts (e.g. lower arms and members) (Emadoddin et al. 2009) due to its poor stretch-flangability (Sugimoto et al. 1999). Also, this steel lacks sufficient performance in bendability and edge formability (De Cooman et al. 2004). Based on the fact that the bainitic steel exhibits an excellent stretch-flangability due to its uniform fine lath structure, (Sugimoto et al. 2000) have developed a new type of high-strength TRIP type bainitic ferrite (TBF) steel. The microstructure of TBF steel is characterized by bainitic ferrite lath matrix and interlath-retained austenite films. TBF steel is characterized by excellent balance between low edge crack susceptibility and high elongation (Sugimoto et al. 2002, 2006). It develops better stretchability compared to TMP steel with the same chemical composition and amount of retained austenite (Bhadeshia et al. 2001). It also exhibits good bendability and edge formability (Sugimoto et al. 2006). For these excellent properties, TBF steel can be applied to the applications that require high localized strain realization. Moreover, TBF steel shows high fatigue strength (Demeyer et al. 1999) and high impact energy (Hojo et al. 2008). Recently, (Sugimoto et al. 2010a) have developed a new type of hot-forged TBF steel with high hardenability. The present work is aimed at producing ultrahigh-strength hot-forged TBF steel with high elongation. Tensile properties of TRIP steel are affected by heat treatment conditions. The specific purpose of this investigation is to study the effect of austempering temperature (T_A) on the microstructure along with its retained austenite characteristics, and tensile properties of the investigated hot-forged TBF steel. The structure-properties relationship is also discussed.

Methods

One hundred kilogramme Y-blocks of the steel alloy were produced in a medium-frequency induction furnace. Chemical composition of the steel is shown in Table 1. The charge was made up of steel scrap and the required Mn, Si, P, Mo and Nb were added as ferroalloys (Fe–80% Mn), (Fe–75% Si), (Fe–28% P), (Fe–70% Mo) and (Fe–65% Nb). The heats were cast into 300 mm × 200 mm × 50 mm Y-blocks. Each leg of Y-blocks was machined and sectioned into 200 mm × 60 mm × 40 mm specimens. The specimens were homogenized at 1250 °C for 2 h in a muffle furnace, and then furnace cooled.

The martensite start temperature was estimated to be 371 °C by the following equation (Tamura 1970):

Table 1 Chemical composition of the investigated steel alloy

Element	C	Si	Mn	P	S	Mo	Ni	Al	Nb
Wt.%	0.35	0.6	1.5	0.02	0.008	0.25	1.5	1.1	0.05

$$M_S(^\circ C) = 550-361x(\%C)-39x(\%Mn)-17x(\%V)$$
$$-20x(\%Cr)-17x(\%Ni)-10x(\%Cu)$$
$$-5x(\%Mo+\%W)-0x(\%Si)+15x(\%Co)$$
$$+30x(\%Al)$$

(1)

Fee-forging as one of the forming technologies is commonly used to refine the microstructure and increase the strength values. The homogenized steel was reheated at 1200 °C for 30 min and then hot forged into rods of 16–18 mm in diameter. The hot forging was performed using Pneumatic Hammer (150 Kp).

Standard tensile specimens ASTM E-8 with 6 mm φ × 36 mm gauge length were prepared. For microstructure investigation, hot-forged round specimens of 6 mm φ were also prepared. In order to obtain TRIP type bainitic ferrite steel, the austempering treatments were performed by austenitizing of the hot-forged tensile and round specimens in an electrically heated furnace, followed by rapid quench into a salt bath at the austempering temperature. The specimens were austempered at 325–475 °C for 20 min, then quenched into oil and cooled to room temperature, as shown in Fig. 1.

The microstructure was identified by optical microscopy, scanning electron microscopy (SEM) and X-ray diffraction measurements. Specimens for optical microscopy were etched with 5% nital and rinsed with water followed by etching in a 10% sodium meta-bisulphite solution. With these etchants, retained austenite appears white, ferrite appears grey and bainite and martensite appear black (Jeong 1994). For microstructure investigation using SEM, samples were etched with 2% nital.

Volume fraction of the retained austenite (V_γ) was estimated by X-ray diffraction using a Cu target at 45 KV and 40 mA. V_γ was calculated using Eq. 2, where I_γ and I_α are the average integrated intensity obtained at the $(200)\gamma$, $(220)\gamma$, $(311)\gamma$ and $(200)\alpha$, $(211)\alpha$ diffraction peaks. K_α and K_γ are the reflection coefficients of the ferrite and austenite phases, respectively (Maruyama 1977).

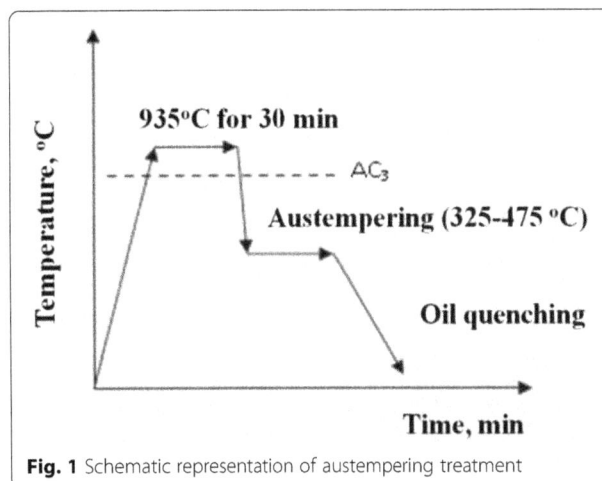

Fig. 1 Schematic representation of austempering treatment

$$V_\gamma = \frac{I_\gamma\,K_\alpha}{I_\gamma\,K_\alpha + I_\alpha\,K_\gamma} \tag{2}$$

Carbon concentration of the retained austenite was estimated from the following equation (Dyson & Holmes 1970). The retained austenite lattice parameter (a_γ) was measured based on the $(220)\gamma$ diffraction peak.

$$a_\gamma = 3.578 + 0.033\left(\text{wt\% } C_\gamma\right) \tag{3}$$

Tensile test was carried out on a universal testing machine (1000 KN) at room temperature with a cross head speed of 0.5 mm min^{-1}.

Results and discussion

Influence of austempering temperature on retained austenite characteristics

Figure 2a–c shows the variation of retained austenite characteristics with T_A in TBF steel austempered for 20 min. As shown in Fig. 2a, a small amount of austenite (3%) is retained at room temperature when the steel is austempered at 325 and 350 °C, lower than the M_s. This is due to transformation of large amount of residual austenite into bainitic ferrite/martensite mixed matrix. With increasing the austempering temperature up to 425 °C, the volume fraction of γ_r is considerably increased showing the maximum value of 16% at 400 °C. Further increase of the austempering temperature up to 475 °C leads to decrease of γ_r due to the transformation of low carbon residual austenite to fresh martensite upon cooling after austempering. At temperatures of 450 °C and higher, carbon begins to precipitate after certain overaging times resulting in a decrease of the amount of γ_r (Hausman 2013). Addition of Mn and Ni to TBF steel was found to increase the amount of γ_r (Kim et al. 2003) and the complex addition of Nb and Mo was found to increase the amount of γ_r as well especially when the TBF steel is austempered at 450–500 °C (Sugimoto et al. 2007). It is well known that the

stability of γ_r is affected by a combination of many factors such as grain size, carbon content and morphology of retained austenite. The properties of the matrix and the alloying elements also influence the stability of γ_r. The addition of 0.05% Nb and 0.2% Mo to TBF steel was found to decrease the carbon concentration in γ_r especially at high austempering temperatures (Sugimoto et al. 2007).

Influence of austempering temperature on the tensile properties

Figure 3 shows the engineering flow curves of the TBF steel austempered at (325–475 °C). As shown in the figure, ultrahigh-strength hot-forged TBF steels exhibiting different tensile properties were produced. These TBF steels exhibit a continuous yielding behaviour. In dual-phase steel, this behaviour is mainly attributed to high dislocation density produced by martensite transformation (Saleh & Priestner 2001). In TRIP steel, the yielding behaviour may be explained not only by dislocation density but also by other factors due to the complex structure of this steel.

It is well known that the discontinuous yielding behaviour is produced in the steel due to the presence of interstitial atoms and the yield point phenomenon in the steel is a function of the composition and heat treatment of the material. In the investigated steel, the carbide formers (Mo and Nb) and nitride formers (Al, Si and Nb) reduce the level of the interstitial atoms of carbon and nitrogen, which are strong lockers of dislocation. Consequently, the continuous yielding is promoted as shown in Fig. 3.

Figure 4a–c shows the variation of the tensile strength (TS), total elongation (TEL) and strength-elongation balance (TSxTEL) with T_A in the TBF steel austempered for 20 min. As shown in Fig. 4a, b, increasing the austempering temperature from 325 up to 400 °C leads to a great decrease of TS and a slight increase of TEL. With increasing T_A to 425 °C, TS is slightly decreased while the TEL and TSxTEL balance are greatly increased to maximum values of 29%

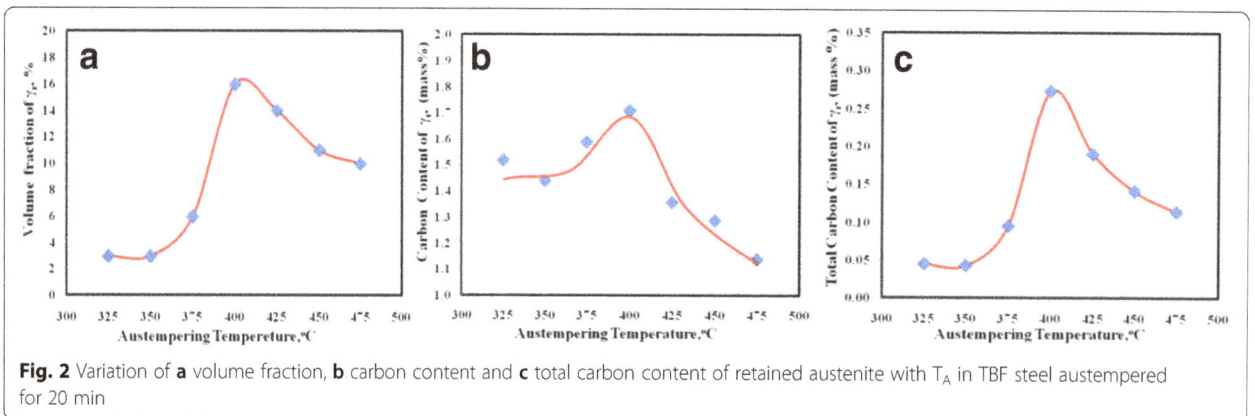

Fig. 2 Variation of **a** volume fraction, **b** carbon content and **c** total carbon content of retained austenite with T_A in TBF steel austempered for 20 min

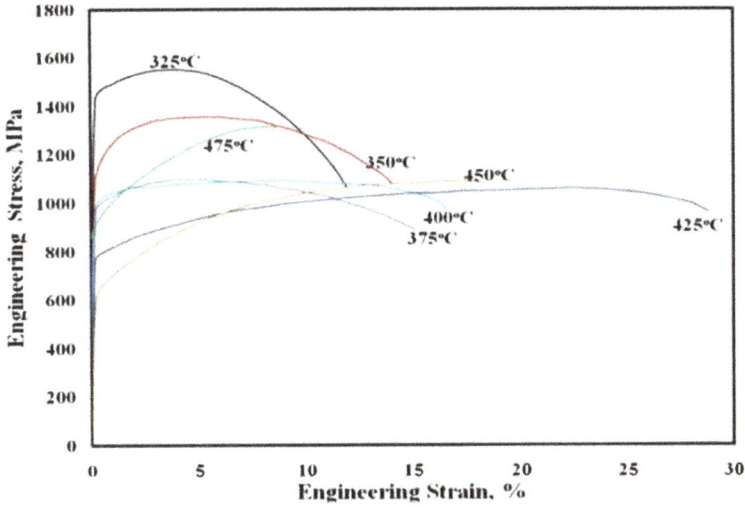

Fig. 3 Engineering stress-strain curves of the TBF steel austempered at different temperatures

and 30 GPa, respectively. Upon further increase of T_A, the TS is increased showing a considerable increase at 475 °C while the TEL and TSxTEL are decreased indicating serious decreases at 475 °C.

Influence of austempering temperature on strain-hardening behaviour

Figure 5 shows the variation of strain-hardening rate with the true strain in TBF steel austempered at different temperatures. The behaviour of strain hardening at high strain levels is related to the transformation rate of γ_r which is greatly affected by the stability of γ_r. Hence, the different transformation rate at high strain levels shown in Fig. 5 may be attributed to the location of the γ_r in the matrix, carbon content and size of the retained austenite grains (Chiang et al. 2011).

Microstructure and tensile properties

Figures 6 and 7 shows typical optical and SEM micrographs of the steel austempered at different temperatures.

It is obvious that the austempering temperature has a great effect on the amount and morphology of the phases present in the steel microstructure.

TBF steel austempered at temperatures lower than M_s

Microstructure of the TBF steel austempered at 325 and 350 °C, lower than M_s, consists of bainitic ferrite/martensite mixed matrix of very fine lath structure and interlath-retained austenite films as shown in Figs. 6a, b and 7a. This fine microstructure clearly shown in Fig. 7a has been attributed to the increase in transformation driving force and suppression of coalescence of fine bainite laths (Bhadeshia et al. 2001). The fine prior austenite grain size shown in Fig. 7 can be attributed to the combined addition of Nb and Mo which has a better refining effect than single Nb addition (Huab et al. 2015). Recently, (Hojo et al. 2010) have shown that the complex addition of Al-Nb-Mo to TBF steel tends to refine the lath structure and γ_r films.

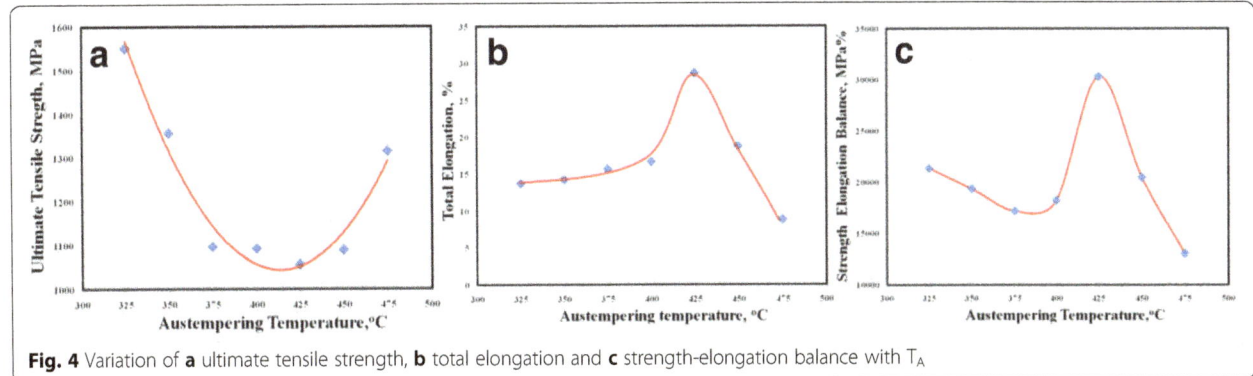

Fig. 4 Variation of **a** ultimate tensile strength, **b** total elongation and **c** strength-elongation balance with T_A

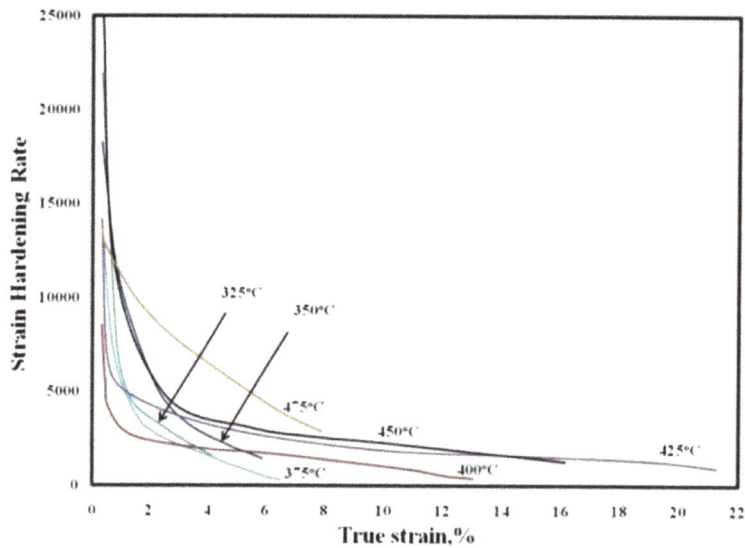

Fig. 5 Variation of strain hardening rate (dσ/dε) with true strain for the steel alloy austempered at different temperatures

TBF steel austempered at temperatures lower than its M_s indicates higher tensile strength "up to 1552 MPa" compared to the steel austempered at temperatures higher than M_s (Fig. 4a). This is due to its finer lath structure and its bainitic ferrite/martensite (α_m) mixed matrix. Tensile strength of this TBF steel is decreased with increasing austempering temperature, from 325 to 350 °C (Fig. 4a) due to the decrease of martensite lathes in the mixed matrix. The strengthening effect of martensite lathes is more dominant in this steel than a small TRIP effect of (3%) γ_r. TBF steel austempered at temperatures lower than M_s exhibits a smaller total elongation compared to that austempered at 425 and 450 °C, higher than the M_s (Fig. 4b). This is due to steep fall of strain hardening at an early stage which may be caused by small internal stress hardening and initial martensite hardening resulting from a small volume fraction of second phase (Sugimoto et al. 2000), followed by great decrease of strain hardening at low strain (Fig. 5) which in turn results from insignificant TRIP effect of small γ_r volume fraction.

TBF steel austempered at temperatures higher than M_s
TBF steel austempered at 400 °C The fine uniform microstructure of TBF steel austempered at 400 °C shown in Fig. 7d contains a large amount of γ_r ($\gamma_r = 16\%$). However, this TBF steel exhibits a smaller total elongation compared to TBF steel austempered at 425 °C (Fig. 4b). This can be attributed to very high stability of γ_r ($C_\gamma = 1.7$ wt%) which prevents high amount of γ_r from transforming during tensile deformation. This results in a considerable decrease of strain hardening at high strain levels (Fig. 5). Therefore, the total elongation

and strength-elongation balance are greatly decreased (Fig. 4b, c). According to (Chiang et al. 2011), the very high levels of carbon in $\gamma_r > 1.8$ wt % C prevent the γ_r from transforming completely.

TBF steel austempered at 425 °C Figures 6e and 7d show the microstructure of TBF steel austempered at 425 °C. This fine uniform microstructure mainly consists of bainitic ferrite matrix of lath structure and interlath γ_r films. It is evident from the figures (Fig. 6d, e and 7c, d) that the matrix structure is changed in some zones from bf lathes to granular bainitic ferrite and the retained austenite films are changed to a few of coarse and island type γ_r. K. Sugimoto et al. (2010a) have also detected and reported these morphologic changes. It is also obvious from these figures that the pro-eutoctoid ferrite (α_{PE}) and upper bainite are developed. The grain refinement through Nb addition leads to promotion of transformation during cooling from austenitization which causes ~20% transformed phase fraction prior to the overaging step (Hausman 2013). Ultrahigh-strength TBF steel austempered at 425 °C exhibits high total elongation. As shown in Fig. 4b, c, the total elongation and strength-elongation balance were considerably increased at 425 °C to the maximum values of 29% and 30 GPa%, respectively. The large total elongation can be attributed to the uniform microstructure and fine lath structure matrix and to the pronounced TRIP effect of a large amount of stable retained austenite films which gradually transforms upon loading to martensite. The strain-induced transformation of γ_r simultaneously relaxes the localized stress concentration at the matrix/second phase interface to suppress the void

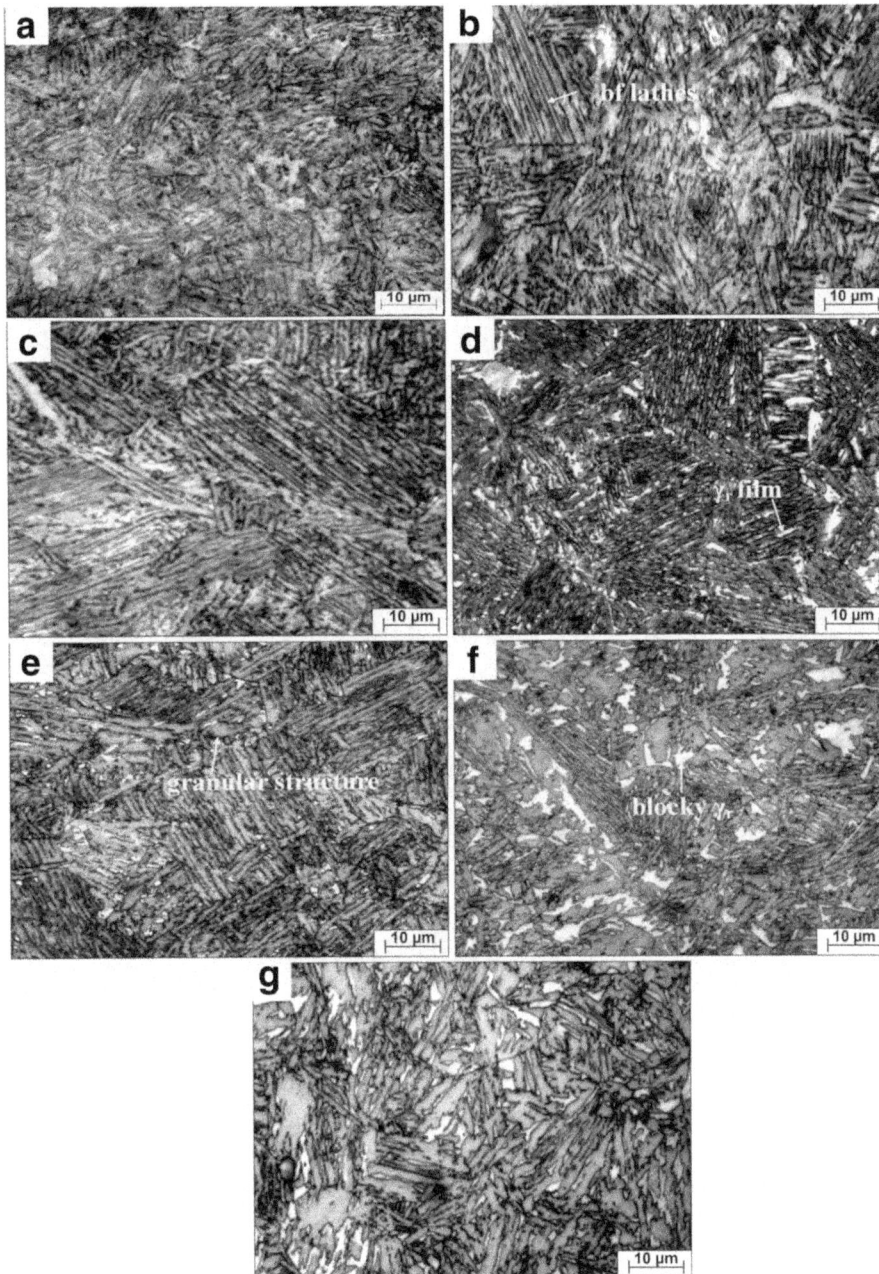

Fig. 6 Optical micrographs of the TBF steels austempered at **a** 325 °C, **b** 350 °C, **c** 375 °C, **d** 400 °C, **e** 425 °C, **f** 450 °C and **g** 475 °C for 20 min

initiation (Sugimoto et al. 2000). Recently, (Zhao et al. 2014) have reported that the film type γ_r transforms into martensite and hardens the bf matrix. The resultant compressive stress on the bf matrix prevents crack propagation. Consequently, the gradual transformation of high amount of stable γ_r films in the TBF steel austempered at 425 °C prevents the rapid decrease of strain hardening rate at low strain and permits for the strain hardening to continue up to a high strain (Fig. 5) which results in a considerable increase of the total elongation (Fig. 4b).

The pro-eutectoid soft phases of ferrite and upper bainite contribute to enhanced elongation. The complex addition of Al and Nb to the investigated ultrahigh-strength TBF steel contributes also to enhanced total elongation (Sugimoto et al. 2010b). The reduced strain induced martensite formation at low strain levels in the steel austempered at 425 °C results in increasing the work hardening at high strain levels (Fig. 5). This result has been also detected and reported by (Hausman 2013).

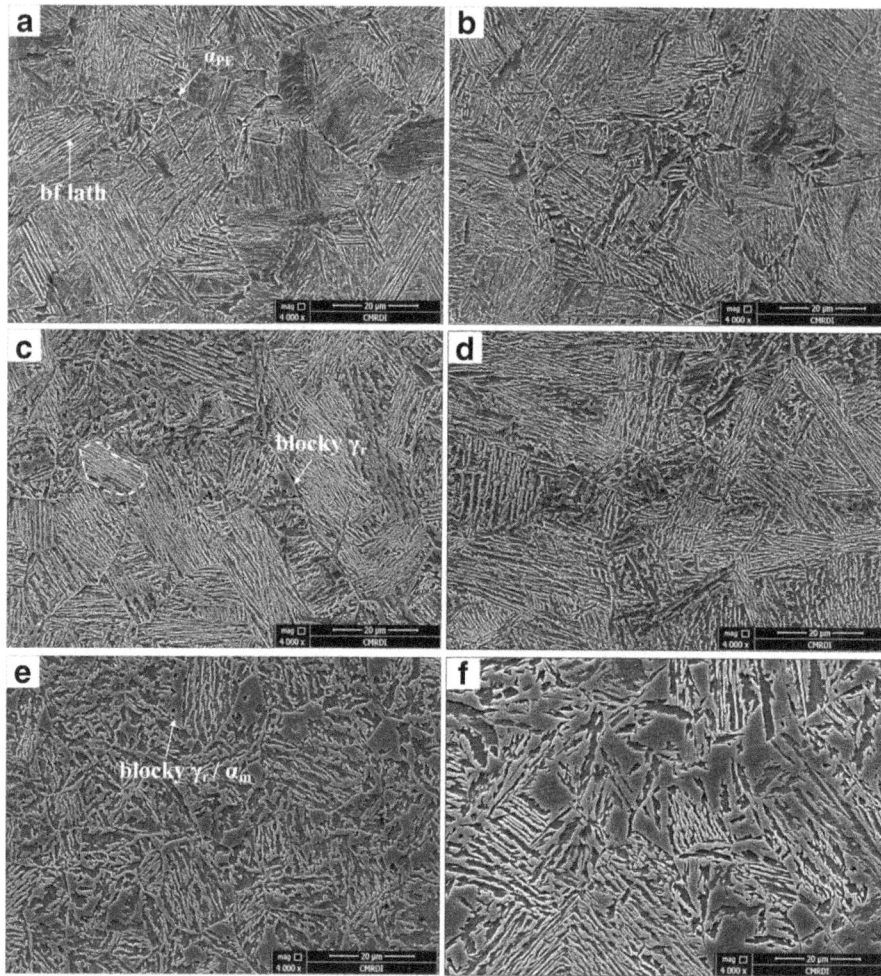

Fig. 7 SEM micrographs of the TBF steels austempered at **a** 350 °C, **b** 375 °C, **c** 400 °C, **d** 425 °C, **e** 450 °C, **f** 475 °C for 20 min

TBF steel austempered at 450 and 475 °C When the steel is austempered at higher temperatures of 450 and 475 °C, the numbers of bainitic ferrite lathes and the fraction of interlath γ_r films are greatly reduced while the amount of granular bainitic ferrite and both of blocky and island type γ_r/α_m are increased due to the morphologic changes (Fig. 6f, g), which can be attributed to Nb addition (Sugimoto et al. 2006; Hausman 2013; Bleck et al. 2001). Increasing austempering temperature leads to an increase in the thickness of bainitic ferrite lathes (Fig. 7e, f) due to the reduced nucleation kinetics of bainite formation at higher T_A. Increase in lath thickness is also attributed to low dislocation density due to the high temperatures (Zhao et al. 2014). Austempering at these higher temperatures leads to a small degree of super cooling that reduces the driving force of bainite formation. Consequently, the numbers of bainitic ferrite lathes are reduced (Fig. 6f, g) as well as the carbon content of the residual austenite. Upon cooling the steel

after austempering, a high amount of lower carbon content residual austenite could not be stabilized at room temperature and both blocky and island type retained austenite/fresh martensite are formed (Figs. 6f, g and 7e, f).

As shown in Fig. 4b, c, the total elongation and strength-elongation balance of TBF steel are decreased with increasing austempering temperature from 425 to 450 °C due to the morphologic change of high amount of γ_r from film type to a less stable blocky one (compare Fig. 6e and f). Presence of high amount of initial martensite and reduced numbers of bainitic ferrite lathes (compare Fig. 7d and e) also lead to the decrease of total elongation of TBF steel austempered at 450 °C.

Enhanced strain hardening rate of TBF steels austempered at higher temperatures (450 and 475 °C) over a small strain range (Fig. 5) is attributed to the high long-range internal stress resulting from both the initial martensite hardening of original martensite and the difference in strength between the granular bainitic ferrite soft matrix and the γ_r hard second phase (Sugimoto et al. 2006). The total elongation and strength-elongation balance of TBF steel austempered at 475 °C are seriously decreased (Fig. 4b, c). The decrease of total elongation is firstly attributed to the rapid transformation of blocky γ_r at low strain which results in a rapid increase of initial strain hardening (Fig. 5). In addition, the coarse microstructure consisting of high amount of large blocks of γ_r/α_m and insufficient numbers of bainitic ferrite lathes (Fig. 6g) results also in the decrease of total elongation. Martensite blocks act as the source of voids initiation and insufficient numbers of bf lathes lead to the decrease of γ_r stability (Sugimoto et al. 2002).

Conclusions

The effect of T_A on the microstructure, retained austenite characteristics and mechanical properties of hot-forged TBF alloyed steel were investigated. The main results are as follows:

- Ultrahigh-strength hot-forged TBF steels with tensile strength ranging from 1058 to 1552 MPa were attained when TBF steel is austempered at 325–475 °C.
- Tensile strength of the hot-forged TBF steel was considerably increased, when austempered at temperature lower than M_s due to its hard fine lath-like bainitic ferrite/martensite mixed matrix. Relatively, small total elongation at these temperatures is due to both the steep fall of strain hardening at an early stage and the great decrease of strain hardening at low strain which resulted from insignificant TRIP effect of small amount of γ_r.
- The considerable decrease of strain hardening at high strain levels and the great decreases of total elongation and strength-elongation balance in TBF steel austempered at 400 °C are due to very high stability of γ_r ($C_\gamma = 1.7$ wt%) which prevents high amount of γ_r from transforming.
- Total elongation and strength-elongation balance of ultrahigh-strength hot-forged TBF steel were greatly increased to 29% and 30 GPa%, respectively, when austempered at 425 °C. Increase of total elongation is due to uniform fine bainitic ferrite lathes, interlath γ_r films and pronounced TRIP

  effect of large amount of γ_r which prevents rapid decrease of strain hardening rate at low strain and leads to a relatively high strain hardening at high strain levels. Pro-eutectoid soft phases and granular bainitic ferrite contribute also to enhanced total elongation.

- Rapid transformation of blocky γ_r at low strain in hot-forged TBF steels austempered at 450 and 475 °C resulted in rapid increase of initial strain hardening. Additionally, the large blocks of martensite and insufficient numbers of bf lathes and γ_r films resulted in serious decreases of the total elongation and strength-elongation balance, especially in TBF steel austempered at 475 °C.

Authors' contributions
All authors read and approved the final manuscript.

Competing interests
The authors declare that they have no competing interests.

Author details
[1]Plastic Deformation Department, Central Metallurgical R& D institute (CMRDI), Cairo, Egypt. [2]Production and Mechanical Design Department, Faculty of Engineering, Tanta University, Tanta, Egypt.

References
Bhadeshia, H. (2001). *Bainite in Steels londone, IOM commercial Ltd.*
Bleck, W. (2002). *International Conference on TRIP-Aided High Strength Ferrous Alloys* (pp. 13–23).
Bleck, W, Frehn, A, Ohlert, J (2001). In Niobium Science and Technology: Proceedings of the International Symposium Niobium 2001, Orlando, Florida, USA (pp. 727–752).
Chiang, J, Lawrence, B, Boyd, JD, Pilkey, AK. (2011). Effect of microstructure on retained austenite stability and work hardening of TRIP steels. *Materials Science and Engineering, A 528*, 4516–4521.
De Cooman, B, Barbé, L, Mahieu, J, Krizan, D, Samek, L, De Meyer, M. (2004). Mechanical properties of low alloy intercritically annealed cold rolled TRIP sheet steel containing retained austenite, *Canadian metallurgical Q, 43*(1), 13–24.
Demeyer, M, Vander, D, DeCoon, B. (1999). The influence of the substitutionof Si by Al on the properties of cold rolled C–Mn–Si TRIP steels. *ISIJ International, 39*, 813–22.
Dyson, D, Holmes, B. (1970). Effect of alloying additions on the lattice parameter austenite. *Journal of Iron Steel Institute, 208*, 469–474.
Emadoddin, E, Asmari1, H, Habibolah Zadeh, A. (2009). Formability evaluation of TRIP-aided steel sheets with different microstructures: polygonal ferrite and bainitic ferrite matrix. International Journal of Material Forming, 781–784.
Hausman, K. (2013). The influence of Nb on transformation behavior and mechanical properties of TRIPassisted bainitic-ferritic sheet steels. ,A 588, Materials Science and Engineering. *Materials Science and Engineering, A588*, 142–150.
Hojo, T, Kobayashi, J, Kajiyama, T, Sugimoto, K. (2010). Effects of alloying elements on impact properties of ultra high-strengthTRIP-aided bainitic ferrite steels. *Jīn Shān Gāo Zhuān Jì Yào*, No. 52, 9–16.
Hojo, T, Sugimoto, K, Mukai, Y, Ikeda, S. (2008). Effects of aluminum on delayed fracture properties of ultra high strength low alloy TRIP-aided steels. *ISIJ International, 48* (No. 6), 824–829.
Huab, H, Xu, G, Wang, L, Xue, Z, Zhang, Y, Liu, G. (2015). The effects of Nb and Mo addition on transformation and properties in low carbon bainitic steels. *Material Science and Design, 84*, 95–99.
Jeong, W. (1994). Proc. of the Symp. on: High-strength Sheet Steels for the Automotive Ind., (p. 267). Baltimore.
Kim, S, Gil Lee, C, Lee, T, Oh, C. (2003). Effect of Cu, Cr and Ni on mechanical properties of 0.15 wt.% C TRIP-aided cold rolled steels. *Scripta Materialia, 48*, 539–544.
Maruyama, H. (1977). X-Ray measurement of retained austenite volume fraction. Journal Japanese Social Heat Treatment. *Journal Japanese Social Heat Treatment, 17*, 198–204.
Matsumura, O, Sakuma, Y, Ishii, Y, Zhao, J. (1992). Effects of Alloying Elements on Impact Properties of Ultra High-Strenght TRIP-Aided Bainitic Ferrite Steels. *ISIJ International, 32*, 1110–1116.

Ojima, Y, Shiroi, Y, Taniguchi, Y, Kato, K, SAE Tech. (1998). Application to Body Parts of High-Strength Steel Sheet Containing Large Volume Fraction of Retained Austenite, SAE Technical Paper 980954. Paper Series, (No. 980954), 39–50.

Saleh, M, Priestner, R. (2001). Retained austenite in dual-phase silicon steels and its effect on mechanical properties. *Journal of Material Processing Technology, 113*, 587–593.

Sugimoto, K, Iida, T, Sakaguchi, J, Kashima, T. (2000). Retained austenite characteristics and tensile properties in a TRIP type bainiric sheet steel. *ISIJ International, 40*(No. 9), 902–908.

Sugimoto, K, Kobayashi, M, Nagasaka, A, Hashimoto, S. (1995). Warm stretch-formability of TRIP-aided Dualphase Sheet Steels. *ISIJ International, 35*(11), 1407–1414.

Sugimoto, K, Kobayashi, M, Hashimoto, S. (1992). Ductility and Strain-Induced Transformation in a High-Strength Transformation-Induced Plasticity-Aided Dual-PhaseSteel. *Metallurgical Transactions, A 23*, 3085–3091.

Sugimoto, K, Muramatsu, T, Hashimoto, S, & Mukaid, Y. (2006). Formability of Nb bearing ultra high-strength TRIP-aided sheet steels. *Journal of Materials Processing Technology, 177*, 390–395.

Sugimoto, K, Murata, M, Muramatsu, T, Mukai, Y. (2007). Formability of C–Si–Mn–Al–Nb–Mo Ultra High-strength TRIP-aided Sheet Steels. *ISIJ International, 47*(No. 9), 1357–1362.

Sugimoto, K, Murata, M, Song, S. (2010). Formability of Al–Nb bearing ultra high-strength TRIP-aided sheet steels with bainitic ferrite and/or martensite matrix. *ISIJ International, 50*(No. 1), 162–168.

Sugimoto, K, Nagasaka, A, Kobayashi, M, Hashimoto, S. (1999). Effects of retained austenite parameters on warm stretch-flangeability in TRIP-aided dual-phase sheet steels. *ISIJ International, 39*, 56–63.

Sugimoto, K, Nakano, K, Song, S, Kashima, T. (2002). Retained austenite characteristics and stretchflangeability of high-strength low-alloy TRIP type bainitic sheet steels. *ISIJ International, 42*(No. 4), 450–455.

Sugimoto, K, Sato, S, Arai, G. (2010). Hot Forging of Ultra High-Strength TRIP-Aided Steel. *Materials Science Forum, 638-642*, 3074–3079.

Sugimoto, K, Sun, X, Kobayashi, M, Haga, T, & Shirasawa, H. (1997). Fatique properties of TRIP-aided dual phase sheet steel *Transactions of the Japan Society of Mechanical Engineering, 63A*, 717.

Tamura, I (1970). *Steel Material Study on the Strength*. (p. 40). Tokyo: Nikkan Kogyo Shinbun Ltd.

Zhao, Z, Yin, H, Zhao, A, Gong, Z, He, J, Tong, T, Ju, H. (2014). The influence of the austempering temperature on the transformation behavior and properties of ultra-high-strength TRIP-aided bainitic ferritic sheet steel. *Materials Science & Engineering, A 613*, 8–16a.

9

Dynamic assessment of direct-current mobility in field-assisted sintered oxide dispersion-strengthened V-4Cr-4Ti alloys

Vinoadh Kumar Krishnan[*] and Kumaran Sinnaeruvadi

Abstract

Background: Vanadium alloy is one of the potential candidate material for structural applications in a commercial fusion reactor. Extended survival of a structural material has a direct consequence on the net energy produced in a fusion reaction, it is important to develop ultra-functional materials with tailored microstructures, to meet the harsh fusion environments. Microstructure of material, indeed depend upon the thermodynamics and kinetics of material processing.

Methods: Aiming to meet the harsh fusion conditions, we have developed oxide dispersion strengthened V-4Cr-4Ti alloys by high energy ball milling and field assisted sintering technique. Possible microstructural, morphological aftermaths observed in ball milled yttria dispersed V-4Cr-4Ti powders is explored.

Results and conclusion: Electron microscopy and laser particle analysis acknowledge that yttria addition aids powder agglomeration during ball milling. Ball milled powder was then consolidated (to a relative density of ~100%) using field assisted sintering technique, under optimal sintering conditions. Densification profile has implied that heterogeneous powder characteristic (apparent particle size and shape of powder) tends to impede the direct-current conductivity across the powder particle during various stages of field assisted sintering. In order to understand the kinetics of the field assisted sintering process on the starting powders, a new method was developed to compute the activation energy required for the direct-current conductivity across the individual powder particles. Relatively higher activation energy (for direct-current conductivity) is required for sintering yttria dispersed V-4Cr-4Ti powder than its V-4Cr-4Ti counterpart.

Keywords: Vanadium, Yttria, Milling, Sintering, Densification, Activation energy

Background

Paybacks of V–4Cr–4Ti alloy as a candidate material in fusion power generation has been extensively discussed elsewhere (Zinkle et al. 2008; Smith et al. 1996; Johnson and Smith 1998; Smith et al. 1995; Muroga et al. 2002). V–4Cr–4Ti alloys have been relentlessly proving themselves as a potential alternative to basic steels and oxide dispersion-strengthened steels owing to their superior performance as a structural material like high temperature mechanical properties, irradiation resistance at elevated temperatures (>400 °C), low activation under fast neutron flux, and high ductility (Chung et al. 1996; Rice and Zinkle

1998; Gelles 1998). Even though the microstructural stability of V–4Cr–4Ti alloys at elevated temperature (>650 °C) is superior to ferritic/martensitic steels, the anticipated power conversion efficiencies expected out of plasma-facing components have increased tremendously (Loomis et al. 1994). This intrigued fusion scientists to look for an ultra-functional V–4Cr–4Ti alloy to withstand harsh conditions inside the fusion reactor, consequently enhancing the power generation efficiency of the same. One such work was attempted in the year 1996, by dispersing ultrafine yttria particles in V–4Cr–4Ti alloy (Yamagata et al. 1996). Shibayama and his co-workers developed the oxide dispersion strengthened (ODS) vanadium alloy by a novel powder metallurgy route, namely mechanical-alloying (MA) (Suryanarayana

* Correspondence: vinoadhkumarkrish@gmail.com
Department of Metallurgical and Materials Engineering, National Institute of Technology, Tiruchirappalli, Tamil Nadu 620015, India

and Norton 1999; Alamo et al. 1992; Miller et al. 2005; Miller et al. 2003; Plant and Steel 1995). Earlier reports have consolidated ODS vanadium alloys using the hot isostatic pressing (HIP) technique, while a recent report suggests that field-assisted sintering (FAST) or spark plasma sintering (SPS) technique offers a variety of microstructural merits that can enhance the high temperature mechanical properties of V–4Cr–4Ti alloys (Krishnan and Sinnaeruvadi 2016). Vital merits offered by the FAST process include self-cleaning of the oxide layer in powder particles before sinter-bonding, retaining the nanostructure, and consolidating the powder particle to full density within a short span of time at lower temperature regimes (Mamedov 2000; Aalund 2008; Yucheng and Zhengyi 2002; Anselmi-tamburini et al. 2005; Chen et al. 2005). But no single literature is available to explain the densification mechanism behind sintering (FAST) V–4Cr–4Ti and yttria-dispersed V–4Cr–4Ti powders. Once the densification mechanism and its associated kinetics behind FAST-processing of the V–4Cr–4Ti powders are understood, tailoring the material properties via FAST process to match fusion-relevant conditions become conceivable (Tong et al. 2009; Thornton 1999; Thornton et al. 2004; Iimura et al. 2009; Yu et al. 1997).

In the present investigation, variations in current consumption while sintering V–4Cr–4Ti and yttria-dispersed V–4Cr–4Ti powders are discussed briefly. Further, the DC (direct-current) conductivity profile while sintering powder was obtained in order to understand the effect of yttria addition on the microstructural and morphological characteristics of ball-milled powders. Finally, the activation energy (E_a) required for the sintering current to traverse through the individual powders was deduced.

Methods
Powder preparation
Elemental powders of vanadium, titanium, and chromium with a purity of 99.9% and particle size equivalent of –325 mesh size were ball-milled using Retsch, PM-400 planetary ball mill. The typical ball-milling conditions used for synthesizing V–4Cr–4Ti and nano-yttria-dispersed V–4Cr–4Ti powders are presented in Table 1. Ascribed powders in the aforementioned atomic proportions were transferred into tungsten carbide (WC) vials inside the glove box under high pure argon environment (99.99% pure).

Powder characterization
Powder morphology, apparent particle size, and chemical composition of ball-milled V–4Cr–4Ti and yttria-dispersed V–4Cr–4Ti powders were studied with the help of scanning electron microscopy images accompanied with energy-dispersive absorptive X-ray spectroscopy (SEM-EDAX). The milled powders were dispersed in ethanol and ultrasonicated for 120 s. The sonicated powders were transferred onto a glass plate and dried in air; subsequently, the powder was transferred onto a carbon tape placed over an aluminum stub and maintained under vacuum 24 h prior to electron microscopy-imaging. From the electron microscopy images, quantitative estimates of apparent particle size (μm) were calculated using ImageJ (version 1.50b) software. Further, to confirm the apparent particle size and its distribution, laser particle size analysis was carried out using Mastersizer 3000 laser diffraction particle size analyzer (Malvern Instruments, UK).

Field-assisted sintering
Ball-milled powders were transferred into the graphite die-punch setup inside the glove box maintained under high pure argon (99.99%) environment. Subsequently, the powder was sintered at 1100 °C for 20 min with a uniaxial pressure of 40 MPa. The relative density of FAST-processed samples was calculated using Archimedes principle. Auxiliary data (recorded every 5 s) with information containing voltage, current, temperature, vacuum, and displacement were recorded during the process. From the auxiliary data, electrical resistivity and conductivity values were computed from the fundamental equations as follows:

$$R = \rho \left(\frac{l}{A} \right) \tag{1}$$

where R is the electrical resistance in ohms calculated from the auxiliary data generated by FAST, ρ is the electrical resistivity in ohm-meter offered by the ball-milled powders during sintering, l is the dynamic sample thickness (linear shrinkage) in millimeter, and A is the cross-sectional area of the sample in square millimeters. Using the dynamic electrical resistivity (ρ) value calculated from Eq. (1), dynamic electrical conductivity σ in mho per meter was calculated from the following equation:

Table 1 Ball-milling conditions employed for synthesizing V–4Cr–4Ti and yttria-dispersed V–4Cr–4Ti powders

Variables	Conditions
Type of ball mill	High-energy planetary ball mill
Milling vessel and media	Tungsten carbide (WC)
Plate rotation	250 rpm
Milling time	10 h
Ball-to-powder ratio	10:1
Ball-milling atmosphere	High purity argon (99.99% pure)
Surfactant	NaCl

$$\sigma = \frac{1}{\rho} \qquad (2)$$

In order to evaluate the rate constant of a chemical reaction on an absolute scale (Kelvin, K), the Arrhenius equation was used to calculate the activation energy for DC conductivity while sintering ball-milled V–4Cr–4Ti and yttria-dispersed V–4Cr–4Ti powders. The typical Arrhenius equation that was used to plot and compute the activation energy profiles is mentioned below:

$$\sigma = A \exp\left(\frac{E_a}{K_b T}\right) \qquad (3)$$

where, A is the pre-exponential factor, E_a is the experimental activation in joules, K_b is Boltzmann's constant $(1.3806 \times 10^{-22} \text{ J K}^{-1})$, and T is the absolute temperature in Kelvin. Mathematically, the slope value from the Arrhenius plot is calculated based on the following equation:

$$\ln(\sigma) = \ln(A) + \text{slope}.\left(\frac{1}{T}\right) \qquad (4)$$

The slope value computed by plotting $\ln(\sigma_{dc})$ in the abscissa and $\Delta(1/T)$ in the ordinate is principally used to calculate the activation energy (E_a) for DC conductivity as shown below:

$$E_a = -K_b \times \text{slope} \qquad (5)$$

Results and discussion

Electron microscopy images accompanied with energy-dispersive absorptive spectroscopy (SEM-EDAX) patterns of unmilled (Fig. 1a), milled (Fig. 1b), and yttria-dispersed (Fig. 1c–e) V–4Cr–4Ti powders are presented in Fig. 1. The unmilled powders exhibit irregular shape with an apparent particle size of ~40 μm. The milled powder has resulted in a narrow particle size distribution with an apparent particle size of ~8 μm (Fig. 1b). A steady increment in the apparent particle size with increasing Y_2O_3 content is noticed. Most importantly, the addition of nano-yttria in the V–4Cr–4Ti powder must have catalyzed the fracturing phenomenon during ball-milling. Upon reaching the limit of comminution, due to very high surface energy, the powder particles must have agglomerated, resulting in a relatively broader apparent particle size distribution. The typical apparent particle sizes of ball-milled powders corresponding to 0.3, 0.6, and 0.9 at.% yttria additions are ~14, ~19, and ~24 μm, respectively (Fig. 1c–e). The apparent particle size (~8 μm) and homogenous shape distribution in the ball-milled V–4Cr–4Ti powders (Fig. 1b) endorse the balance between fracturing and cold-welding phenomena during milling (Yu and Standish 1993). For ball-milling V–4Cr–4Ti powders with varying

nano-yttria content, the exponential supremacy of cold-welding (Fig. 1c–e) is evident leading to the agglomeration of powders. Thus, yttria addition could have possibly suppressed the role of a surfactant (NaCl) during ball-milling, leading to excessive cold-welding and the subsequent agglomeration of powders. The EDAX spectra corresponding to the unmilled and milled V–4Cr–4Ti powders reveal the presence of Na and Cl content (Fig. 1b) while the EDAX spectra representing the nano-yttria-dispersed powder show the absence of Na and Cl spectra (Fig. 1c–e). The absence of Na and Cl spectra in the EDAX patterns (Fig. 1c–e) principally confirms the due absence of NaCl's role as a surfactant/lubricant in the later stages of ball-milling. The absence of surface-active agents (NaCl) must have ramped up the surface tension of powder particles during milling, thereby promoting powder agglomeration (Adams 1994; Mills et al. 2000; Mullier et al. 1991).

Typical outcomes corresponding to the laser particle analysis of V–4Cr–4Ti powders with varying yttria content are presented in Fig. 2 and Table 2. Upon ball-milling, substantial size reduction in the powders is evident and the same can be inferred from the left-sided shift and relatively narrow size distribution profile. Most importantly, while ball-milling the nano-yttria-dispersed V–4Cr–4Ti powders, significant right-sided shift and steady growth in the width of the distribution profile can be observed. Both right-sided shift and growth in the size distribution profile (with yttria addition in the V–4Cr–4Ti powders) approve the clustering or agglomeration of powder particles during milling (Fig. 2 and Table 2). As a point of mention, sintering of powders with a relatively broader particle size distribution can yield relatively dense compacts (close to theoretical density) for a less total linear shrinkage. Ascribed discussion pertaining to the agglomeration of powders with yttria addition strongly concedes with the electron microscopy studies.

Figure 3 reveals the DC conductivity profiles deduced from the auxiliary data generated by the FAST apparatus. DC conductivity profiles clearly demonstrate that yttria addition in the V–4Cr–4Ti powder suppresses the current conductivity across the individual powder particles. Increase in the electrical resistance with varying yttria content can be inferred from the upward shift in the DC conductivity profiles as a function of sintering temperature. Gradual agglomeration of powders and growing complexity in the particle morphology with yttria addition should be the possible reasons behind the upward shift in the DC conductivity profiles.

The activation energy values computed from the DC conductivity profile is shown in Fig. 4. Yttria addition clearly shoots up the E_a values for current conduction (DC conductivity) across the powder particles during FAST-processing. The gradual increment in the E_a values

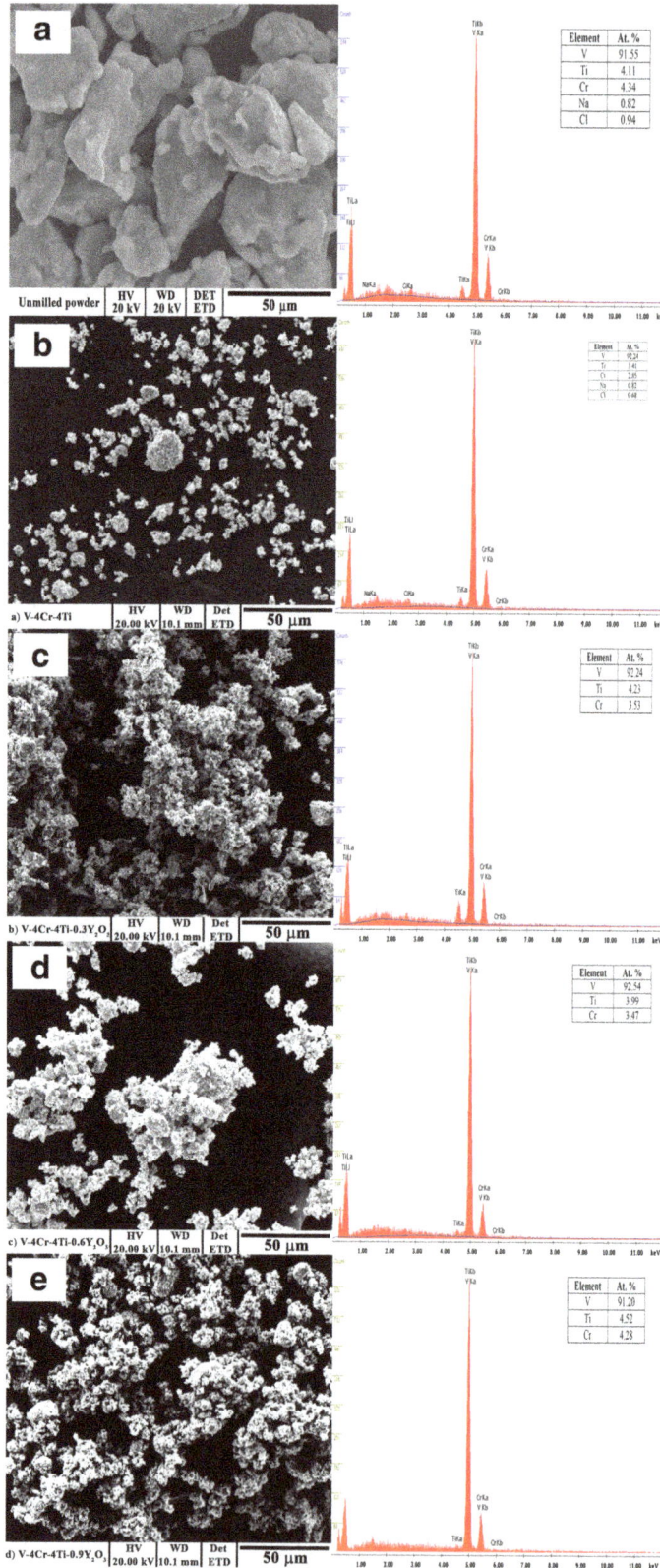

Fig. 1 Electron microscopy images accompanied with energy-dispersive absorptive spectroscopy patterns of **a** unmilled V–4Cr–4Ti powder, **b** milled V–4Cr–4Ti powder, and V–4Cr–4Ti powder ball-milled with varying yttria content: **c** 0.3% yttria, **d** 0.6% yttria, and **e** 0.9% yttria

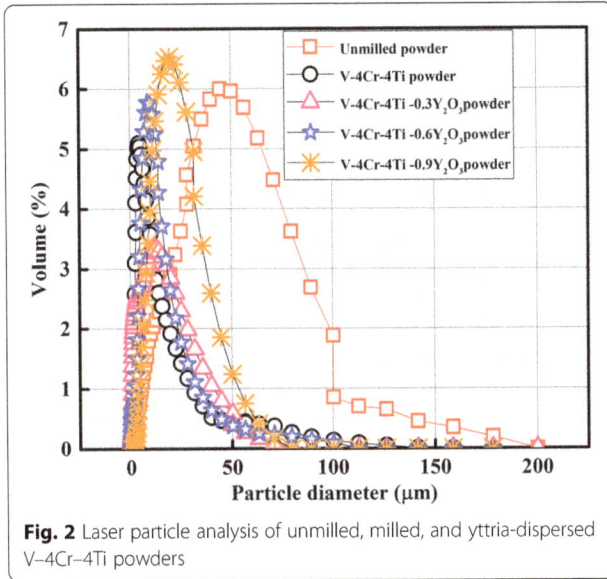

Fig. 2 Laser particle analysis of unmilled, milled, and yttria-dispersed V–4Cr–4Ti powders

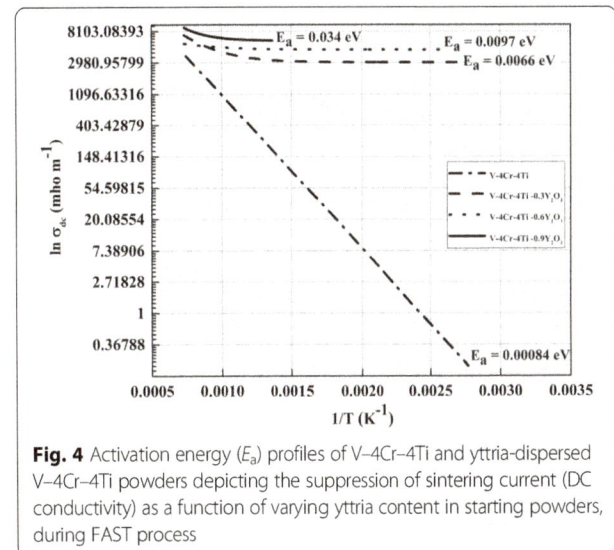

Fig. 3 DC conductivity profiles computed from the auxiliary data generated from the FAST setup while sintering V–4Cr–4Ti and yttria-dispersed V–4Cr–4Ti powders

with increasing yttria content (from 0.3 to 0.9 at.%) clearly indicates the sluggish electromigration phenomenon across the individual powder particles (ODS vanadium powders) during sintering. Apart from the particle size and shape effect, yttria addition must have principally increased the population of metal-to-ceramic interfaces. Since the electrical conductivity across the metal-to-metal and the metal-to-ceramic interface is different altogether during sintering, the population of metal-to-ceramic interfaces could have possibly delayed (or suppressed) the subsequent stages of sintering, viz. spark discharge, formation of ionization column, vaporization, localized neck formation, plastic deformation, mass transport via ionization column (surface and volume diffusion), and solidification. Ascribed stages of sintering solely depend on the DC current conduction and its associated joule-heating phenomenon across the powders. Even if one stage gets delayed, overlapping of the subsequent stages of sintering can lead to poor sintering kinetics henceforth. The poorer the sintering kinetics, the lower will be the driving force for electromigration across the powders. If we compare the activation energy (E_a) values for DC

current conduction of V–4Cr–4Ti (0.00084 eV) and V–4Cr–4Ti–0.9Y$_2$O$_3$ (0.034 eV) powders, by adding 0.9 at.% of yttria in V–4Cr–4Ti, the activation barrier required for DC current conduction across the powder particles has increased by 40 times. The solid line (see Fig. 4) indicating the activation energy profile of the V–4Cr–4Ti powder is relatively much steeper than the activation profiles of ODS vanadium powders. A steeper activation profile (V–4Cr–4Ti powder) indicates excellent DC current conduction across the powder particles during sintering. On the other hand, when the yttria content is gradually increased from 0.3 to 0.9 at.%, the activation profile becomes significantly flat indicating a higher activation barrier (or poor sintering kinetics). Even though the increase in yttria content has led to an inhomogeneous and larger particle size, the samples have achieved relative

Table 2 Comparison of particle size outcomes obtained from the SEM and laser particle analysis

Composition	d_{SEM} (μm)	d_{LPA} (μm)
Unmilled V–4Cr–4Ti powder	40 ± 15	54 ± 9
Milled V–4Cr–4Ti	8 ± 2	12 ± 2.4
V–4Cr–4Ti–0.3Y$_2$O$_3$	14 ± 4	15 ± 1.4
V–4Cr–4Ti–0.6Y$_2$O$_3$	19 ± 7	21 ± 2.4
V–4Cr–4Ti–0.9Y$_2$O$_3$	24 ± 8	27 ± 8.2

d_{LPA} (μm) mean particle size measured using laser particle analyzer, d_{SEM} (μm) mean particle size measured using scanning electron microscopy

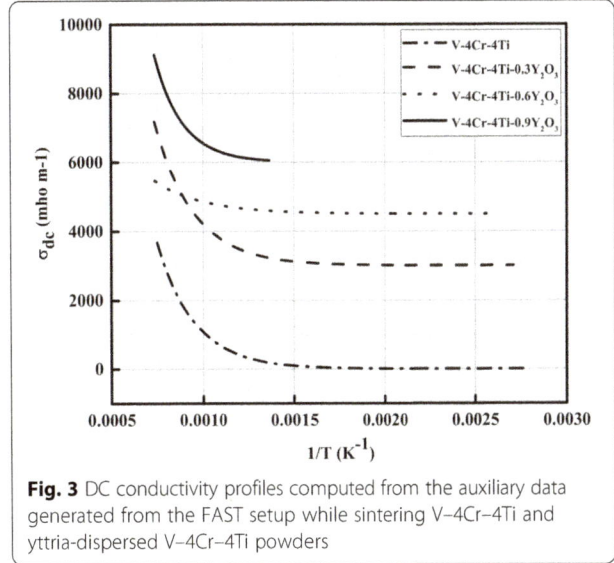

Fig. 4 Activation energy (E_a) profiles of V–4Cr–4Ti and yttria-dispersed V–4Cr–4Ti powders depicting the suppression of sintering current (DC conductivity) as a function of varying yttria content in starting powders, during FAST process

density close to the theoretical density with relatively lesser shrinkage/densification. The apparent relative densities of FAST-processed samples, viz. V–4Cr–4Ti, V–4Cr–4Ti–0.3Y_2O_3, V–4Cr–4Ti–0.6Y_2O_3, and V–4Cr–4Ti–0.9Y_2O_3, sintered at identical sintering conditions are found to be ~98%.

Conclusions

V–4Cr–4Ti and yttria-dispersed V–4Cr–4Ti powders have been synthesized by ball-milling and consolidated using FAST process. Yttria addition has principally suppressed the surfactant's role as a process control agent during milling, resulting in powder agglomeration and the subsequent increase in the apparent particle size distribution. Electron microscopy imaging and laser particle size analysis of ball-milled powders do endorse the increase in the apparent particle size with varying yttria content in V–4Cr–4Ti powder. The steady increment in the apparent particle size distribution of ball-milled powders (with varying yttria content) has attributed an exponential increase in the current consumption during the FAST process. Hence, the magnitude of sintering current consumption (for sintering the discussed powders) during the FAST process can be seen as a function of yttria content in V–4Cr–4Ti powders. For the given sintering condition, FAST-processing of V–4Cr–4Ti powders consume a relatively low order of current when compared with its yttria-dispersed V–4Cr–4Ti counterpart. Activation energy values computed from the DC current conductivity profile also confirms the fact that yttria addition (0.3 to 0.9 at.%) steadily increases the insulation resistance for DC current conductivity during FAST process.

Highlights

- Novel method developed for computing activation energy during sintering
- Effect of yttria-doping (in V–4Cr–4Ti) on DC conductivity during sintering
- Yttria addition in V–4Cr–4Ti powder leads to powder agglomeration
- Powder characteristics affect dynamic current mobility during sintering

Authors' contributions

VKK planned and carried out the experiments. VKK analyzed the metadata and generated the useful inferences presented in the manuscript. VKK drafted this manuscript along with SK. Significant efforts related to the proof-reading of the manuscript were done by SK along with VKK. The manuscript was uploaded by VKK, and the revisions will be attended by both the authors, collectively. Both authors read and approved the final manuscript.

Competing interests

The authors declare that they have no competing interests.

References

Aalund, R. (2008). Spark plasma sintering. Ceramic Industry magazine. *158*, 607–610.

Adams, MJ. (1994). Agglomerate compression strength measurement test using a uniaxial confined. *78*, 5–13.

Alamo, A, Regle, H, Pons, G, & Béchade, JL (1992). Microstructure and textures of ODS ferritic alloys obtained by mechanical alloying. In Materials Science Forum (Vol. 88, pp. 183-190). Trans Tech Publications.

Anselmi-tamburini, U, Gennari, S, Garay, JE, & Munir, ZA (2005). Fundamental investigations on the spark plasma sintering/synthesis process II. Modeling of current and temperature distributions. *394*, 139–148.

Chen, W, Anselmi-Tamburini, U, Garay, JE, Groza, JR, & Munir, ZA. (2005). Fundamental investigations on the spark plasma sintering/synthesis process. *Materials Science and Engineering: A, 394*(1-2), 132–138.

Chung, HM, Loomis, BA, & Smith, DL (1996). Development and testing of vanadium alloys for fusion applications. *239*, 139–156.

Gelles, DS (1998). Microstructural examination of irradiated V ± (4 ± 5%) Cr ± (4 ± 5%) Ti. *263*, 1380–1385.

Iimura, K, Suzuki, M, Hirota, M, & Higashitani, K. (2009). Higashitani, Simulation of dispersion of agglomerates in gas phase – acceleration field and impact on cylindrical obstacle, Advanced Powder Technology, *20*(2), 210–5.

Johnson, WR, & Smith, JP (1998). Fabrication of a 1200 kg ingot of V ± 4Cr ± 4Ti alloy for the DIII ± D radiative divertor program. *263*, 1425–1430.

Krishnan, VK, Sinnaeruvadi, K, NU SC, RMHM (2016). http://www.sciencedirect.com/science/article/pii/S0263436815302900.

Loomis, BA, Chung, HM, Nowicki, LJ, & Smith, DL (1994). !!!! I. *3115*(94), 2–6.

Mamedov, V. (2000). *Spark plasma sintering as advanced PM sintering method.*

Miller, MK, Kenik, EA, Russell, KF, Heatherly, L, Hoelzer, DT, & Maziasz, PJ (2003). Atom probe tomography of nanoscale particles in ODS ferritic alloys. *353*, 140–145.

Miller, MK, Hoelzer, DT, Kenik, EA, & Russell, KF (2005). Stability of ferritic MA/ODS alloys at high temperatures. *13*, 387–392.

Mills, PJT, Seville, JPK, Knight, PC, & Adams, MJ (2000). The effect of binder viscosity on particle agglomeration in a low shear mixer granulator. 140-147. http://www.sciencedirect.com/science/article/pii/S0032591000002242.

Mullier, MA, Seville, JPK, & Adams, MJ (1991). The effect of agglomerate on attrition during processing. *65*, 321–333.

Muroga, T, Nagasaka, T, Abe, K, Chernov, VM, Matsui, H, Smith, DL, Xu, ZY, & Zinkle, SJ. (2002). Vanadium alloys—overview and recent results. *Journal of Nuclear Materials, 307–311*(1 SUPPL), 547–554.

Plant, CL, & Steel, K (1995). Dispersion behaviour of oxide particles in mechanically alloyed ODS steel. *14*, 1600–1603.

Rice, PM, & Zinkle, SJ (1998). Temperature dependence of the radiation damage microstructure in V ± 4Cr ± 4Ti neutron irradiated to low dose. *263*, 1414-1419.

Smith, DL, Chung, HM, Loomis, BA, Matsui, H, Votinov, S, Witzenburg, WV (1995). Fusion engineering and design development of vanadium-base alloys for fusion first-wall-blanket applications, *3796* (94).

Smith, DL, Chung, HM, & Loomis, BA (1996). b ! fY, *15*(96). http://www.sciencedirect.com/science/article/pii/S0022311596002310.

Suryanarayana, C, & Norton, MG. (1999). *Book review X-ray diffraction : a practical approach* (pp. 13–15). https://books.google.co.in/books?hl=en&lr=&id=RRfrBwAAQBAJ&oi=fnd&pg=PA3&dq=Xray+diffraction+:+a+practical+approach&ots=NwGLKtau7e&sig=jTAiYITe9dP0eJSA4-70K77jSmk#v=onepage&q=X-ray%20diffraction%20%3A%20a%20practical%20approach&f=false.

Thornton, C. (1999). *Numerical simulations of agglomerate impact breakage* (pp. 74–82).

Thornton, C, Ciomocos, MT, & Adams, MJ. (2004). Numerical simulations of diametrical compression tests on agglomerates. *140*, 258–267.

Tong, ZB, Yang, RY, Yu, AB, Adi, S, & Chan, HK. (2009). Numerical modelling of the breakage of loose agglomerates of fine particles. *Powder Technology, 196*(2), 213–221.

Yamagata, I, Kurishita, H, & Kayano, H (1996). Development of oxide dispersion strengthened vanadium alloy and its properties. *239*, 162–169.

Yu, A. B., Bridgwater, J., & Burbidge, A (1997). On the modelling of the packing of fine particles. *Powder technology, 92* (3), 185–194.

Yu, AB, & Standish, N (1993). Characterisation of non-spherical particles from their packing behaviour. *74*, 205–213.

Yucheng, W, & Zhengyi, F (2002). Study of temperature field in spark plasma sintering. *90*, 34–37.

10

Improved shear correction factors for deflection of simply supported very thick rectangular auxetic plates

T. C. Lim

Abstract

Background: The first-order shear deformation theory (FSDT) for plates requires a shear correction factor due to the assumption of constant shear strain and shear stress across the thickness; hence, the shear correction factor strongly influences the accuracy of the deflection solution; the third-order shear deformation theory (TSDT) does not require a correction factor because it facilitates the change in shear strain across the plate thickness.

Methods: This paper obtains an improved shear correction factor for simply supported very thick rectangular plates by matching the deflection of the Mindlin plate (FSDT) with that of the Reddy plate (TSDT).

Results: As a consequence, the use of the exact shear correction factor for the Mindlin plate gives solutions that are exactly the same as for the Reddy plate.

Conclusions: The customary adoption of 5/6 shear correction factor is a lower bound, and the exact shear correction factor is higher for the following: (a) very thick plates, (b) narrow or long plates, (c) high Poisson's ratio plate material, and (d) highly patterned loads, while the commonly used shear correction factor of 5/6 is still valid for the following: (i) marginally thick plates, (ii) square plates, (iii) negative Poisson's ratio materials, and (d) uniformly distributed loadings.

Keywords: Aspect ratio, Auxetic materials, Incompressible materials, Shear deformation, Thick plates

Background

It is well known that the shear correction factors of plates are simpler than those for beams (Dong et al., 2010; Puchegger et al., 2003; Hlavacek and Chleboun, 2000; Pai and Schultz, 1999; Popescu and Hodges, 2000; Yu and Hodges, 2004; Chan et al., 2011; Pai et al., 2000; Hutchinson, 1980; Hutchinson, 2001; Han et al., 1999); this is due to the cross-sectional geometry in beams being more varied than for plates. For plates, the commonly adopted shear correction factor is typically 5/6; in some instances, Poisson's ratio is taken into account (e.g., Rössle, 1999; Lee et al., 2002). Exact shear correction factors for vibrating Mindlin plates have been proposed by Stephen (1997) and Hull (2005, 2006). In this paper, exact shear correction factors for simply supported very thick rectangular Mindlin plates are derived by comparing its deflection against that of Reddy plates. The Mindlin plate,

which adopts the first-order shear deformation theory (FSDT), requires a correction factor due to its assumption of uniform shear across the plate thickness while the Reddy plate, which adopts the third-order shear deformation theory (TSDT), does not require any correction as it caters for the varying shear strain across the plate thickness. The rigor of the Reddy plate, therefore, forms the justification for its use as a benchmark for evaluating the accuracy of Mindlin plate deflection—this has been done for triangular plates (Lim, 2016a, b). Following a recent preliminary analysis (Lim, 2016c) to evaluate the ratio of maximum deflection of Reddy plate to that of Kirchhoff plate or the classical plate theory (CPT), the TSDT is now being employed for extracting the exact shear correction factor of rectangular plates in the FSDT.

Methods

General consideration

Figure 1 illustrates a simply supported thick rectangular plate of sides a and b, measured along the x and y axes,

Correspondence: alan_tc_lim@yahoo.com
School of Science and Technology, SIM University, Singapore, Singapore

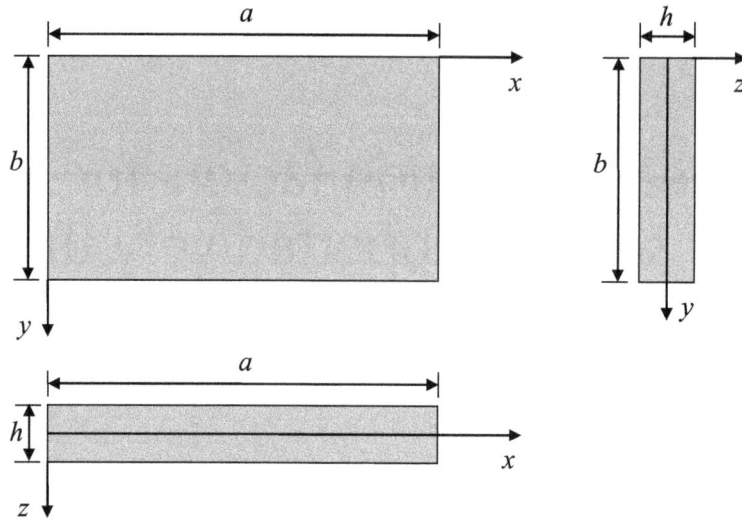

Fig. 1 Geometrical nomenclature adopted for analysis

respectively, while the thickness h is measured along the z axis. Recall that the Mindlin plate deflection w^M is related to the Kirchhoff plate deflection w^K as

$$w^M = w^K + \frac{M^K}{\kappa Gh} \tag{1a}$$

where the moment sum, or Marcus moment, is defined as

$$M^K = -D\nabla^2 w^K \tag{1b}$$

and D is the plate flexural rigidity, while the Reddy plate deflection w^R is related to the Kirchhoff plate deflection as

$$w^R = w^K + \frac{1}{Gh}\left(\alpha C_1 D\nabla^2 w^R + C_4 M^K\right) \tag{2a}$$

where $280C_1 = 3h^2 = 4/\alpha$, $C_4 = 17/14$, and

$$\lambda_0^2 = \frac{Gh}{\alpha C_1 D} = \frac{70Gh}{D}, \tag{2b}$$

with G being the shear modulus of the plate.

With reference to Eqs.(1) and (2), it is useful to describe the plate flexural ridigity

$$D = \frac{Eh^3}{12(1-\nu^2)} \tag{3a}$$

and the shear modulus

$$G = \frac{E}{2(1+\nu)}, \tag{3b}$$

where E is Young's modulus, as the ratio

$$\frac{D}{G} = \frac{h^3}{6(1-\nu)} \tag{3c}$$

so that Eqs.(1) and (2) can be expressed as

$$w^M = w^K - \frac{h^2}{6\kappa(1-\nu)}\nabla^2 w^K \tag{4a}$$

and

$$w^R - \frac{h^2}{420(1-\nu)}\nabla^2 w^R = w^K - \frac{17h^2}{84(1-\nu)}\nabla^2 w^K \tag{4b}$$

respectively, to facilitate comparison. Equating the Mindlin and Reddy plate deflections gives the following general relationship under the same boundary condition:

$$\frac{1}{\kappa} = \frac{17}{14} - \frac{1}{70}\frac{\Delta w^R}{\Delta w^K} \tag{5a}$$

where

$$\Delta = \nabla^2 = \frac{\partial^2}{\partial x^2} + \frac{\partial^2}{\partial y^2}. \tag{5b}$$

Perusal to Eq. (5a) suggests that a meaningful exact shear correction factor can be obtained if both the Reddy plate and Kirchhoff plate deflections are known. Neglecting the higher order term in Eq. (5a) gives a shear correction factor of $\kappa = 14/17$. The two constant shear correction factors of 5/6 and 14/17 have been discussed by Wang et al. (2000).

Uniform load

As the Kirchhoff plate deflection for a simply supported rectangular plate under uniform load $q = q_0$ is

$$w^K = \frac{16q_0}{\pi^6 D} \sum_{m=1}^{\infty} \sum_{n=1}^{\infty} \frac{\sin\frac{m\pi x}{a}\sin\frac{n\pi y}{b}}{mn\left(\frac{m^2}{a^2}+\frac{n^2}{b^2}\right)^2} \tag{6}$$

with $m, n = 1, 3, 5, \ldots$, we adopt a similar profile deflection for the Reddy plate

$$w^R = A^R \sum_{m=1}^{\infty} \sum_{n=1}^{\infty} \frac{\sin\frac{m\pi x}{a}\sin\frac{n\pi y}{b}}{mn\left(\frac{m^2}{a^2}+\frac{n^2}{b^2}\right)^2}, \tag{7}$$

where A^R is the amplitude term of the Reddy plate. Substituting the deflection profiles of Kirchhoff and Reddy plates into the relationships described by Eq. (4) leads to

$$w^M = \frac{16q_0}{\pi^6 D} \left\{ \sum_{m=1}^{\infty} \sum_{n=1}^{\infty} \frac{\sin\frac{m\pi x}{a}\sin\frac{n\pi y}{b}}{mn\left(\frac{m^2}{a^2}+\frac{n^2}{b^2}\right)^2} \right. \tag{8}$$
$$\left. + \frac{\pi^2 h^2}{6\kappa(1-v)} \sum_{m=1}^{\infty} \sum_{n=1}^{\infty} \frac{\sin\frac{m\pi x}{a}\sin\frac{n\pi y}{b}}{mn\left(\frac{m^2}{a^2}+\frac{n^2}{b^2}\right)} \right\}$$

and

$$A^R = \frac{16q_0}{\pi^6 D} \frac{\displaystyle\sum_{m=1}^{\infty}\sum_{n=1}^{\infty}\frac{\sin\frac{m\pi x}{a}\sin\frac{n\pi y}{b}}{mn\left(\frac{m^2}{a^2}+\frac{n^2}{b^2}\right)^2} + \frac{17\pi^2 h^2}{84(1-v)}\sum_{m=1}^{\infty}\sum_{n=1}^{\infty}\frac{\sin\frac{m\pi x}{a}\sin\frac{n\pi y}{b}}{mn\left(\frac{m^2}{a^2}+\frac{n^2}{b^2}\right)}}{\displaystyle\sum_{m=1}^{\infty}\sum_{n=1}^{\infty}\frac{\sin\frac{m\pi x}{a}\sin\frac{n\pi y}{b}}{mn\left(\frac{m^2}{a^2}+\frac{n^2}{b^2}\right)^2} + \frac{\pi^2 h^2}{420(1-v)}\sum_{m=1}^{\infty}\sum_{n=1}^{\infty}\frac{\sin\frac{m\pi x}{a}\sin\frac{n\pi y}{b}}{mn\left(\frac{m^2}{a^2}+\frac{n^2}{b^2}\right)}} \tag{9}$$

Introducing the function

$$f(a,b,x,y) = \frac{\displaystyle\sum_{m=1}^{\infty}\sum_{n=1}^{\infty}\frac{\sin\frac{m\pi x}{a}\sin\frac{n\pi y}{b}}{mn\left(\frac{b}{a}m^2+\frac{a}{b}n^2\right)}}{\displaystyle\sum_{m=1}^{\infty}\sum_{n=1}^{\infty}\frac{\sin\frac{m\pi x}{a}\sin\frac{n\pi y}{b}}{mn\left(\frac{b}{a}m^2+\frac{a}{b}n^2\right)^2}} \tag{10}$$

allows Eq. (9) to be contracted as

$$A^R = \frac{16q_0}{\pi^6 D} \frac{1 + \frac{17\pi^2 h^2}{84(1-v)ab}f(a,b,x,y)}{1 + \frac{\pi^2 h^2}{420(1-v)ab}f(a,b,x,y)}. \tag{11}$$

The terms a/b in Eq. (10) and $h/\sqrt{ab}$ in Eq. (11) indicate the plate aspect ratio and its relative thickness, respectively. For a square plate, these reduce to $a/b = 1$ and h/a. Using Eq. (11) and equating the Mindlin and Reddy plate deflection gives

$$1 + \frac{\pi^2 h^2}{420(1-v)ab}f(a,b,x,y) + \frac{\pi^2 h^2}{6\kappa(1-v)ab}f(a,b,x,y)$$
$$+ \frac{\pi^4 h^4}{2520\kappa(1-v)^2 a^2 b^2}(f(a,b,x,y))^2$$
$$= 1 + \frac{17\pi^2 h^2}{84(1-v)ab}f(a,b,x,y). \tag{12}$$

Hence, the usual shear correction factor of

$$\kappa = \frac{5}{6} \tag{13}$$

is obtained from Eq. (12) if the highest order term is neglected. Taking into account the highest order term, we have the exact shear correction factor

$$\kappa = \frac{5}{6}\left[1 + \frac{\pi^2}{420(1-v)}\frac{h^2}{ab}f(a,b,x,y)\right]. \tag{14a}$$

Since the maximum deflection takes place at the plate center, it is practical to consider the shear correction factor there, i.e.,

$$\kappa = \frac{5}{6}\left[1 + \frac{\pi^2}{420(1-v)}\frac{h^2}{ab}f\left(a,b,\frac{a}{2},\frac{b}{2}\right)\right] \tag{14b}$$

where the function described by Eq. (10) becomes

$$f\left(a,b,\frac{a}{2},\frac{b}{2}\right) = \frac{\displaystyle\sum_{m=1}^{\infty}\sum_{n=1}^{\infty}\frac{(-1)^{\frac{m+n}{2}-1}}{mn\left(\frac{b}{a}m^2+\frac{a}{b}n^2\right)}}{\displaystyle\sum_{m=1}^{\infty}\sum_{n=1}^{\infty}\frac{(-1)^{\frac{m+n}{2}-1}}{mn\left(\frac{b}{a}m^2+\frac{a}{b}n^2\right)^2}} \tag{15}$$

at the plate center. For an extremely long and narrow plate, Eq. (15) reduces to

$$\lim_{\frac{a}{b}\to\infty} f\left(a,b,\frac{a}{2},\frac{b}{2}\right) = \frac{a}{b} \tag{16}$$

We note that both the numerator and denominator of Eq. (15) are dependent on the plate aspect ratio a/b. Setting $a \geq b$ for the uniformly loaded plate, Table 1 lists the denominator and numerator of Eq. (15) by performing double series summation. The summation was performed up to $m = n = 41$ in order to obtain sufficient numerical accuracy.

A simple curve fit based on Table 1 gives the function

$$f\left(a,b,\frac{a}{2},\frac{b}{2}\right) = -0.0062\left(\frac{a}{b}\right)^3 + 0.1281\left(\frac{a}{b}\right)^2$$
$$+ 0.1956\left(\frac{a}{b}\right)$$
$$+ 1.4598 \quad ; \quad 1 \leq a/b \leq 10, \tag{17}$$

with a statistical accuracy of $R^2 = 0.9998$.

Table 1 Computed results of Eq. (15)

Plate aspect ratio $\frac{a}{b}$	Numerator $\sum\limits_{m=1}^{\infty}\sum\limits_{n=1}^{\infty}\dfrac{(-1)^{\frac{m+n}{2}}-1}{mn\left(\frac{b}{a}m^2+\frac{a}{b}n^2\right)}$	Denominator $\sum\limits_{m=1}^{\infty}\sum\limits_{n=1}^{\infty}\dfrac{(-1)^{\frac{m+n}{2}}-1}{mn\left(\frac{b}{a}m^2+\frac{a}{b}n^2\right)^2}$	Ratio $f\left(a,b,\frac{a}{2},\frac{b}{2}\right)$
1.0	0.44895	0.24409	1.839281
1.2	0.44059	0.23578	1.868649
1.5	0.40575	0.20627	1.967082
2.0	0.34749	0.15215	2.283865
3.0	0.24622	0.08167	3.014816
5.0	0.15405	0.031175	4.941460
7.5	0.10389	0.013907	7.470339
10.0	0.07859	0.007824	10.04473

Sinusoidal load

Suppose the load distribution takes a sinusoidal form

$$q = q_0 \sin\frac{m\pi x}{a}\sin\frac{n\pi y}{b} \tag{18}$$

instead of being uniformly distributed—whereby m and n quantify the load waviness along the x and y axes, respectively—then Eqs.(10) and (14a) reduce to

$$f(a,b,x,y) = \frac{b}{a}m^2 + \frac{a}{b}n^2 \tag{19}$$

and

$$\kappa = \frac{5}{6}\left[1 + \frac{\pi^2}{420(1-\nu)}\frac{h^2}{ab}\left(\frac{b}{a}m^2 + \frac{a}{b}n^2\right)\right] \tag{20a}$$

respectively. Although Eq. (20a) can be written in a simpler way as

$$\kappa = \frac{5}{6}\left[1 + \frac{\pi^2 h^2}{420(1-\nu)}\left(\frac{m^2}{a^2} + \frac{n^2}{b^2}\right)\right], \tag{20b}$$

the former is instructive for showing the effect of relative plate thickness $h/\sqrt{ab}$ and aspect ratio a/b, in addition to Poisson's ratio. Unlike the previous section on uniform load, this section on sinusoidal load allows one to observe the interlacing effect of load waviness pattern and plate aspect ratio on the shear correction factor.

For the special case of square plate, perusal to Table 2 shows that load waviness increases the shear correction factor. Reference to the same table also shows that waviness is strongly influenced by the aspect ratio of the plate; the effect of load waviness along the longer side

diminishes as the plate becomes long or very narrow, i.e.,

$$\begin{aligned}
\kappa &\approx \frac{5}{6}\left[1 + \frac{n^2\pi^2}{420(1-\nu)}\frac{h^2}{b^2}\right] \quad ; \quad a \gg b \\
\kappa &\approx \frac{5}{6}\left[1 + \frac{m^2\pi^2}{420(1-\nu)}\frac{h^2}{a^2}\right] \quad ; \quad a \ll b
\end{aligned} \tag{21}$$

and consequently, the relative thickness is governed by the ratio of the plate thickness to its shorter side.

Results and discussion

In determining the range of relative thickness that is applicable for the shear deformation theories, one may refer to Steele and Balch (2009) who classified the plate thickness into four categories: (i) $a/h > 100$, (ii) $20 < a/h < 100$, (iii) $3 < a/h < 20$, and (iv) $a/h < 3$. This implies that one may then adopt the membrane theory for $h/a < 0.01$, CPT for $h/a < 0.05$, shear deformation theories for $h/a < 0.3333$, and elasticity theory for $h/a > 0.3333$. It therefore follows that the TSDT-based shear correction factor for FSDT problems are therefore applicable for relative thickness range of $h/a < 0.3333$. As such, the following results were computed for relative thickness up to 0.2 since shear deformation theories are not applicable for relative thickness of 1/3 and above. As with the CPT and FSDT, the TSDT is applicable for auxetic materials since the development of these theories are not confined to cases where Poisson's ratio is positive.

Uniform load

Figure 2(a) and (b) shows the effect of relative plate thickness and plate aspect ratio on the shear correction

Table 2 Shear correction factor expressions for special cases of rectangular plates under sinusoidal loads

	Simple sinusoidal load distribution ($m = n = 1$)	General sinusoidal load distributions ($m, n \geq 1$)
Square plates $a = b$	$\kappa = \frac{5}{6}\left[1 + \frac{\pi^2}{210(1-\nu)}\frac{h^2}{a^2}\right]$	$\kappa = \frac{5}{6}\left[1 + \frac{\pi^2}{420(1-\nu)}\frac{h^2}{a^2}(m^2+n^2)\right]$
Rectangular plates $a \neq b$	$\kappa = \frac{5}{6}\left[1 + \frac{\pi^2}{420(1-\nu)}\frac{h^2}{ab}\left(\frac{b}{a}+\frac{a}{b}\right)\right]$	$\kappa = \frac{5}{6}\left[1 + \frac{\pi^2}{420(1-\nu)}\frac{h^2}{ab}\left(\frac{b}{a}m^2+\frac{a}{b}n^2\right)\right]$

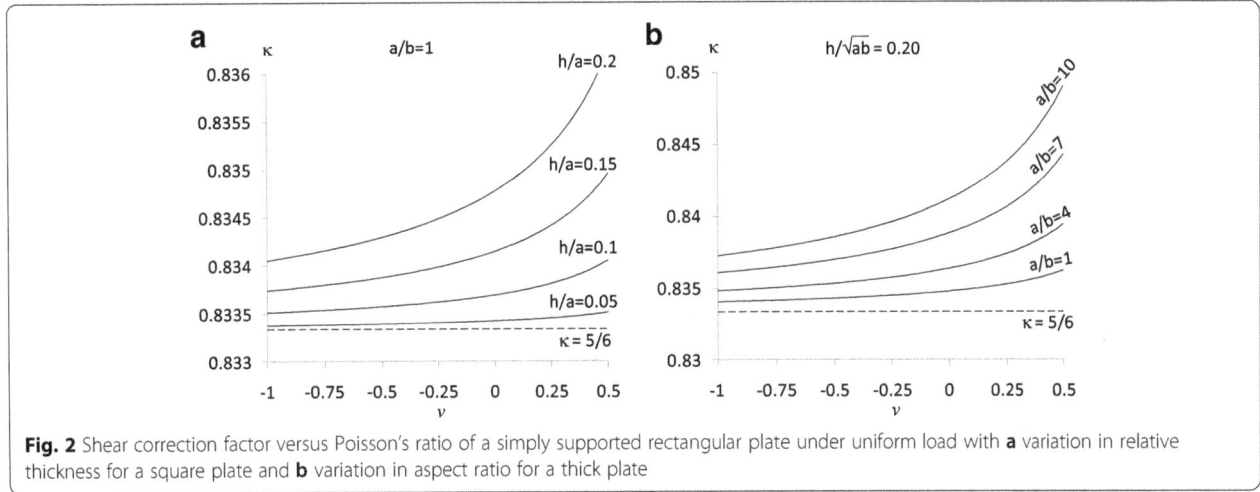

Fig. 2 Shear correction factor versus Poisson's ratio of a simply supported rectangular plate under uniform load with **a** variation in relative thickness for a square plate and **b** variation in aspect ratio for a thick plate

factor of a uniformly loaded plate for the entire range of Poisson's ratio. Specifically, the shear correction factor increases when (i) the plate becomes thicker, (ii) the plate becomes longer or narrower, and (iii) Poisson's ratio of the plate material is greater. The curves of the shear correction factors are plotted for Poisson's ratio of the range $-1 \leq \nu \leq 0.5$. This range is applicable for isotropic materials, in which solids of negative Poisson's ratio are termed "auxetic" materials (Lim, 2010, 2015a, 2016d); no bounds exist for Poisson's ratio of anisotropic ones (Ting, 2005; Lim, 2015b; Boldrin et al., 2016). The dashed lines in this and subsequent figures indicate the lower bound for the shear correction factor, i.e., $\kappa = 5/6$, for comparison. The influence of the plate geometry, in terms of the in-plane aspect ratio and the relative thickness, on the shear correction factor is plotted in Fig. 3 for $\nu = 0.3$.

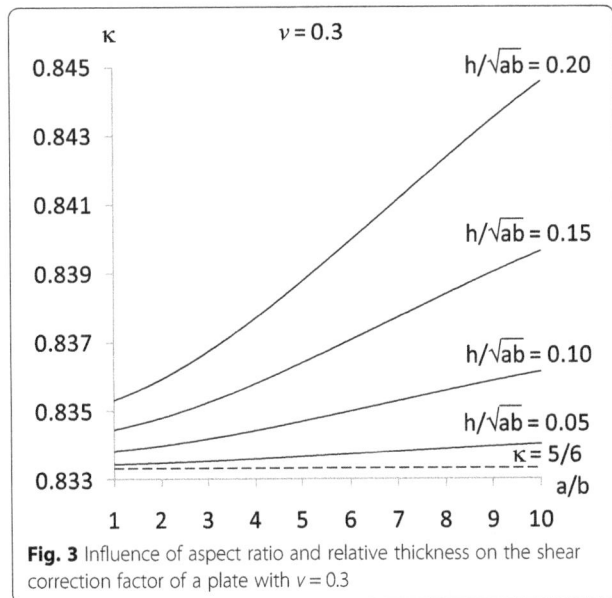

Fig. 3 Influence of aspect ratio and relative thickness on the shear correction factor of a plate with $\nu = 0.3$

Sinusoidal load

In the case of sinusoidal load, there is a qualitatively comparable trend in the effect of plate geometry (aspect ratio and relative thickness) and Poisson's ratio on uniform load. In addition, the waviness of the transverse static load increases the shear correction factor, as evidenced in Fig. 4 for square plates.

In the special case of square plates, the shear correction factor is unchanged when the load waviness changes direction. For example, the shear correction factor for $(m, n) = (3, 1)$ is similar to that for $(m, n) = (1, 3)$; likewise, the shear correction factor for $(m, n) = (5, 1)$ is similar to that for $(m, n) = (1, 5)$. This observation, however, does not hold for rectangular plates. Perusal to Eqs.(20) or (21) shows that for very long or very narrow plates, the load waviness measured along the shorter side has greater influence than that along the longer side, as shown in Fig. 5.

Comparison with other cases

This section makes two types of comparisons, i.e., (a) with plates of other shapes but with similar boundary condition and (b) with similar plates but other boundary conditions. To put into perspective the current results with other plates under similar boundary conditions, a comparison is made with some recently improved shear correction factors of very thick plates. Table 3 summarizes the improved shear correction factors of very thick plates evaluated at the plate centroid for three different Poisson's ratio within isotropic solids: extremely auxetic ($\nu = -1$), typical solids ($\nu = 0.3$), and incompressible solids ($\nu = 0.5$). The plates considered for comparison are simply supported isosceles right triangular plate (Lim, 2016a), equilateral triangular plate (Lim, 2016b), square plate, and rectangular plate of aspect ratio 4 under uniform load and possess the dimensionless plate thickness of $h/a = 0.2$.

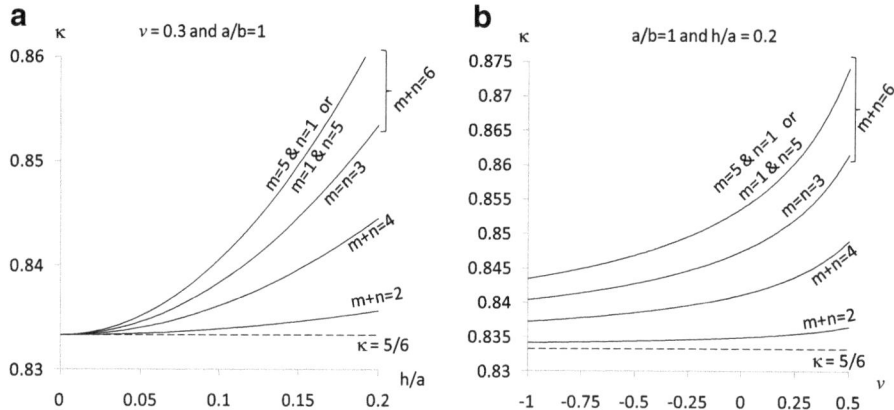

Fig. 4 Effect of load waviness on the shear correction factor of a sinusoidally loaded square plate with varying **a** relative thickness and **b** Poisson's ratio

The descriptions of shear correction factors developed herein apply only for rectangular plates of simply supported boundary condition, and are therefore not applicable for thick rectangular plates of clamped and/or free edges. Nevertheless, it is worthy to note that the FSDT for Levy plates, as reviewed by Wang et al. (2000), adopts a shear correction factor of 5/6. Therefore, on this basis and on the basis of the results obtained from this paper, it can be said that the shear correction factor of 5/6 could well continue to be a tight lower bound and that Poisson's ratio, alongside the plate's relative thickness, exerts influence on the shear correction factor.

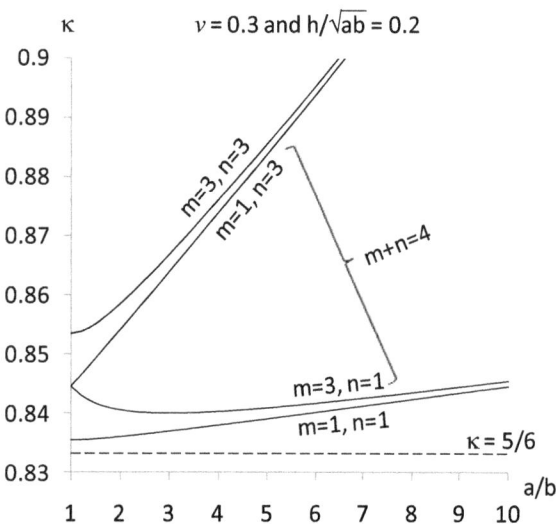

Fig. 5 Asymmetric effect of load waviness on shear correction factor of a rectangular plate with Poisson's ratio 0.3 and relative thickness 0.2

Conclusions

Exact shear correction factors for simply supported very thick rectangular plates under static loads have been developed herein for the case of uniform and sinusoidal loads using the Mindlin-Kirchhoff relationship and the Reddy-Kirchhoff relationship. Results obtained herein for uniform and sinusoidal loads show that the exact shear correction factor is higher than the commonly used shear correction factor of 5/6 under the following conditions:

(a) Very thick plates
(b) Very long or narrow plates
(c) Plates made from large Poisson's ratio (especially incompressible materials)
(d) Highly patterned loading pattern or sinusoidal load with high load waviness

However, the use of the lower bound shear correction factor of $\kappa = 5/6$ is valid under the following conditions:

(a) Marginally thick plates
(b) Square or almost square plates
(c) Plates made from auxetic materials, and
(d) Less wavy load pattern, especially uniform loads.

Nomenclature

A = Amplitude of plate deflection
a, b = In-plane dimensions of rectangular plate along x, y axes
D = Plate flexural rigidity
E = Young's modulus
G = Shear modulus
h = Plate thickness
M = Marcus moment
m, n = Load waviness along x, y axes
q = Load intensity

Table 3 Summary of shear correction factors of simply supported and uniformly loaded square and triangular plates with $h/a = 0.2$

Schematics of plates	Extremely auxetic material ($v = -1$)	Typical conventional material ($v = 0.3$)	Incompressible material ($v = 1/2$)	Remarks
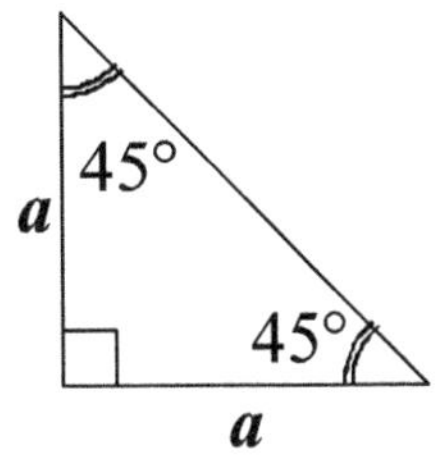	$\kappa = 0.8351$	$\kappa = 0.8384$	$\kappa = 0.8405$	Isosceles right triangular plate (Lim 2016a)
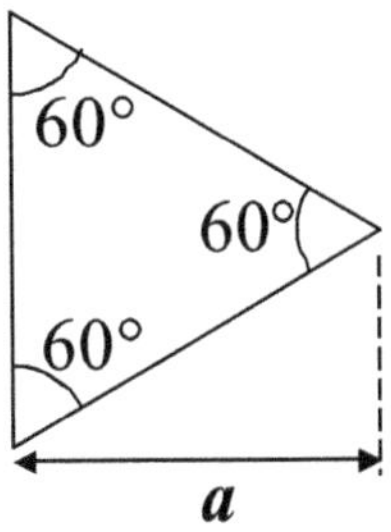	$\kappa = 0.8344$	$\kappa = 0.8364$	$\kappa = 0.8376$	Equilateral triangular plate (Lim 2016b)
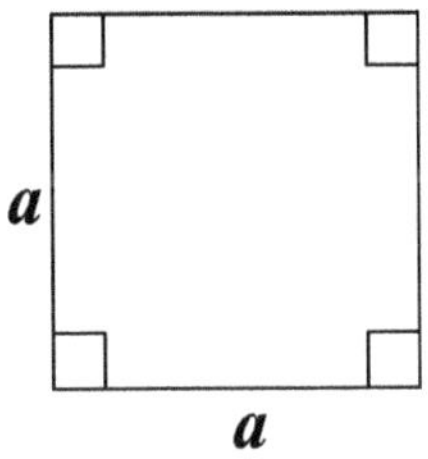	$\kappa = 0.8341$	$\kappa = 0.8354$	$\kappa = 0.8362$	Square plate (this paper)
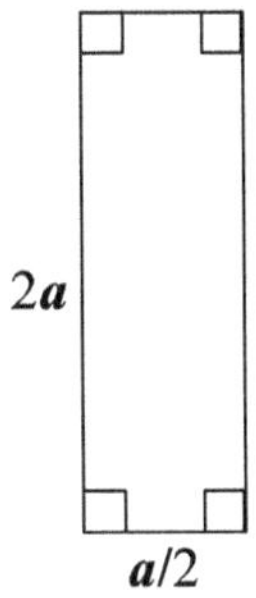	$\kappa = 0.8349$	$\kappa = 0.8377$	$\kappa = 0.8394$	Rectangular plate with aspect ratio of 4 (this paper)

R^2 = Coefficient of determination
w = Plate deflection
x, y = In-plane Cartesian coordinates
κ = Shear correction factor
v = Poisson's ratio

Superscripts
K = Kirchhoff plate
M = Mindlin plate
R = Reddy plate

Subscript
0 = Maximum intensity for load.

Competing interests
The author declares that he has no competing interests.

References
Boldrin, L., Hummel, S., Scarpa, F., Di Maio, D., Lira, C., Ruzzene, M., Remillat, C. D. L., Lim, T. C., Rajasekaran, R., & Patsias, S. (2016). Dynamic behaviour of auxetic gradient composite hexagonal honeycombs. *Composites Structures*, *149*, 114–124.

Chan, K. T., Lai, K. F., Stephen, N. G., & Young, K. (2011). A new method to determine the shear coefficient of Timoshenko beam theory. *Journal of Sound and Vibration, 330*(14), 3488–3497.

Dong, S. B., Alpdogan, C., & Taciroglu, E. (2010). Much ado about shear correction factors in Timoshenko beam theory. *International Journal of Mechanical Sciences, 47*(13), 1651–1655.

Han, S. M., Benaroya, H., & Wei, T. (1999). Dynamics of transversely vibrating beams using four engineering theories. *Journal of Sound and Vibration, 225*(5), 935–988.

Hlavacek, I., & Chleboun, J. (2000). Reliable analysis of transverse vibrations of Timoshenko-Mindlin beams with respect to uncertain shear correction factor. *Computer Methods in Applied Mechanics and Engineering, 190*(8-10), 903–918.

Hull, A. J. (2005). An exact analytical expression of the shear coefficient in the Mindlin plate equation. *Journal of the Acoustical Society of America, 117*(4), 2601–2601.

Hull, A. J. (2006). Mindlin shear coefficient determination using model comparison. *Journal of Sound and Vibration, 294*(1&2), 125–130.

Hutchinson, J. R. (1980). Vibrations of solid cylinders. *ASME Journal of Applied Mechanics, 47*(4), 901–907.

Hutchinson, J. R. (2001). Shear coefficients for Timoshenko beam theory. *ASME Journal of Applied Mechanics, 68*(1), 87–92.

Lee, K. H., Lim, G. T., & Wang, C. M. (2002). Thick Levy plates re-visited. *International Journal of Solids and Structures, 39*(1), 127–144.

Lim, T. C. (2010). In-plane stiffness of semiauxetic laminates. *ASCE Journal of Engineering Mechanics, 136*(9), 1176–1180.

Lim, T. C. (2015a). Shear deformation in beams with negative Poisson's ratio. *IMechE Journal of Materials Design and Applications, 229*(6), 447–454.

Lim, T. C. (2015b). *Auxetic Materials and Structures*. Singapore: Springer.

Lim, T. C. (2016a). Refined shear correction factor for very thick simply supported and uniformly loaded isosceles right triangular auxetic plates. *Smart Materials and Structures, 25*(5), 054001.

Lim, T. C. (2016b). Higher order shear deformation of very thick simply-supported equilateral triangular plates under uniform load. *Mechanics Based Design of Structures and Machines, 44*(4), 514–522.

Lim, T. C. (2016c). Combined effect of load waviness and auxeticity on the shear deformation in a class of rectangular plates. *IOP Conference Series: Materials Science and Engineering, 157*, 012011.

Lim, T. C. (2016d). Large deflection of circular auxetic membranes under uniform load. *ASME Journal of Engineering Materials and Technology, 138*(4), 041011.

Pai, P. F., Anderson, T. J., & Wheater, E. A. (2000). Large-deformation tests and total-Lagrangian finite-element analyses of flexible beams. *International Journal of Solids and Structures, 37*(21), 2951–2980.

Pai, P. F., & Schultz, M. J. (1999). Shear correction factors and an energy-consistent beam theory. *International Journal of Solids and Structures, 36*(10), 1523–1540.

Popescu, B., & Hodges, D. H. (2000). On asymptotically correct Timoshenko-like anisotropic beam theory. *International Journal of Solids and Structures, 37*(3), 535–558.

Puchegger, S., Bauer, S., Loidl, D., Kromp, K., & Peterlik, H. (2003). Experimental validation of the shear correction factor. *Journal of Sound and Vibration, 261*(1), 177–184.

Rössle, A. (1999). On the derivation of an asymptotically correct shear correction factor for the Reissner-Mindlin plate mode. *Comptes Rendus de l'Académie des Sciences - Series I – Mathematics, 328*(3), 269–274.

C. R. Steele, C. D. Balch (2009) Introduction to the Theory of Plates. Stamford University. http://web.stanford.edu/~chasst/Course%20Notes/Introduction%20to%20the%20Theory%20of%20Plates.pdf. Accessed 24 October 2016.

Stephen, N. G. (1997). Mindlin plate theory: best shear coefficient and higher spectra validity. *Journal of Sound and Vibration, 202*(4), 539–553.

Ting, T. C. T. (2005). Poisson's ratio for anisotropic elastic materials can have no bounds. *Quarterly Journal of Mechanics and Applied Mathematics, 58*(1), 73–82.

Wang, C. M., Reddy, J. N., & Lee, K. H. (2000). *Shear Deformable Beams and Plates*. Oxford: Elsevier.

Yu, W., & Hodges, D. H. (2004). Elasticity solutions versus asymptotic sectional analysis of homogeneous, isotropic, prismatic beams. *ASME Journal of Applied Mechanics, 71*(1), 15–23.

11

The effect of pin profiles on the microstructure and mechanical properties of underwater friction stir welded AA2519-T87 aluminium alloy

S. Sree Sabari*, S. Malarvizhi and V. Balasubramanian

Abstract

Background: AA2519-T87 is a new armour grade aluminium alloy employed in the fabrication of light combat military vehicles. Joining of this material using fusion welding processes, results in the formation of solidification defects like porosity, alloy segregation and hot cracking. In order to overcome the solidification related problems, solid state welding processes such as friction stir welding (FSW) can be used. Though the joining takes place below the melting temperature of the material, the thermal cycle experienced by the thermo-mechanical-affected zone (TMAZ) and heat-affected zone (HAZ) is causing grain coarsening and precipitates dissolution in the age-hardenable aluminium alloys, which deteriorate the joint properties. To get rid of this problem, underwater friction stir welding (UWFSW) process can be employed. The water cooling reduces the heat, and thus, the thermal softening required taking place in TMAZ and HAZ. Therefore, the material flow is entirely different in FSW and UWFSW.

Methods: In this investigation, a comparative study is made to understand the influence of tool pin profiles, namely straight cylindrical (STC), taper cylindrical (TAC), straight threaded cylindrical (THC) and taper threaded cylindrical (TTC) on stir zone characteristics and the resultant tensile properties of both FSW and UWFSW joints.

Results: From this investigation, it is found that the joint made by taper threaded pin profiled tool underwater cooling medium exhibited higher tensile properties of 345 MPa and joint efficiency of 76 %.

Conclusion: The enhancement in the strength is attributed to the precipitation hardening, grain boundary strengthening and narrowing of lower hardness distribution region (LHDR).

Keywords: Friction stir welding, Underwater friction stir welding, Pin profiles, Microstructure, Tensile properties, Microhardness

Background

In recent days, the lighter material is used to construct the light combat vehicles to improve its mobility (Børvik et al. 2011). Among the lighter materials, copper-containing high strength aluminium alloy is the potential candidate which exhibits high toughness and high specific strength to fit into the armour applications. AA2519-T87 is a new generation age-hardenable armour grade aluminium alloy which has good ballistic properties. It is presently well demonstrated

that friction stir welding (FSW) process can join aluminium plates with better joint properties over other welding processes (Barcellona et al. 2006; Gachi et al. 2011). The low heat input during FSW avoids the solidification defects and also avoids deteriorated metallurgical transformation that occur at the elevated temperatures (Shukla and Baeslack 2007). Despite, the heat generated during FSW can alter the precipitation behaviour of the aluminium material. The strength and hardness of the age-hardenable aluminium alloys mainly rely on the precipitate type, size and its distribution (Frigaard et al. 2011). During FSW, the thermal condition prevails in the thermo-mechanical-affected zone (TMAZ) and heat-affected zone (HAZ) causes precipitate

* Correspondence: sreesabaridec2006@yahoo.co.in
Centre for Materials Joining and Research (CEMAJOR), Department of Manufacturing Engineering, Annamalai University, Annamalai Nagar, Chidambaram 608 002, Tamil Nadu, India

coarsening and dissolution of precipitates. In addition, the grain coarsening is also occurred in above said zones. Hence, these regions attain lower hardness (softening) and so termed as the lowest hardness distribution region (LHDR) (Fonda and Bingert 2006; Fu et al. 2013).

Zhang et al. studied the microhardness variations in the friction stir welded AA2219 aluminium alloy joint and reported that HAZ recorded lower hardness of 78 HV and hence tensile fracture occurred in this LHDR (Zhang et al. 2011). Similarly, Fonda et al. investigated on the FSW of AA2519 aluminium alloy to study the hardness variations (Fonda and Bingert 2014). Lower hardness values of 80–85 HV were observed in the TMAZ and HAZ, and it was also observed that the tensile fracture occurred in the TMAZ/HAZ interface. In both the above investigations, the hardness values reported in the LHDR were nearly 50 % lower than the hardness of the parent metal. Moreover, they opined that the tensile fracture occurred exactly in the LHDR. Hence, enhancing the hardness of the LHDR is mandatory to improve the joint properties of the FSW joints.

Underwater friction stir welding (UWFSW) is a variant of the FSW process in which the water cooling is utilized to regulate the thermal cycles prevailing in the joint. In FSW process, the joints are fabricated in the air medium (open atmosphere) whereas in UWFSW process, the joints are made in the water medium (submerged in water). During UWFSW, the high heat dissipation capacity of water controls the conduction of heat to TMAZ and HAZ. The low heat prevailing in the TMAZ and HAZ will not be sufficient to coarsen or dissolute the precipitates. In addition, the width of the TMAZ and HAZ can also be minimized by limiting the heat and plastic deformation by UWFSW. Zhang et al. reported that the performance of UWFSW joints of AA2219 Al alloy was enhanced by narrowing the width of the HAZ. Moreover, the controlling of thermal cycles using water cooling improves the hardness from 78 to 98 HV in the LHDR (Zhang et al. 2012). Liu et al. made an investigation on the tensile properties of FSW and UWFSW of AA2219 Al alloy. The study reported that the tensile strength of the UWFSW joint is 5 % higher than the FSW joint (Liu et al. 2010). In an another investigation, Liu et al. reported that the UWFSW AA2219 Al joints showed increase in tensile strength with increase in welding speed, but drastically decreased on further increasing the welding speed and end up with defect formation at the higher welding speed (Liu et al. 2011). Zhang et al. reported that the water cooling improved the tensile strength of FSW AA2219 Al alloy joints at lower welding speed and showed no obvious effect in the higher welding speeds (Zhang et al. 2014).

Though the UWFSW process will yield joints with superior properties, the process parameters and tool geometry has to be selected in such a way to attain defect free, sound joints. The material flow behaviour mainly decides the quality of the FSW joints. The material flow behaviour in FSW depends on the process parameters such as tool rotation speed, tool traverse speed, axial force, tool tilt angle, tool shoulder diameter, tool shoulder profile and tool pin profile. Among the parameters, tool pin profile plays an important role in stirring and extruding the material around the tool pin. The pin profile plays a primary role in controlling the rotary and transverse material flow, though the other parameters are supplementing the material flow by supplying the sufficient heat and force. The flow behaviour of materials in FSW and UWFSW process is entirely different, because of the difference in heat dissipation capacity of the cooling mediums. Hence, the thermal softening in the preheat zone, stir zone and TMAZ differs which in turn causes the difference in material flow behaviour.

Many research works (Chen et al. 2013; Colegrove and Shercliff 2005; Zhao et al. 2014) were previously carried out to understand the flow behaviour of materials in FSW process, but limited research works have been reported so far related to the material flow behaviour of UWFSW process. Hence, in this investigation, an attempt has been made to study the effect of pin profiles on stir zone characteristics and the resultant tensile properties of the joints made by FSW and UWFSW processes.

Methods

Rolled plates of AA2519-T87 aluminium alloy were used as the parent material in this investigation. The chemical composition of the parent metal was quantified using spectro-chemical analysis, and the composition is presented in Table 1. The joint configuration of 150 × 150 × 6 mm was used, and the welding was done normal to the rolling direction using the tools with four different pin profiles, namely straight cylindrical (STC), straight threaded cylindrical (THC), taper cylindrical (TAC) and taper threaded cylindrical (TTC). The dimensions of the four pin profiles are shown in the Fig. 1. The joints were fabricated under air and water cooling mediums, and they are designated as FSW and UWFSW joints, respectively. The process parameters and the welding conditions used to in FSW and UWFSW process are presented in Table 2. These welding parameters were selected based on trial experiments to attain defect free, sound joints. The specimens were extracted from the joints to test and characterize the FSW and UWFSW.

Table 1 Chemical composition (wt%) of AA 2519-T87 aluminium alloy

Cu	Mg	Mn	Fe	V	Si	Ti	Al
5.71	0.47	0.27	0.1	0.05	0.04	0.02	Balance

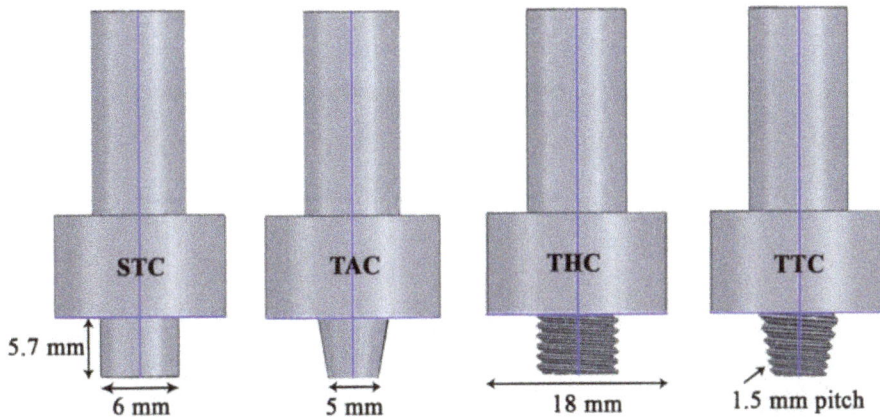

Fig. 1 Dimensions of the tool pin profiles

Metallographic procedures were followed to reveal the microstructural characteristics of the welds by optical microscopy (OM). The OM specimens were polished using water emery papers and etched using Keller's reagent for 10 s to reveal the microstructure. The grain size of stir zone is measured in the pin influenced region (PIR). Similarly, the grain size of TMAZ and HAZ is measured in the mid-thickness region.

One hundred kilo Newton servo controlled universal testing machine (Make: FIE–BLUESTAR, India, Model: UNITEK 94100) was employed to evaluate the transverse tensile properties of the FSW and UWFSW joints. The tensile specimens were extracted, machined and tested as per the ASTM E8M guidelines. Before testing, the samples were flattened to ensure the equal cross-sectional area along the entire gauge length of the specimen. The tensile properties such as yield strength, ultimate tensile strength and elongation were evaluated. The cross section of the tensile tested samples were polished and etched to reveal the entire fracture path using optical microscope. The tensile fracture path was identified by the macro-structural analysis. Scanning electron microscope (SEM) was employed to characterize the fracture surfaces. Hydraulic-controlled

Table 2 Welding parameters and tool dimensions used in this investigation

Process parameters	Values
Tool rotational speed (rpm)	1300
Welding speed (mm/min)	30
Pin length (mm)	5.7
Tool shoulder diameter (mm)	18
Pin diameter (mm)	5–6
Tool tilt angle, degree	2°
Pin profile	Taper threaded pin profile
Tool material	Hardened super high speed steel

Vickers microhardness tester (Make: Shimadzu and model: HMV-2T) was used to measure the microhardness along the cross section of the weld joint. The indentations were made under the load of 4.9 N for a dwell time of 15 s. The single hardness profile was obtained along the mid-thickness region with the indentation of 1 mm spacing. The correlation between the entire fracture path and the LHDR cannot be agreed from the single hardness profile. Hence, the hardness distribution maps were obtained by indenting along five test lines which are 1 mm spacing along the thickness direction. In each test lines, 25 indentations were made and a total of 125 indentations were made to obtain the hardness distribution map.

Transmission electron microscope (TEM) was employed to characterize the microstructure of the LHDR. Thin sample of 1-mm thick was extracted from the weld joints using wire-cut electro-discharge machining (WEDM) process. The samples were polished to 100-μm thick, and then a 3-mm diameter of sample was extracted from the LHDR for further polishing. The samples were reduced to 10-μm thick using ion milling process to reveal the microstructure under TEM.

Results

Macrographs

Table 3 shows the appearance of the top surface and the cross-sectional macrograph of the FSW and UWFSW joints fabricated using different pin profiles. The surfaces of all the joints are free from surface defects. The weld surface is smooth and composed of closely spaced ripples in all the joints. However, a distinct band of white region is observed next to the weld region on both sides of the UWFSW joints.

The macro features of the stir zone exhibit different material flow behaviour. From the macrograph, the stir zone can be divided into upper shoulder influenced region (SIR), middle pin influenced region (PIR) and lower

Table 3 Effect of tool pin profile on top surface and cross-sectional macrographs

Name of the joint	Top surface	Cross section	Observation
FSW-STC			Defect-free top surface but tunnel defect is observed at the advancing side
FSW-TAC			Defect-free top surface but tunnel defect is observed at the advancing side
FSW-THC			Defect-free stir zone at both top surface and cross section
FSW-TTC			Defect-free stir zone at both top surface and cross section
UWFSW-STC			Defect-free top surface but tunnel defect is observed at the advancing side
UWFSW-TAC			Defect-free top surface but tunnel defect is observed at the advancing side

Table 3 Effect of tool pin profile on top surface and cross-sectional macrographs *(Continued)*

UWFSW-THC	White band		Defect-free stir zone at both top surface and cross section
UWFSW-TTC	White band		Defect-free stir zone at both top surface and cross section

vortex region (VOR). In both air and water cooling medium, the tunnel defects are observed in the advancing side-PIR of the joints fabricated using STC and TAC profiled tools. But the joints fabricated using THC and TTC profiled tools yielded defect free stir zones in both air and water cooling medium. The defective joints are not considered further analysis, and the defect-free THC and TTC joints alone are considered.

Microstructure

Figure 2 shows the optical micrograph of the parent metal. It is characterized by the presence of elongated grains oriented towards the rolling direction. Figures 3a, b and 4a, b show the stir zone micrographs of the joints fabricated using THC and TTC tools in both air and water cooling medium. It can be observed that dynamic recrystallization has occurred during the FSW and

Fig. 2 Optical micrograph of parent metal

UWFSW process. It can also be noticed that grains are fine and equi-axially oriented in the SZ irrespective of the cooling medium. The average grain diameter at various regions were quantified and presented in Table 4. In stir zone, the grain diameter of FSW joints is higher than the UWFSW joints. The joint made using TTC tool shows lower grain size than THC tool under both the cooling mediums and the average grain diameter is 15 and 3.3 µm for FSW and UWFSW joints, respectively. The average grain diameter of the stir zone of the joint made by THC tool is measured as 17.5 and 5.2 µm for FSW and UWFSW joints, respectively.

Figure 3c–f shows the TMAZ micrographs of the joints fabricated using THC in both air and water cooling medium. The advancing side-thermo-mechanically-affected zone (AS-TMAZ) of joint made by THC profiled tool shows coarse and severely deformed elongated grains at the interface. The retreating side-thermo-mechanically-affected zone (RS-TMAZ) micrograph is characterized by elongated and upward oriented grains whereas the deformation is gradually reduced from the interface. Figure 4c–f show the TMAZ micrographs of the joints fabricated using TTC in both air and water cooling medium. The joint fabricated using TTC tool exhibit symmetric material flow in both the AS-TMAZ and RS-TMAZ in both the air cooling and water cooling conditions. The grain size and extent of deformation is more or less similar in both the sides.

The average grain diameter of AS-TMAZ and RS-TMAZ are equal and measured as 54 and 50 µm for the THC and TTC joint, respectively for FSW joints. Similarly, the average grain diameter of AS-TMAZ and RS-TMAZ are equal and measured as 85 and 82 µm for the THC and TTC joint, respectively, for FSW joints. In comparison, the grain diameter of TMAZ of UWFSW joint is

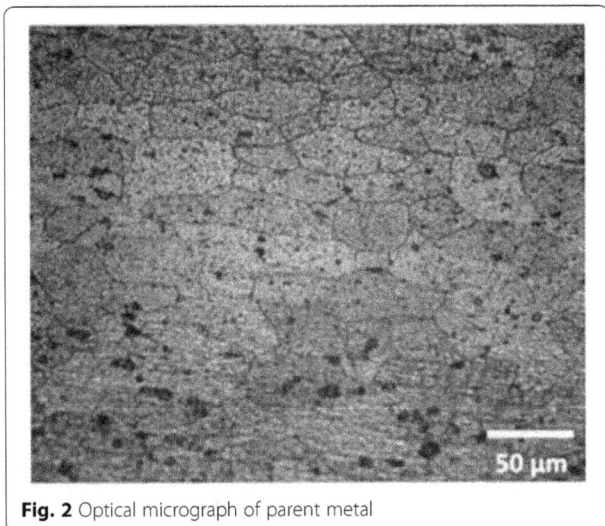

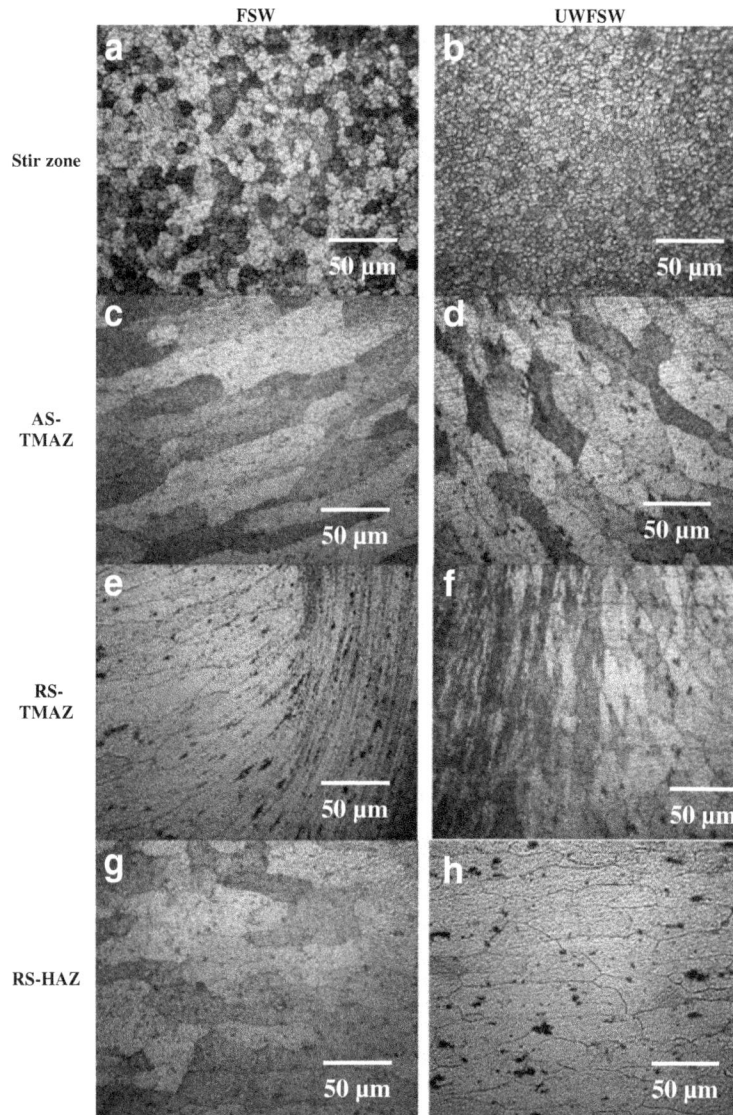

Fig. 3 Optical micrograph of the various regions of THC joint **a**. Stir zone of FSW joint. **b**. Stir zone of UWFSW joints. **c**. AS-TMAZ of FSW joints. **d**. AS-TMAZ of UWFSW joints. **e**. RS-TMAZ of FSW joints. **f**. RS-TMAZ of UWFSW joints **g**. RS-HAZ of FSW joints. **h**. RS-HAZ of UWFSW joints

40 % higher than the FSW joints is respective of pin profiles. It is noticed that the TMAZ micrograph of UWFSW joints shows the interface microstructure from SZ to TMAZ. But at the same magnification level, only a part of TMAZ is seen for the FSW joints. This suggests that the width of the TMAZ region is much wider in FSW joints.

Figures 3g, f and 4g, f show the HAZ micrographs of the joints It is observed that no mechanically induced deformation took place but grain coarsening occurred in all the joints. The HAZ micrograph of the joint made by THC profiled tool reveals larger grains of 52 μm, but TTC joint shows grains of 49 μm in water cooling condition. The HAZ grain size of UWFSW-THC joint is higher than the grain size of PM (of 49 μm) whereas it is equal to the HAZ grain size of UWFSW-TTC joint. In air cooling condition,

the HAZ micrograph of the joint made by THC profiled tool reveals larger size of 64 μm, but TTC joint shows grains of 60 μm. The FSW joints contain coarser grains in HAZ than the UWFSW joints.

Figure 5 shows the TEM micrographs of parent metal and LHDR of all the joints. Figure 5a shows the parent metal micrograph which is characterized by the presence of fine, dense and uniformly distributed θ' ($CuAl_2$) precipitates. The precipitates are oriented in two directions which is normal to each other. The LHDR is characterized by the presence of precipitate free zone (PFZ) and coarsened precipitates. The LHDR of the joints contains lower volume fraction of precipitates than the parent metal. The FSW joints fabricated using THC and TTC profiled pins show more or less identical precipitation behaviour; however, the size of the precipitates of

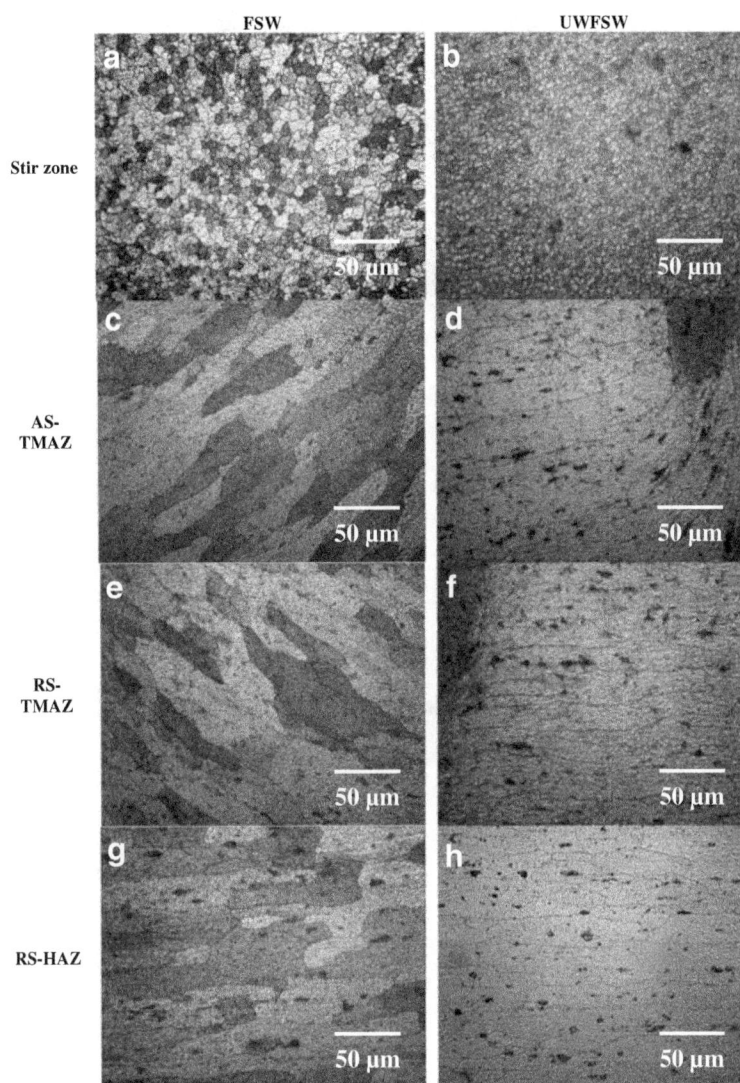

Fig. 4 Optical micrograph of the various regions of TTC joint **a**. Stir zone of FSW joint. **b**. Stir zone of UWFSW joints. **c**. AS-TMAZ of FSW joints. **d**. AS-TMAZ of UWFSW joints. **e**. RS-TMAZ of FSW joints. **f**. RS-TMAZ of UWFSW joints **g**. RS-HAZ of FSW joints. **h**. RS-HAZ of UWFSW joints

TTC joint is appreciably lower than the THC joints. The LHDR of FSW joints are characterized by the presence of few coarse θ precipitates and larger PFZ. The dissolution of precipitate is relatively lower in LHDR of UWFSW joints, and it is composed of dense coarse stable θ precipitates. From the TEM micrographs, it is evident that the volume fraction of precipitates is higher in UWFSW joints than in FSW joints.

Microhardness

Figure 6 shows the microhardness measurement across the mid-thickness region of the joints. In all the joints, typical W-shaped hardness plots were recorded. Among the various regions, the TMAZ on both the AS and RS of the joints recorded lower hardness. In air cooling condition, FSW-THC joint recorded lowest hardness of 78 HV whereas FSW-TTC joint recorded 80 HV in the RS. In the water

Table 4 Average grain diameter of various regions

Pin profile	SZ (μm)	AS-TMAZ (μm)	RS-TMAZ (μm)	RS-HAZ (μm)	PM (μm)
FSW-THC	17.5	85	85	64	49
FSW-TTC	15	82	82	60	
UWFSW-THC	5.2	54	54	52	
UWFSW-TTC	3.3	50	50	49	

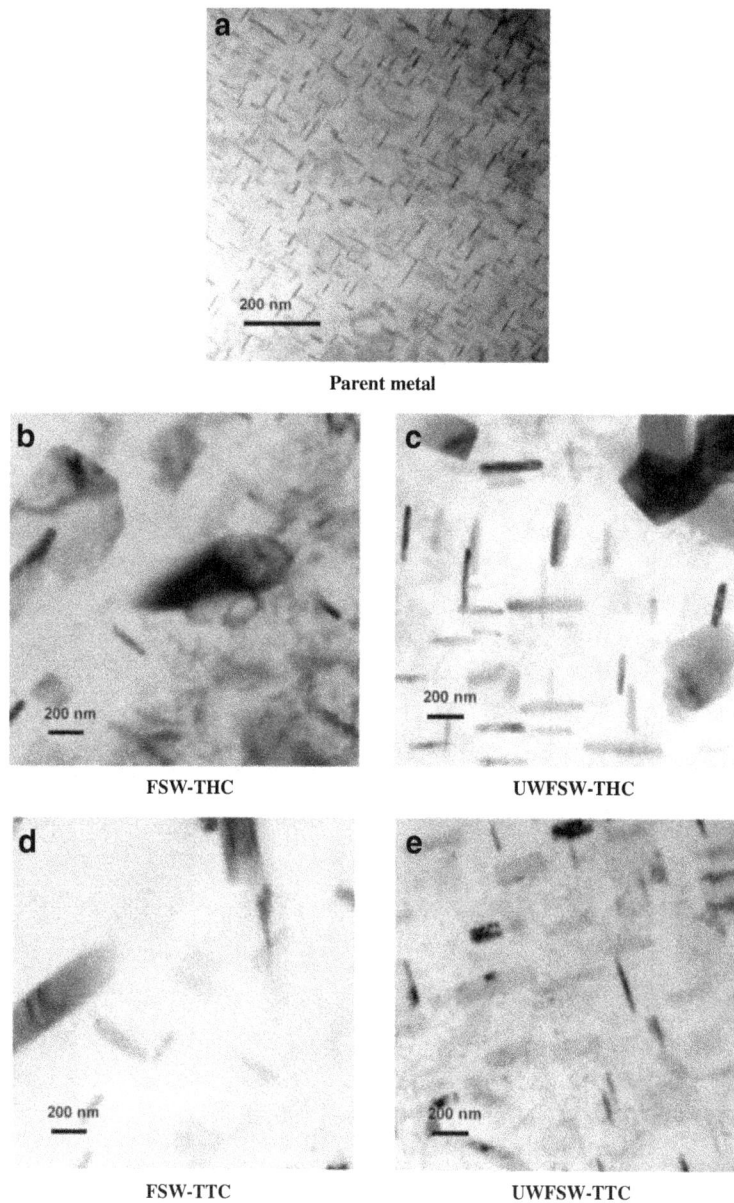

Fig. 5 TEM images of parent metal region and LHDR **a**. Parent metal. **b**. LHDR of FSW-THC. **c**. LHDR of UWFSW-THC. **d**. LHDR of FSW-TTC. **e**. LHDR of UWFSW-TTC

cooling condition, UWFSW-THC joint recorded lowest hardness of 82 HV whereas UWFSW-TTC joint recorded 93 HV in the RS. It is observed that the location of the LHDR is closer to the weld centre in UWFSW joints, but it is marginally away from the weld center in FSW joints. It is also observed that the LHDR is wider in FSW joints and narrower in the UWFSW joints.

In all the joints, the hardness of the SZ are higher than the TMAZ. SZ of UWFSW-TTC joint recorded higher hardness of 105 HV whereas lower hardness of 86 HV was recorded in FSW-THC joint. The hardness of the HAZ are lower; however, it is higher than the TMAZ. It is observed that there is an increase in hardness from TMAZ to PM region. The joint

fabricated using water cooling medium recorded marginally higher hardness in all the regions than the joint fabricated under air cooling medium. Among the pin profiles, the joint fabricated using TTC pin profiled tool in water cooling medium recorded higher hardness than its counterparts.

Tensile properties

Figure 7 shows the stress strain curves of the FSW and UWFSW joints fabricated using THC and TTC tools. The transverse tensile properties like yield strength, ultimate tensile strength and elongation are derived from the stress strain curves and presented in the Table 5. The unwelded parent metal (PM) showed

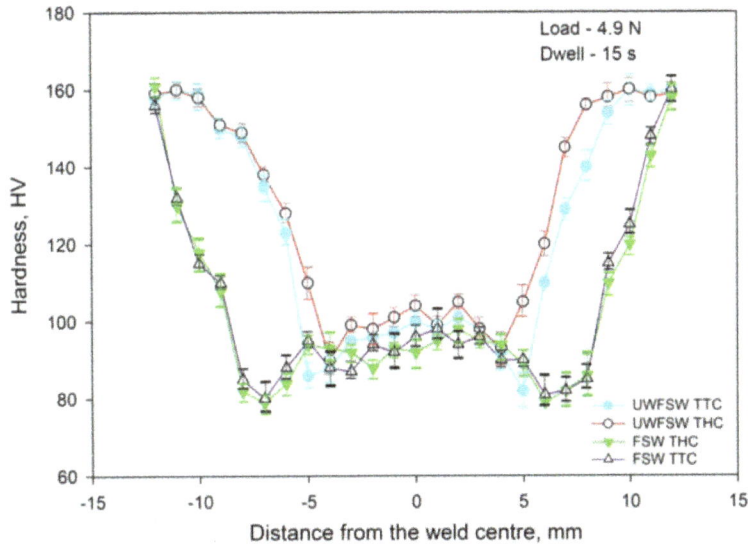

Fig. 6 Microhardness plot

tensile strength of 452 MPa with an elongation of 11.2 %. The UWFSW-THC joint exhibited tensile strength of 327 MPa which is 30 % lower than PM, and it showed an elongation of 7.94 % which is also 30 % lower than PM. The UWFSW-TTC joint yielded tensile strength of 345 MPa which is 25 % lower than PM, and it showed an elongation of 9.17 % which is 20 % lower than PM. FSW-TTC joint exhibited tensile strength of 267 MPa and joint efficiency of 59 % which is 9 % higher than the FSW-THC joint. Of the four joints, UWFSW-TTC joint showed higher joint efficiency of 76 % which is 5 % higher than UWFSW-THC joint, 29 % higher than FSW-THC joint and 22 % higher than FSW-TTC joint.

Fracture surface

Table 6 shows the fracture location of the tensile tested specimen. The cross-sectional macrograph of the fractured specimen reveals the entire fracture path. It can be clearly seen that the fracture is observed in the TMAZ in all the joints. The FSW joints show the regular fracture path which is 45°

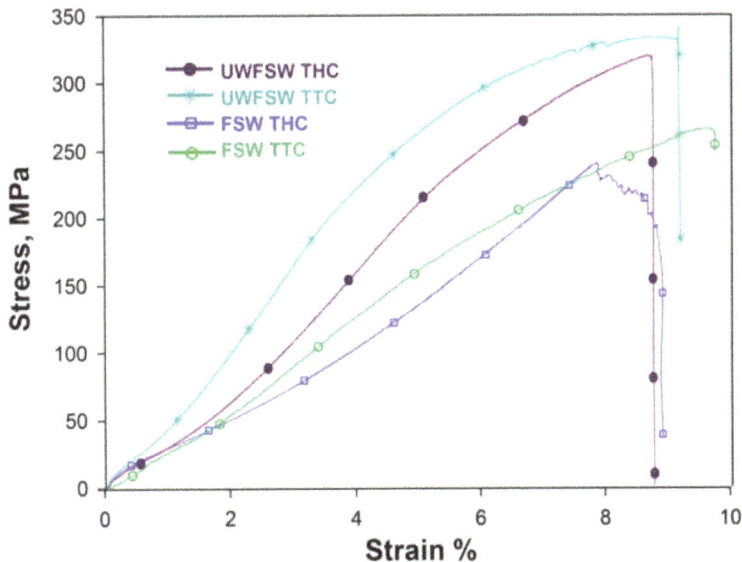

Fig. 7 Stress strain curves

Table 5 Transverse tensile properties of the joints

	0.2 % Yield strength (MPa)	Tensile strength (MPa)	Elongation in 50-mm gauge length (%)	Joint efficiency (%)
Parent metal	427	452	11.2	–
UWFSW-THC	301	327	7.94	72
UWFSW-TTC	322	345	9.17	76
FSW-THC	218	244	9.2	54
FSW-TTC	230	267	9.85	59

inclined to the tensile loading direction. But the UWFSW joints show an irregular fracture path along the outer periphery of the stir zone. From the hardness map, it is yet again confirmed that the LHDR is present at both the AS and RS. The dotted lines imposed on the hardness map describe fracture location and the fracture path. The fracture path is having close agreement with the LHDR of the hardness maps of all the joints.

Figure 8 shows the typical fracture surface at macro- and microlevel magnifications. The joints fabricated using THC and TTC pin profile in water cooling medium shows two different patterns on fracture surface, one at the SIR and another at PIR. But the entire fracture region of FSW-THC and FSW-TTC joints exhibit only one fracture pattern. At higher magnification, all the joints show fine populated dimples in the fracture surface, irrespective of the pin profiles and cooling

Table 6 Effect of tool pin profile on fracture path

Name of the joint	Micrograph map	Fracture path	Observation
FSW-THC			Fracture is occurred in the RS-TMAZ which is 45° inclined to the loading direction
FSW-TTC			Fracture is occurred in the RS-TMAZ which is 45° inclined to the loading direction
UWFSW-THC			Fracture is occurred in the TMAZ in which the fracture is irregular
UWFSW-TTC			Fracture is occurred in the TMAZ in which the fracture is irregular

78 90 100 110 120 130 140 150 160

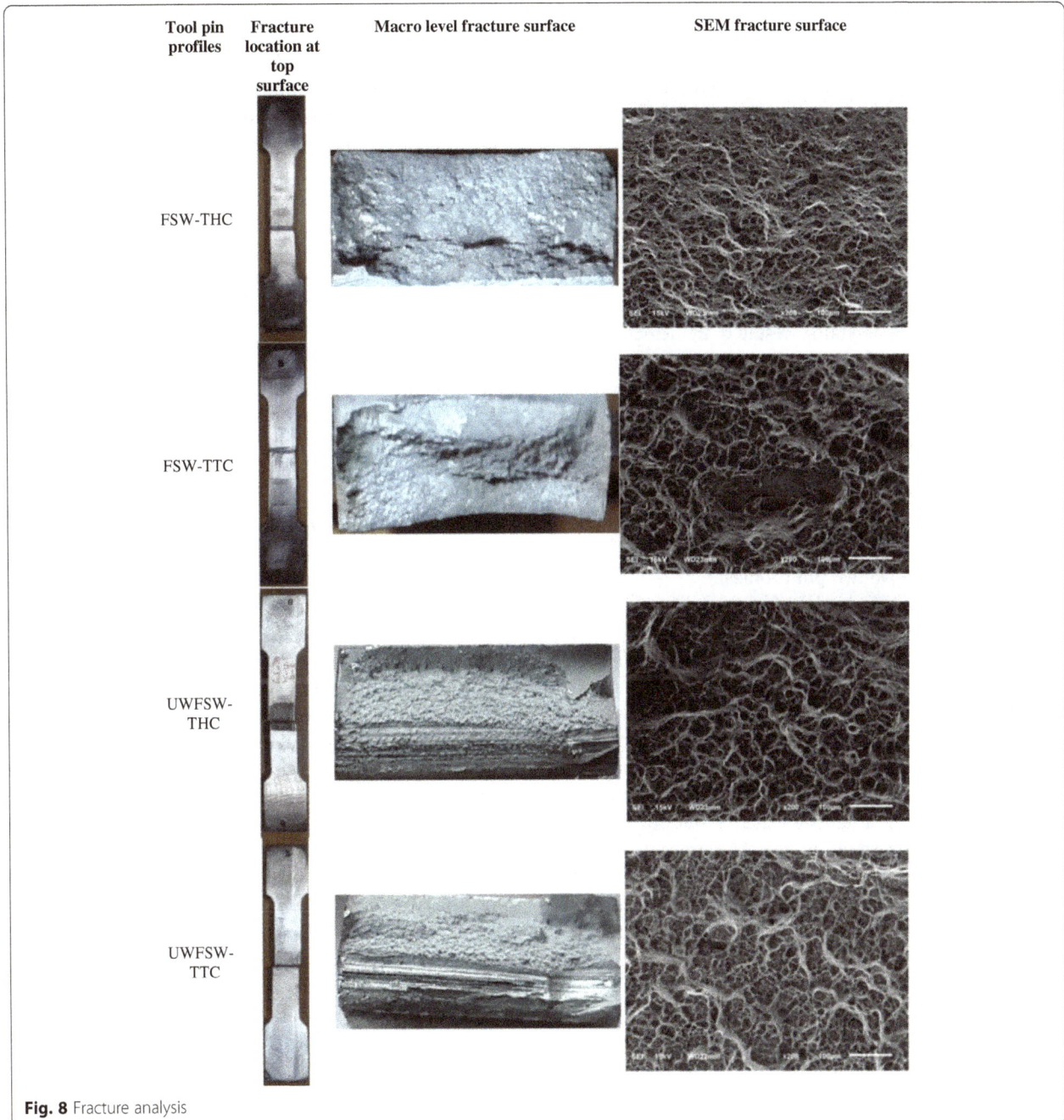

Tool pin profiles	Fracture location at top surface	Macro level fracture surface	SEM fracture surface
FSW-THC			
FSW-TTC			
UWFSW-THC			
UWFSW-TTC			

Fig. 8 Fracture analysis

medium. The dimples are oriented towards the loading direction and the presence of dimples suggests that the joints are failed predominantly in the ductile mode.

Discussion

The effect of tool pin profiles on the joint quality

In UWFSW process, the soundness of joint is decided by the heat generation and the material flow behaviour (Mishra & Ma 2005). The heat generation and material flow are classified into three states, namely insufficient state, balance state and excess state. The tool pin profile is one of the predominant parameters controlling these states. The defect-free joints can be made, if the proper tool pin profile is used for making FSW joints (Suresha et al. 2011; Zhao et al. 2005; Kumar et al. 2011).

In this investigation, the defect-free joints were achieved while using THC and TTC pin profiled tools in both air and water cooling medium. During each rotation of the tool, the threaded profile extrudes cylindrical sheets of material around the tool pin from AS to RS and from RS to AS. The peaks and valleys of threaded pin profile offer more friction over the plasticized material which promotes sticking condition. Thus, the extent of heat generation and plasticization is sufficient to attain the balance state of heat and material flow. In general, the defect is formed in the weld periphery, i.e. the transformed layers around the tool pin. So, the pin profile should capable of forming sufficient transforming layers. By creating enough transforming layers by plasticization and shearing of materials, the threaded pin profiles (THC and TTC) are capable of resulting sound joints.

The joints fabricated using STC and TAC pin profile, using both cooling medium, resulted in defects at the stir zone. The plain, featureless surface area offers lower friction over the plasticized material which promotes slipping condition. In addition, the featureless surface cannot direct the material flow intensively. Palanivel et al. reported that the plain taper surface results drop in frictional heat because of lower heat generation during sliding over sticking condition (Palanivel et al. 2012). Thus, an insufficient heat state is prevailing in the SZ while using the above two pin profiles.

The STC pin profile is plain and straight, and thus, it exhibits a regular material flow during welding. The material transport from AS to RS and from RS to AS is uniform from SIR to PIR, i.e. top to bottom of the joint. Usually, it is not preferable because the heat generation is varied from SIR to PIR. Thus, the optimum combination of heat input and material flow is not met throughout the stir zone. The material flow should vary as like heat input to attain the balance state of heat and material flow. The size of the defect is larger in air cooling condition than the water cooling condition. The size of the defect should be reduced as approaching from insufficient heat state to balance state. Thus, it is inferred that the STC pin profile cannot create defect-free stir zone, since the plain and straight profile cannot able to support and direct the material to flow. Thus, in STC joints, the defects are formed primarily due to the attainment of insufficient material flow state rather due to the heat state.

In TAC pin profile, the tapers are beneficial for aiding upwards and downward flow; however, the plain, featureless surface area creates lower friction which promotes slipping condition. In underwater condition, the heat generated is low, and thus, insufficient heat state is attained. But, during the air cooling condition, the heat generated is high and it is approaching the balance heat state. As the heat generated increases, the size of the defect is decreased. Thus, in TAC joints, the defects are formed due to the insufficient heat and material flow state.

Because of the insufficient heat generation and poor material flow, the plastic deformation around the pin is

limited, and thus, the formation of transforming layer is limited. Thus, the joints fabricated using STC and TAC tool exhibited defects in the stir zone. The wider transforming layer created around the tool pin with the help of threads is the prime reason for the defect-free stir zone formation. By this way, the threaded pin profile tools create the balance state of heat and material flow to form the defect-free stir zone.

It is also inferred that the change of cooling medium does not significantly influence the mechanism of defect formation. This is because the FSW process is carried out by localize application of frictional heat and pressure. The heat and pressure experienced is almost similar in FSW and UWFSW joints. The effect of cooling is higher in the TMAZ and HAZ region and minimal in the SZ. However, the extent of heat loss from SZ is different which varies the defect size, grain size and resultant mechanical properties.

The effect of tool pin profiles on microstructure

The heat generation and the plastic deformation during FSW and UWFSW process drive the recrystallization process in the stir zone (Liu et al. 2010; Zhang et al. 2014). Thus, a new set of fine grains are observed in the stir zones of the joints fabricated using THC and TTC tools in both the cooling conditions. In the TMAZ, the heat input and the deformation is not sufficient to recrystallize the grains (Yoon et al. 2015). Thus, elongated coarse grains are observed in both the joints. The HAZ only experiences the heat and no deformation is occurred; hence, this region exhibit undeformed coarse grains. The high heat dissipation capacity of water, readily convect the heat from the SZ, TMAZ and HAZ. Thus, the heat availability in UWFSW in various regions is lower than the FSW joints. During the FSW process, the loss of heat due to air cooling is minimal. Because of high heat input and slow cooling rate, the joint fabricated using air cooling medium reveals marginally larger grains and wider stir zone than the joint fabricated using water cooling medium. In FSW joint, the width of TMAZ is wider and it is located away from the weld centerline. This is attributed to the presence of wider stir zone and occurrence of extensive deformation in the TMAZ. In addition to the change of the cooling medium, the change of pin profile has effect in heat generation. The THC profiled tool exhibit larger frictional contact area to create higher frictional heat than TTC profiled tool pin. Thus, the THC joint exhibit high heat input and slow cooling rate than the TTC joint.

Zhang et al. (Fu et al. 2013) referred SZ as the reprecipitation zone, SZ-TMAZ interface as the dissolution zone and TMAZ to HAZ as the overaging zone. During FSW and UWFSW process, the heat generated is high to solutionize the precipitates in the SZ. On cooling cycle, the heat is utilized to reprecipitate in the SZ. In the SZ-TMAZ interface, the heat is sufficient to solutionize the precipitates, but unable to reprecipitate because of the high cooling rate. The heat prevails in the regions from TMAZ to HAZ is not sufficient to solutionize and so coarsening of precipitate is happening in the above said regions. Thus, these regions are termed as overaged zone. Because of this reason, TMAZ undergone overaging during FSW and UWFSW and coarse precipitates are observed. However, both coarsening and the dissolution of precipitates are observed in the LHDR because of the heterogeneous precipitation. Increase in heat input increases the precipitate size and dissolution of precipitates. Hence, the high input FSW-THC joint exhibits low volume fraction of coarse precipitates in the LHDR.

The effect of tool pin profiles on the mechanical properties of the joints

The age-hardenable materials are mainly strengthened due to the presence of fine θ' precipitates. During loading, these precipitates act as the obstacle for the dislocation motion. The coarse stable θ precipitates cannot provide the hindrance effect due to low coherency. From the Hall-Petch relation, it was understood that the hardness or strength decreases with an increase in the grain size (Xu et al. 2012). The TMAZ micrographs exhibit coarser grains and so reduced grain boundary strengthening is observed in this region. Thus, the lower hardness of LHDR is attributed to the low grain boundary strengthening and precipitate hardening.

The heat input governs the above said strengthening mechanism. Decrease in heat input will increase the strengthening effects. Thus, the UWFSW joint recorded higher hardness in the entire region than the FSW joint and therefor the UWFSW joint exhibited higher tensile strength than the FSW joint. The higher tensile strength is attributed to the presence of relatively fine grains, high volume fraction of precipitates and appreciably narrow LHDR. On comparing the four joint conditions (FSW-THC, FSW-TTC, UWFSW-THC and UWFSW-TTC), the UWFSW-TTC joint shows enhanced tensile and hardness properties because this joint experience the lowest temperature compared to its counterparts. So, the pin profile capable of formation of the defect-free stir zone at the minimum heat input is appreciable.

During tensile loading, the load will concentrate on the weakest zone in the joint. The TMAZ is identified as the LHDR, and thus, the load is concentrated on it and the failure occurred in this region. This is consistent with the microhardness map and the fracture locations of both the joints (Table 6). Thus, the fracture is occurring in the weakest region, i.e. at TMAZ. The FSW joints exhibit wider LHDR, and the UWFSW joints exhibit narrow LHDR near to the weld periphery. Because of wider

LHDR, the fracture path is 45° to the loading direction and the failure is occurred by simple shearing. But in UWFSW joints, due to narrowed LHDR, the fracture path is restricted near to the weld periphery, and thus, the shape of the fracture path is similar to the shape of the weld periphery. In addition, the grain orientation difference at the interface offers resistance to the tensile fracture, and thus, the fracture surface is not smooth.

From the tensile test results, it was observed that the weld joint undergone reduction in the ductility property (Table 5). The measures of ductility, i.e. elongation of the joints were lower than the parent metal. The poor precipitation strengthening and grain boundary strengthening of TMAZ offers less resistance to tensile load. Therefore, the load was accommodated in TMAZ which cause yielding of TMAZ. The load concentration phenomenon is called strain localization (Fu et al. 2011). Because of strain localization, the TMAZ alone contributes to elongate during tensile loading. Therefore, a reduced elongation value was observed in the joints compare to the parent metal. The elongation of joints are almost similar; however, the UWFSW joints exhibit lower elongation than the FSW joints. This was attributed to the narrow TMAZ of UWFSW joints which undergone high extent of strain localization than the FSW joints.

Conclusions

The effect of tool pin profiles on the stir zone characteristics and tensile properties of friction stir welded and underwater friction stir welded AA2519-T87 aluminium alloy joints were investigated and the following conclusions are derived:

1) Of the four tool pin profiles used in this investigation, straight threaded cylindrical (THC) pin profile and taper threaded cylindrical (TTC) pin profile yielded defect-free weld surface and stir zone formation in both air cooling and water cooling medium. It is attributed to the attainment of balance state of heat generation and material flow during stirring.

2) The UWFSW-TTC joint fabricated using taper threaded pin profile with water cooling exhibited tensile strength of 345 MPa and joint efficiency of 76 %, which is 5 % higher than UWFSW-THC joint, 29 % higher than FSW-THC joint and 22 % higher than FSW-TTC joint.

3) The presence of relatively finer grains in the stir zone, higher volume fraction of precipitates, marginally higher hardness of stir zone and appreciably lower width of lower hardness distribution region are the main reasons for the better performance of UWFSW-TTC joints than its counterparts.

Competing interests
The authors declare that they have no competing interests.

Authors' contributions
SM and VB participated in the sequence alignment and drafted the manuscript. All authors read and approved the final manuscript.

Acknowledgements
The authors gratefully acknowledge the financial support of the Directorate of Extramural Research & Intellectual property Rights (ER&IPR), Defense Research Development Organization (DRDO), New Delhi through a R&D project no. DRDO-ERIPER/ERIP/ER/0903821/M/01/1404. The authors also wish to record the sincere thanks to M/S Aleris Aluminium, Germany, for supplying the material to carry out this investigation.

References
Barcellona, A, Buffa, G, Fratini, L, & Palmeri, D. (2006). On microstructural phenomena occurring in friction stir welding of aluminium alloys. *Journal of Materials Processing Technology, 177*, 340–343.

Børvik, T, Olovsson, L, Dey, S, & Langseth, M. (2011). Normal and oblique impact of small arms bullets on AA6082-T4 aluminium protective plates. *International Journal of Impact Engineering, 38*, 577–589.

Chen, G, Shi, Q, Li, Y, Sun, Y, Dai, Q, Jia, J, Zhu, Y, & Wu, J. (2013). Computational fluid dynamics studies on heat generation during friction stir welding of aluminum alloy. *Computational Materials Science, 79*, 540–546.

Colegrove, PA, & Shercliff, HR. (2005). 3-Dimensional CFD modelling of flow round a threaded friction stir welding tool profile. *Journal of Materials Processing Technology, 169*, 320–327.

Fonda, RW, & Bingert, JF. (2006). Precipitation and grain refinement in a 2195 Al friction stir weld. *Metallurgical and Materials Transactions A, 37*(12), 3593–3604.

Fonda, RW, & Bingert, JF. (2014). Microstructural evolution in the heat-affected zone of a friction stir weld. *Metallurgical and Materials Transactions A, 35A*, 1487–1499.

Frigaard, Ø, Grong, Ø, & Midling, OT. (2011). A process model for friction stir welding of age hardening aluminum alloys. *Metallurgical and Materials Transactions A, 32A*, 1189–1200.

Fu, R, Sun, Z, Sun, R, Li, Y, Liu, H, & Liu, L. (2011). Improvement of weld temperature distribution and mechanical properties of 7050 aluminum alloy butt joints by submerged friction stir welding. *Materials and Design, 32*, 4825–4831.

Fu, R, Zhang, J, Li, Y, Kang, J, Liu, H, & Zhang, F. (2013). Effect of welding heat input and post-welding natural aging on hardness of stir zone for friction stir-welded 2024-T3 aluminum alloy thin-sheet. *Materials Science and Engineering A, 559*, 319–324.

Gachi, S, Boubenider, F, & Belahcene, F. (2011). Residual stress, microstructure and microhardness measurements in AA7075-T6 FSW welded sheets. *Nondestructive of Testing and Evaluation, 26*, 1–11.

Kumar, K, Satish Kailas, V, & Srivatsan, TS. (2011). The role of tool design in influencing the mechanism for the formation of friction stir welds in aluminum alloy 7020. *Materials and Manufacturing Processes, 26*, 915–921.

Liu, HJ, Zhang, HJ, Huang, YX, & Yu, L. (2010). Mechanical properties of underwater friction stir welded 2219 aluminum alloy. *Transactions of the Nonferrous Metals Society of China, 20*, 1387–1391.

Liu, HJ, Zhang, HJ, & Yu, L. (2011). Effect of welding speed on microstructures and mechanical properties of underwater friction stir welded 2219 aluminum alloy. *Materials and Design, 32*, 1548–1553.

Mishra, RS, & Ma, ZY. (2005). Friction stir welding and processing. *Materials Science and Engineering R, 50*, 1–78.

Palanivel, R, Koshy Mathews, P, Murugan, N, & Dinaharan, I. (2012). Effect of tool rotational speed and pin profile on microstructure and tensile strength of dissimilar friction stir welded AA5083-H111 and AA6351-T6 aluminum alloys. *Materials and Design, 40*, 7–16.

Shukla, AX, & Baeslack, WA. (2007). Study of microstructural evolution in friction-stir welded thin-sheet Al–Cu–Li alloy using transmission-electron microscopy. *Scripta Materialia, 56*, 513–516.

Suresha, CN, Rajaprakash, BM, & Upadhya, S. (2011). A study of the effect of tool pin profiles on tensile strength of welded joints produced using friction stir welding process. *Materials and Manufacturing Processes, 26*, 1111–1116.

Xu, WF, Liu, JH, Chen, DL, Luan, GH, & Yao, JS. (2012). Improvements of strength and ductility in aluminum alloy joints via rapid cooling during friction stir welding. *Materials Science and Engineering A, 548*, 89–98.

Yoon, S, Ueji, R, & Fuji, H. (2015). Effect of rotation rate on microstructure and texture evolution during friction stir welding of Ti-6Al-4V plates. *Materials Science and Engineering, 10*, 103–107.

Zhang, HJ, Liu, HJ, & Yu, L. (2011). Microstructure and mechanical properties as a function of rotation speed in underwater friction stir welded aluminum alloy joints. *Materials and Design, 32*, 4402–4407.

Zhang, HJ, Liu, HJ, & Yu, L. (2012). Effect of water cooling on the performances of friction stir welding heat-affected zone. *Journal of Materials Engineering and Performance, 21*, 1182–1187.

Zhang, Z, Xiao, BL, & Ma, ZY. (2014). Influence of water cooling on microstructure and mechanical properties of friction stir welded 2014Al-T6 joints. *Materials Science and Engineering A, 614*, 6–15.

Zhao, Y, Lin, S, Wu, L, & Qu, F. (2005). The influence of pin geometry on bonding and mechanical properties in friction stir weld 2014 Al alloy. *Materials Letters, 59*, 2948–2952.

Zhao, Y, Wang, Q, Chen, H, & Yan, K. (2014). Microstructure and mechanical properties of spray formed 7055 aluminum alloy by underwater friction stir welding. *Materials and Design, 56*, 725–730.

12

Effect of nonlinear thermal radiation on double-diffusive mixed convection boundary layer flow of viscoelastic nanofluid over a stretching sheet

. Ganesh Kumar[1], B. J. Gireesha[1], S. Manjunatha[2*] and N. G. Rudraswamy[1]

Abstract

Background: The present exploration deliberates the effect of nonlinear thermal radiation on double diffusive free convective boundary layer flow of a viscoelastic nanofluid over a stretching sheet. Fluid is assumed to be electrically conducting in the presence of applied magnetic field. In this model, the Brownian motion and thermophoresis are classified as the main mechanisms which are responsible for the enhancement of convection features of the nanofluid. Entire different concept of nonlinear thermal radiation is utilized in the heat transfer process.

Methods: Appropriate similarity transformations reduce the nonlinear partial differential system to ordinary differential system which is then solved numerically by using the Runge–Kutta–Fehlberg method with the help of shooting technique. Validation of the current method is proved by having compared with the preexisting results with limiting solution.

Results: The effect of pertinent parameters on the velocity, temperature, solute concentration and nano particles concentration profiles are depicted graphically with some relevant discussion and tabulated result.

Conclusions: It is found that the effect of nanoparticle volume fraction and nonlinear thermal radiation stabilizes the thermal boundary layer growth. Also it was found that as the Brownian motion parameter increases, the local Nusselt number decreases, while the local friction factor coefficient and local Sherwood number increase.

Keywords: Double diffusion, Mixed convection, Nanofluid, Viscoelastic fluid, Nonlinear thermal radiation

Background

In the modern day, a great deal of interest has been created on heat and mass transfer of the boundary layer flow over a stretching sheet, in view of its numerous applications in various fields such as polymer processing industry in manufacturing processes. Sakiadis (1961a; Sakiadis, 1961b) first studied the boundary layer problem assuming velocity of a bounding surface as constant. Crane (1970) computed an exact similarity solution for the boundary layer flow of a Newtonian fluid toward an elastic sheet which is stretched with velocity proportional to the distance from the origin. It is noteworthy to mention that both of these studies are regarding Newtonian fluid. Subsequently, the pioneering works of Sakiadis and Crane have been extended by

including various effects such as suction/injection, porosity, magnetic field, variable material properties, thermal radiation, heat source/sink, and slip boundary, for either a Newtonian or non-Newtonian fluids. In recent years, researches on boundary layer flow and heat transfer in nanofluids have received amplified devotion due to their growing importance in numerous industrial and biomedical applications. Nanofluid is dilute suspension of nanoparticles with at least one of their principal dimensions smaller than 100 nm and the base fluid. Nanofluids are very stable and free from extra issues of sedimentation, erosion, additional pressure drops, and rheological characteristics. This is due to the tiny size and the low-volume fraction of nanoelements. Thus, nanofluids have principal advantage about enhancement in thermal conductivity and the convective heat transfer coefficient when compared with the customary base fluids which comprise water, oil, and ethylene glycol. The nanomaterials are more operative in terms of heat

* Correspondence: manjubhushana@gmail.com
[2]Faculty of Engineering, Department of Engineering Mathematics, Christ University, Mysore Road, Bengaluru 560074, India
Full list of author information is available at the end of the article

exchange performance in micro/nanoelectro mechanical devices, for growing demands of modern technology, including power station, chemical production, and microelectronics. Especially, the magneto nanofluids are significant in applications like optical modulators, magneto-optical wavelength filters, tunable optical fiber filters, and optical switches. Choi (1995) was first to use the term nanofluids to refer to the fluid with suspended nanoparticles. Thermophysical properties of nanofluids such as thermal conductivity, diffusivity, and viscosity have been studied by Kang et al. (2006). The theoretical/experimental investigations on nanofluid flow and heat transfer were conducted by Rudyak et al. (2010), Wang et al. (1999), Eastman et al. (2001), and Rudraswamy et al. (2016; 2015) initiated an incompressible nanofluid over an impermeable stretching sheet and considered the uniform magnetic field. To make such investigations, the following studies are quite useful (Kai-Long Hsiao (2016; Hsiao 2014; Hsiao 2017a; Hsiao 2017b; Hsiao 2017b) Buongiorno (2006), Khan and Pop (2010), and Kuznetsov and Nield (2010)). They concluded that the nanofluids possess the novel property of enhanced thermal conductivity of the working fluids.

Researchers are paying their attention to investigate the properties of non-Newtonian fluids because of their numerous technological applications, including manufacturing of plastic sheets, performance of lubricants, and movement of biological fluids. Nadeem et al. (2014) initiated the boundary-layer flow and heat transfer of a Maxwell fluid past a stretching sheet. Shehzad et al. (2015) have observed the influence of nanoparticles in MHD flow of Jeffrey fluid over a stretched surface, and they considered the thermal and nanoparticle concentration convective boundary conditions. Khan and Gorla (2011) studied heat and mass transfer in non-Newtonian nanofluids over a stretching surface with prescribed wall temperature and surface nanoparticle concentration. Recently, Rizwan et al. (2014), Ganesh et al. (2017), and Rudraswamy et al. (2017) also discussed the nanofluid in the presence of Newtonian heating and viscous dissipation over a stretching sheet (Fig. 1).

During the past four decades, the investigators have devoted to the double-diffusive phenomena because of their various applications in chemical engineering, solid-state physics, oceanography, geophysics, liquid gas storage, production of pure medication, oceanography, high-quality crystal production, solidification of molten alloys, and geothermally heated lakes and magmas, etc. Khan and Aziz (2011) analyzed the double-diffusive natural convective boundary layer flow in a porous medium saturated with a nanofluid over a vertical plate with prescribed surface heat, solute, and nanoparticle fluxes. Nield and Kuznetsov (2011) presented the double-diffusive nanofluid convection in a porous medium using analytical method. Numerous models and methods have been proposed by

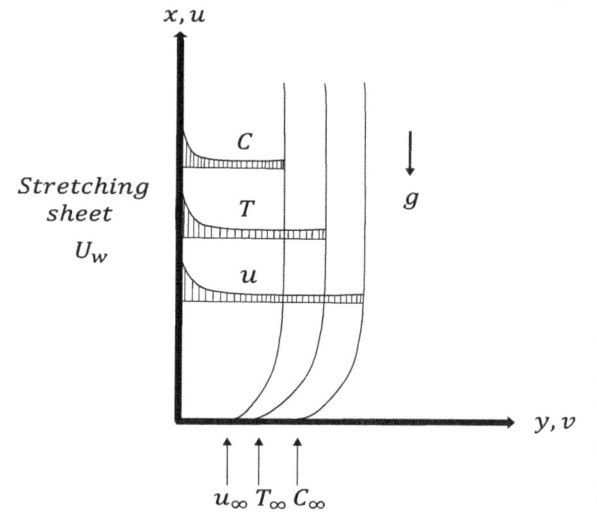

Fig. 1 Geometry of the problem

many researchers and academicians to advance their studies relating to problems involving various parameters. To make such investigations, the following studies are quite useful (Sharma et al. (2012), Hayat et al. (2014), Beg et al. (2014), Goyal and Bhargava (2014), Gaikwad et al. (2007), Wang (1989), and Gorla and Sidawi (1994)).

Motivated by the aforementioned researchers, we have considered the level of species concentration is moderately high so that the thermal diffusion (Soret) and diffusion-thermo (Dufour) effects cannot be neglected. In view of the above discussion, the necessity of Soret and Dufour effects and inclusion of nonlinear thermal radiation in the heat transfer and first-order chemical reaction not only enrich the present analysis but also complement the earlier studies. The inclusion of these phenomena gives rise to additional parameters such as viscoelastic parameter, Brownian motion parameter, thermophoresis parameter, modified Dufour parameter, regular double-diffusive buoyancy ratio, nanofluid buoyancy ratio, mixed convection parameter, radiation parameter, Dufour Lewis number, nanofluid Lewis number, and regular Lewis number. Thus, inclusion of additional parameters contributes to the complexity of the mathematical model representing the flow, heat, and mass transfer phenomena. To the best of author's acquaintance, no work has been so far reported considering above phenomenal together.

Mathematical formulation

Consider a steady two-dimensional laminar flow of an incompressible and electrically conducting viscoelastic nanofluid over a stretching surface. The flow models for elastic-viscous and double-diffusive are considered. The Brownian motion and thermophoresis are taken into consideration. It is also assumed that in its plane, the surface of the sheet is stretched with the velocity $u_w = ax$

Table 1 Comparison of the result for Nusselt number $-\theta'(0)$ when Nd = Le = Ld = $a = \lambda = 0$, Bi = 1, $R = 1$, $\theta_w = 1$, $M = 1.5$, Nc = 0.1, Ln = 0, Nb = Nt $\rightarrow \infty$

Pr	Khan and Pop (2010)	Wang (1989)	Gorla and Sidawi (1994)	Goyal, and Bhargava (2014)	Present result (RKF-45 method)
0.7	0.4539	0.4539	0.5349	0.4539	0.46257
2	0.9113	0.9114	0.9114	0.9113	0.91135
7	1.8954	1.8954	1.8905	1.8954	1.89539
20	3.3539	3.3539	3.3539	3.3539	3.35387
70	6.4621	6.4622	6.4622	6.4621	6.46209

(where $a > 0$ is the constant acceleration parameter). The x-axis is taken vertically upwards, and flow is confind in the region at $y = 0$. T_w, C_w, and N_w denote the constant values of the temperature, solutal concentration, and the nanoparticle concentration at the boundary and at the large distance from the sheet $(y \rightarrow \infty)$, the temperature, solutal concentration, and the nanoparticle concentration are represented by T_∞, C_∞, and N_∞, respectively. It is worth mentioning that $T_w > T_\infty$, $C_w > C_\infty$, and $N_w > N_\infty$ as a results of these conditions, the momentum, thermal, solutal, and nanoparticle concentration boundary layers are formed near the solid surface. The sheet is saturated in a medium which is saturated by a binary fluid with dissolved solutal and containing nanoparticles in suspension. The fluid phase and nanoparticles both are assumed to be in thermal equilibrium state. The thermo physical properties of the nanofluid are assumed to be constant. Under the usual boundary layer approximation, the five governing equations were derived by Buongiorno (2006), Khan and Pop (2010), and Kuznetsov and Nield (2010), which represent equations of conservation of mass, momentum, thermal energy, solute, and nanoparticles.

$$\frac{\partial u}{\partial x} + \frac{\partial v}{\partial y} = 0, \tag{1}$$

$$u\frac{\partial u}{\partial x} + v\frac{\partial u}{\partial y} = \nu\frac{\partial^2 u}{\partial y^2} + \frac{\alpha}{\rho}\left[\frac{\partial u}{\partial x}\frac{\partial^2 u}{\partial y^2} + u\frac{\partial^3 u}{\partial x \partial y^2} + \frac{\partial u}{\partial y}\frac{\partial^2 u}{\partial y^2} + v\frac{\partial^3 u}{\partial y^3}\right] +$$

$$\left[\begin{array}{c} -g\left(\rho_p - \rho_f\right)(N - N_\infty) + \\ (1 - N_\infty)\rho_f\left[g\beta_T(T - T_\infty) + g\beta_c(C - C_\infty)\right] \end{array}\right] - \frac{\sigma B_0^2}{\rho}u, \tag{2}$$

$$u\frac{\partial T}{\partial x} + v\frac{\partial T}{\partial y} = \alpha_m\frac{\partial^2 T}{\partial y^2} + \tau\left[D_B\frac{\partial N}{\partial y}\frac{\partial T}{\partial y} + \frac{D_T}{T_\infty}\left(\frac{\partial T}{\partial y}\right)^2\right]$$

$$+ D_{TC}\frac{\partial^2 C}{\partial y^2} + \frac{\partial q_r}{\partial y}, \tag{3}$$

$$u\frac{\partial C}{\partial x} + v\frac{\partial C}{\partial y} = D_S\frac{\partial^2 C}{\partial y^2} + D_{CT}\frac{\partial^2 T}{\partial y^2}, \tag{4}$$

$$u\frac{\partial N}{\partial x} + v\frac{\partial N}{\partial y} = D_B\frac{\partial^2 N}{\partial y^2} + \frac{D_T}{T_\infty}\frac{\partial^2 T}{\partial y^2}. \tag{5}$$

The boundary conditions to the problem are given as follows:

$$u = u_w, \quad v = 0, \quad -k\frac{\partial T}{\partial y} = h_f\left(T_f - T\right), C = C_w, \quad N = N_w \text{at } y = 0,$$

$$u = v = 0, \quad T \rightarrow T_\infty, \quad C \rightarrow C_\infty, \quad N \rightarrow N_\infty \text{as } y \rightarrow \infty, \tag{6}$$

where u and v are the components of velocity along the x and y directions, ρ_f is the density of base fluid, ρ_p is the nanoparticle density, μ is the absolute viscosity of the base fluid, v is the kinematic viscosity of the base fluid, σ is the electrical conductivity of the base fluid, α is the material fluid parameter, T is the fluid temperature,

Table 2 Comparison of the results for the Nusselt number $(-\theta'(0))$ when Ln = Pr = 10 and Nd = Le = Ld = $a = \lambda = 0$, Bi = 1, $M = 1.5$, Nc = 0.1, $R = 1$, $\theta_w = 1$

Nt	$-\theta'(0)$											
	Nb = 0.1			Nb = 0.2			Nb = 0.3			Nb = 0.4		
	Khan and Pop	Goyal and Bhargava	Present result	Khan and Pop	Goyal and Bhargava	Present result	Khan and Pop	Goyal and Bhargava	Present result	Khan and Pop	Goyal and Bhargava	Present result
0.1	0.9524	0.95244	0.95234	0.5056	0.50561	0.50556	0.2522	0.25218	0.25215	0.1194	0.11940	0.11940
0.2	0.6932	0.69318	0.69317	0.3654	0.36536	0.36536	0.1816	0.18159	0.18159	0.0859	0.08588	0.08590
0.3	0.5201	0.52025	0.52009	0.2731	0.27313	0.27311	0.1355	0.13564	0.13552	0.0641	0.06424	0.06408
0.4	0.4026	0.40260	0.40261	0.2110	0.21100	0.21100	0.1046	0.10461	0.10462	0.0495	0.04962	0.04947
0.5	0.3211	0.32105	0.32109	0.1681	0.16811	0.16810	0.0833	0.08342	0.08331	0.0394	0.03932	0.03939

Table 3 Comparison of the results for the nanoparticle Sherwood number ($-\varphi'(0)$) when Ln = Pr = 10 and Nd = Le = Ld = $a = \lambda = 0$, Bi = 1, $M = 1.5$, Nc = 0.1, $R = 1$, $\theta_w = 1$

Nt	$-\varphi'(0)$											
	Nb = 0.1			Nb = 0.2			Nb = 0.3			Nb = 0.4		
	Khan and Pop	Goyal and Bhargava	Present result	Khan and Pop	Goyal and Bhargava	Present result	Khan and Pop	Goyal and Bhargava	Present result	Khan and Pop	Goyal and Bhargava	Present result
0.1	2.1294	2.12949	2.1290	2.3819	2.38186	2.3816	2.4100	2.41009	2.4098	2.3997	2.39970	2.3994
0.2	2.2740	2.27401	2.2735	2.5152	2.51537	2.5148	2.5150	2.51501	2.5147	2.4807	2.48066	2.4804
0.3	2.5286	2.52855	2.5284	2.6555	2.65550	2.6550	2.6088	2.60876	2.6084	2.5486	2.54848	2.5483
0.4	2.7952	2.79520	2.7949	2.7818	2.78181	2.7812	2.6876	2.68758	2.6871	2.6038	2.60380	2.6034
0.5	3.0351	3.03511	3.0334	2.8883	2.88830	2.8876	2.7519	2.75190	2.7513	2.6483	2.64831	2.6478

C is the solutal concentration, N is the nanoparticle volume fraction, $\alpha_m = \frac{k}{(\rho c)_f}$ is the thermal diffusivity of the fluid, $\tau = \frac{(\rho c)_p}{(\rho c)_f}$ is the ratio of effective heat capacity of nanoparticle material to heat capacity of fluid, T_f is the hot fluid at temperature, D_{TC} and D_{CT} are the Dufour and Soret type diffusivity, D_S is the solutal diffusivity, D_B and D_T are the Brownian diffusion coefficient and thermophoresis diffusion coefficient, β_C is the volumetric solutal expansion coefficient of the fluid, β_T the is volumetric thermal expansion coefficient of the fluid, h_f is the heat transfer coefficient and g and k are the acceleration due to gravity and thermal conductivity of the fluid, respectively.

The Rosseland diffusion approximation for radiation heat flux q_r is given by

$$q_r = -\frac{16\sigma^*}{3k^*} T^3 \frac{\partial T}{\partial y}, \tag{7}$$

where σ^* and k^* are the Stefan–Boltzmann constant and the mean absorption coefficient, respectively.

In view of Eq. (7), Eq. (3) reduces to

$$u\frac{\partial T}{\partial x} + v\frac{\partial T}{\partial y} = \frac{\partial}{\partial y}\left[\left(\alpha + \frac{16\sigma^* T^3}{3k^*}\right)\frac{\partial T}{\partial y}\right]$$

$$+ \tau\left[D_B\frac{\partial T}{\partial y}\frac{\partial C}{\partial y} + \frac{D_T}{T_\infty}\left(\frac{\partial T}{\partial y}\right)^2\right] + D_{TC}\frac{\partial^2 C}{\partial y^2}. \tag{8}$$

Yang et al. [38] have shown that a similarity relation is not possible for flows in stably stratified media. Similarity transformation is possible only for the case of temperature decreasing with height, which is physically unstable. Such is the present case, and hence, Eqs. (2) to (5) subjected to boundary conditions (6) admit self-similar solution in terms of the similarity function $f(\eta)$, $\theta(\eta)$, $\gamma(\eta)$, and $\phi(\eta)$ and the similarity variable η which is defined as follows:

$$u = axf'(\eta), v = -\sqrt{av}f(\eta), \eta = \sqrt{\frac{a}{v}}y,$$

$$T = T_\infty(1 + (\theta_w - 1)\theta), \gamma(\eta) = \frac{C - C_\infty}{C_w - C_\infty}, \phi(\eta) = \frac{N - N_\infty}{N_w - N_\infty}, \tag{9}$$

where $\theta_w = \frac{T_f}{T_\infty}$, $\theta_w > 1$ being the temperature ratio parameter.

Substituting the expressions in Eq. (9) into Eqs. (2)–(5), we obtain the following transformed similarity equations:

$$f''' + ff'' - f'^2 - \alpha(f''^2 - 2ff'f''' + ff'''') \tag{10}$$

$$+ \lambda(\theta + Nc\,\gamma - Nr\phi) - Mf' = 0,$$

$$\left(1 + R\frac{d}{d\eta}(1 + (1 + \theta_w)\theta)^3\right)\theta''(\eta)$$

$$+ \Pr(f\theta' + Nb\theta'f' + Nt\theta'^2 + Nd\gamma'') = 0, \tag{11}$$

$$\gamma'' + Lef'\gamma' + Ld\theta'' = 0, \tag{12}$$

$$\phi'' + Lnf\phi' + \frac{Nt}{Nb}\theta'' = 0. \tag{13}$$

The transformed boundary conditions are

$$f(0) = 0, \ f'(0) = 1, \ \theta'(0) = -Bi(1 - \theta(0)), \ \gamma(0) = 1, \ \phi(0) = 1 \text{ at } \eta = 0$$
$$f'(\infty) \to 0, \ \theta(\infty) \to 0, \ \gamma(\infty) \to 0, \ \phi(\infty) \to 0 \text{ at } \eta \to \infty \tag{14}$$

where $\alpha = \frac{\alpha_1 a}{\mu}$ is the viscoelastic parameter, $Nb = \frac{(\rho C)_p D_B (N_w - N_\infty)}{(\rho C)_f v}$ is the Brownian motion parameter, $Nt = \frac{(\rho C)_p D_T (T_f - T_\infty)}{(\rho C)_f T_\infty v}$ is the thermophoresis parameter, $Nd = \frac{D_{TC}(T_f - T_\infty)}{D_S(C_w - C_\infty)}$ is the modified Dufour parameter, $Nc = \frac{\beta_C(C_w - C_\infty)}{\beta_T(T_f - T_\infty)}$ is the regular double-diffusive buoyancy ratio, $Nr = \frac{(\rho_p - \rho_f)(N_w - N_\infty)}{\rho_f\beta_T(T_f - T_\infty)(1 - N_\infty)}$ is the nanofluid buoyancy

Table 4 The numerical of skin friction coefficient with $a = 0$ and $a = 0.5$

Bi	R	Pr	θ_w	Nd	Ln	Le	Ld	Nb	Nt	λ	M	Nc	Skin friction coefficient	
													$a = 0$	$a = 0.5$
0.2													1.43205	2.96707
0.4													1.36583	2.85068
0.6													1.32221	2.77284
	0												1.36251	2.85839
	0.5												1.34192	2.80810
	1												1.32089	2.75671
		3											1.32497	2.76526
		4											1.34192	2.80810
		5											1.35193	2.83360
			1										1.34876	2.82184
			1.4										1.33371	2.79157
			1.8										1.31274	2.74874
				0.1									1.26834	2.56823
				0.2									1.17796	2.29936
				0.3									1.07752	2.02579
					3								1.34801	2.81638
					4								1.34192	2.80810
					5								1.33795	2.80305
						5							1.34192	2.80810
						10							1.34160	2.80936
						15							1.33928	2.80541
							0.5						1.34192	2.80810
							1						1.34085	2.80503
							1.5						1.33978	2.80196
								0.2					1.36515	2.85138
								0.4					1.31797	2.76266
								0.6					1.26959	2.66933
									0.2				1.34910	2.82314
									0.4				1.33441	2.79227
									0.6				1.31846	2.75816
										0			1.58113	3.22750
										2			1.12634	2.44198
										4			0.73642	1.80525
											1		1.17104	2.47904
											2		1.49715	3.10837
											3		1.77371	3.64713
												1	1.09804	2.44336
												2	0.83753	2.06279
												3	0.58582	1.70306

Table 5 Numerical values of $-f''(0)$, $-\gamma'(0)$, $-\theta'(0)$ and $-\varphi'(0)$ for different physical parameters

Bi	R	Pr	θ_w	Nd	Ln	Le	Ld	Nb	Nt	α	λ	M	Nc	$-f''(0)$	$-\gamma'(0)$	$-\varphi'(0)$	$-\theta'(0)$
0.2														2.96707	1.50463	1.28818	0.26757
0.4														2.85068	1.50341	1.27248	0.41719
0.6														2.77284	1.50424	1.26552	0.50867
	0													2.85839	1.50345	1.27238	0.25574
	0.5													2.80810	1.50372	1.26835	0.46802
	1													2.75671	1.50957	1.27520	0.28380
		3												2.76526	1.50769	1.27218	0.45607
		4												2.80810	1.50372	1.26835	0.46802
		5												2.83360	1.50321	1.26976	0.47224
			1											2.82184	1.50071	1.26348	0.47644
			1.4											2.79157	1.50740	1.27434	0.45783
			1.8											2.74874	1.51675	1.28943	0.43212
				0.1										2.56823	1.57167	1.38537	0.30838
				0.2										2.29936	1.65304	1.52823	0.11343
				0.3										2.02579	1.74113	1.68545	−0.10123
					3									2.81638	1.49844	1.02955	0.47983
					4									2.80810	1.50372	1.26835	0.46802
					5									2.80305	1.50759	1.47537	0.45967
						5								2.80810	1.50372	1.26835	0.46802
						10								2.80936	2.26285	1.27719	0.45463
						15								2.80541	2.84139	1.28580	0.44357
							0.5							2.80810	1.50372	1.26835	0.46802
							1							2.80503	1.46854	1.26896	0.46828
							1.5							2.80196	1.43361	1.26957	0.46854
								0.2						2.85138	1.48668	1.18357	0.50788
								0.4						2.76266	1.52079	1.31270	0.42742
								0.6						2.66933	1.55364	1.35999	0.34719
									0.2					2.82314	1.49937	1.29030	0.47871
									0.4					2.79227	1.50818	1.25051	0.45699
									0.6					2.75816	1.51741	1.22730	0.43399
										1				3.99182	1.53308	1.30072	0.47469
										2				5.89049	1.57107	1.34290	0.48317
										3				7.43629	1.59502	1.36948	0.48839
											0			3.22750	1.46544	1.22729	0.45927
											2			2.44198	1.53371	1.29994	0.47455
											4			1.80525	1.58046	1.34835	0.48425
												1		2.47904	1.53351	1.30079	0.47471
												2		3.10837	1.47670	1.23908	0.46183
												3		3.64713	1.18758	1.18758	0.45060
													1	2.44336	1.52792	1.29302	0.47309
													2	2.06279	1.55163	1.31701	0.47793
													3	1.70306	1.57281	1.33831	0.48215

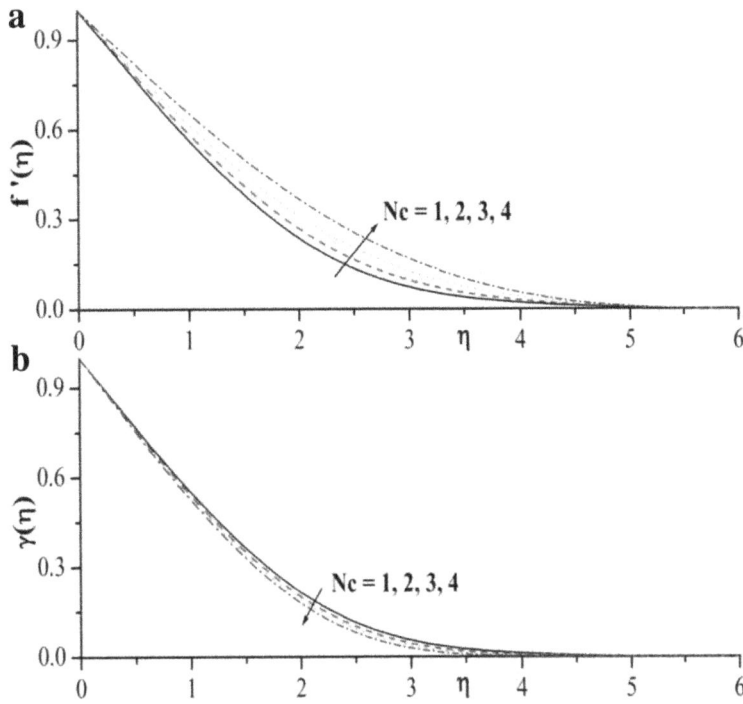

Fig. 2 a Velocity and **b** solute concentration profiles for variable values of Nc

ratio, $\lambda = \frac{(1-N_\infty)(\rho_f)g\beta_T(T_f-T_\infty)}{a\,u_w}$ is the mixed convection parameter, $R = \frac{16\sigma^* T_\infty^3}{3k^* k}$ is the radiation parameter, $Ld = \frac{D_{CT}(T_f-T_\infty)}{D_S(C_w-C_\infty)}$ Dufour Lewis number, $Ln = \frac{\nu}{D_B}$ is the nanofluid Lewis number, $Le = \frac{\nu}{D_s}$ is the regular Lewis number, $M = \frac{\sigma B_0^2}{\rho a}$ is the magnetic parameter, and $Pr = \frac{\nu}{\alpha_m}$ is Prandtl number and $Bi = \frac{h_f}{k}\sqrt{\frac{\nu}{a}}$ is Biot number.

The physical quantities with practical interest in the study are skin friction coefficient, Nusselt number, regular Sherwood number, and nanoparticle Sherwood number.

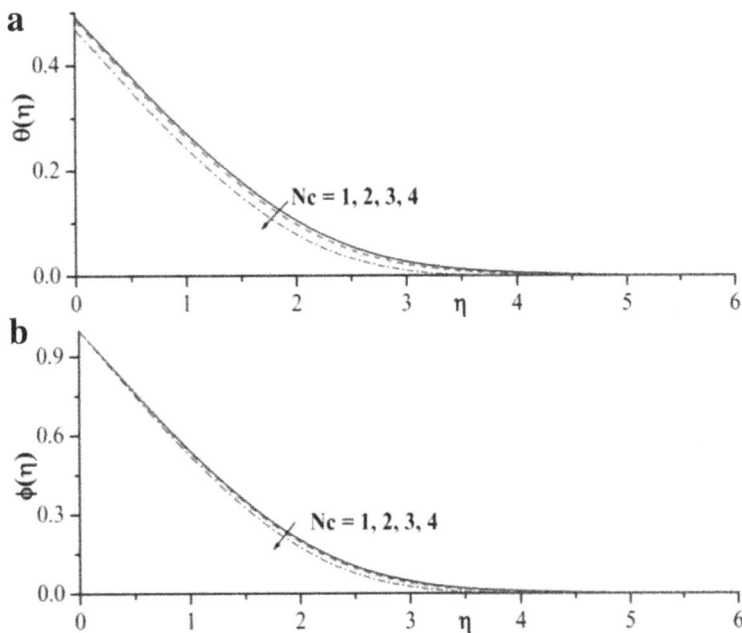

Fig. 3 a Temperature and **b** nanoparticle concentration profiles for variable values of Nc

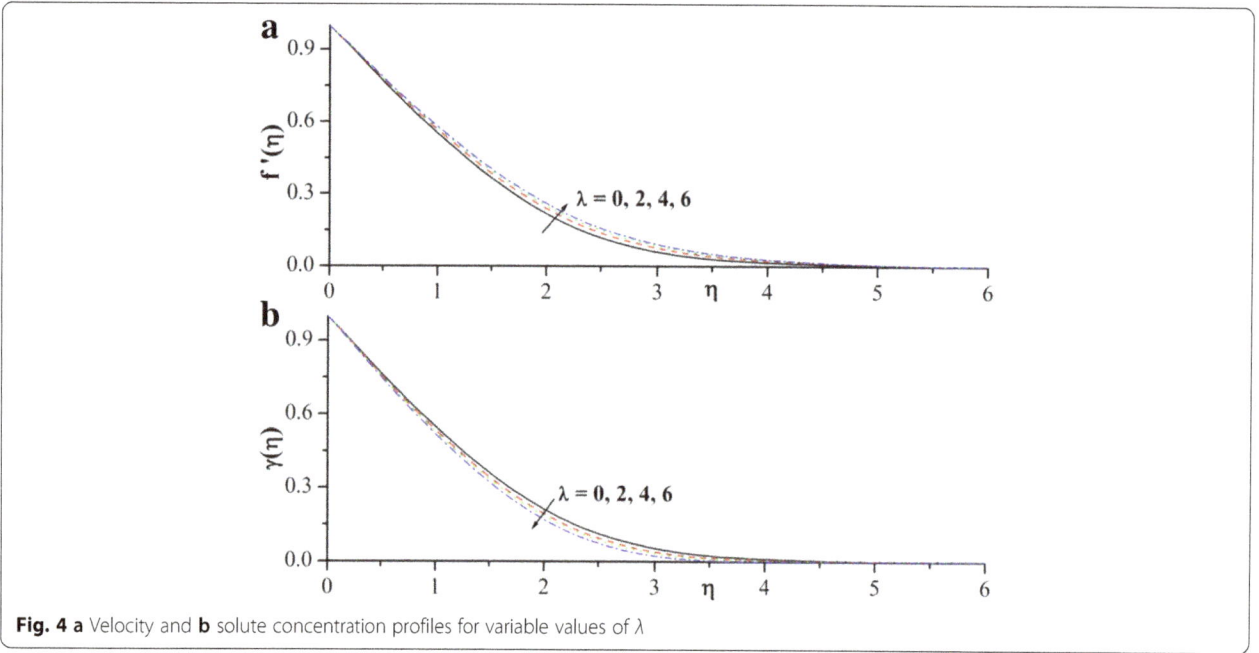

Fig. 4 a Velocity and **b** solute concentration profiles for variable values of λ

These parameters characterize surface drag, wall heat transfer, regular mass transfer, and nanoparticle mass transfer rates, respectively. These quantities are defined as

$$C_f = \frac{\tau_w}{\rho u_w^2/2}, Nu_x = \frac{xq_w}{k\left(T_f - T_\infty\right)}, Sh_x$$
$$= \frac{xq_m}{D_S\left(C_w - C_\infty\right)}, Sh_{xn} = \frac{xq_{np}}{D_B\left(N_w - N_\infty\right)}, \quad (15)$$

where τ_w is the wall skin friction, q_w is the wall heat flux, and q_m and q_{np} are the solutal wall mass flux and

nanoparticle wall mass flux at the surface of the sheet, respectively, which are given by the following expressions:

$$\tau_w = \mu\left(\frac{\partial u}{\partial y}\right)_{y=0} + \frac{\alpha}{\rho}\left(u\frac{\partial^2 u}{\partial x \partial y} v\frac{\partial^2 u}{\partial y^2} - 2\frac{\partial u}{\partial y}\frac{\partial v}{\partial y}\right)_{y=0}, q_w = -k\left(\frac{\partial T}{\partial y}\right) + q_{r_{y=0}},$$

$$q_m = -D_S\left(\frac{\partial C}{\partial y}\right)_{y=0} \text{ and } q_{np} = -D_B\left(\frac{\partial N}{\partial y}\right)_{y=0} \quad (16)$$

Following Kuznetsov and Nield (2010), the reduced local skin friction, local Nusselt number, reduced local

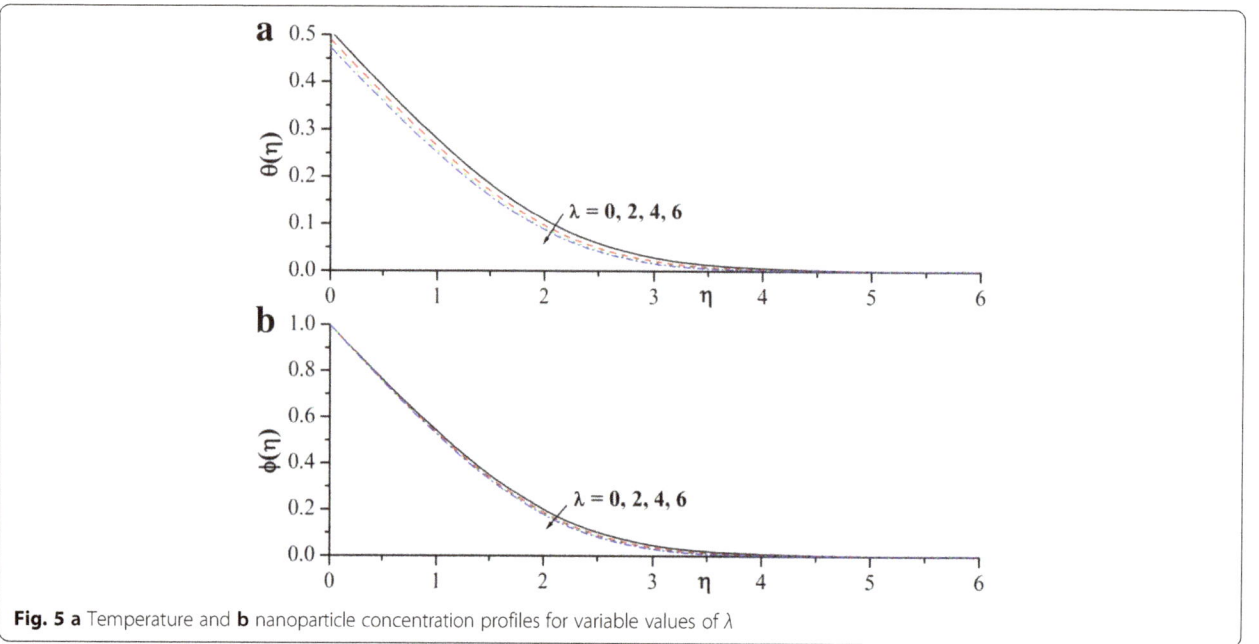

Fig. 5 a Temperature and **b** nanoparticle concentration profiles for variable values of λ

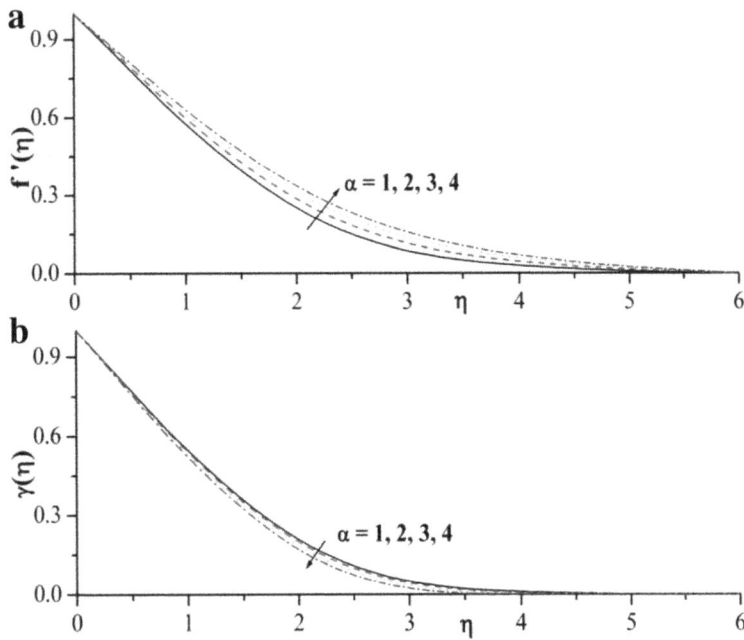

Fig. 6 a Velocity and **b** solute concentration profiles for variable values of a

Sherwood number, and reduced nanofluid Sherwood number can be introduced and represented as

$$C_f Re_x^{\frac{1}{2}} = (1+K)f''(0), Nu_x Re_x^{-\frac{1}{2}} = -(1+R\theta_w^3)\theta'(0),$$

$$Sh_x Re_x^{-\frac{1}{2}} = -\gamma'(0), \text{ and } Sh_{xn} Re_x^{-\frac{1}{2}} = -\phi'(0),$$

$$(17)$$

where $Re_x = \frac{u_w(x)x}{\nu}$ is the local Reynolds number.

Methods

The reduced set of coupled similarity Eqs. (10)–(13) subject to boundary condition (14) are highly nonlinear in nature; thus, it is very difficult possess a closed form analytical solution. Therefore, it has been solved numerically by fourth–fifth order Runge–Kutta–Fehlberg integration scheme with the help of algebraic software Maple. The algorithm in Maple has been well tested for its accuracy and robustness. Thus, this has been used to solve a

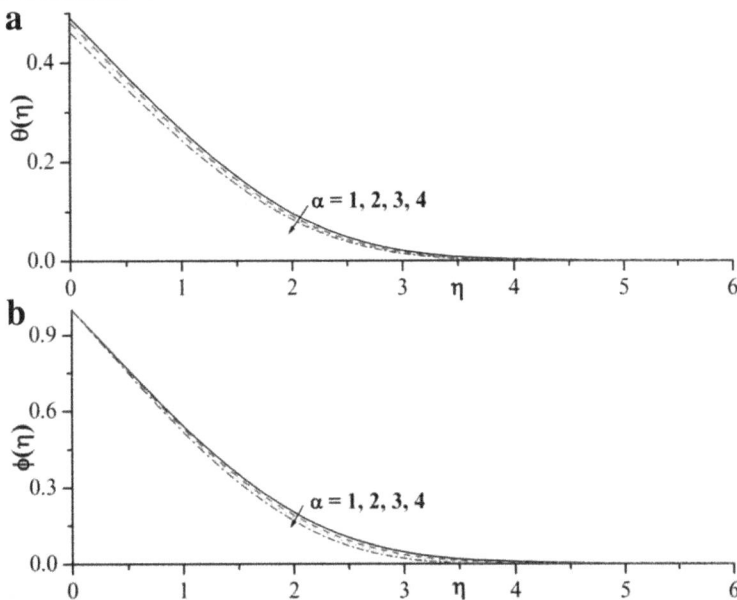

Fig. 7 a Temperature and **b** nanoparticle concentration profiles for variable values of a

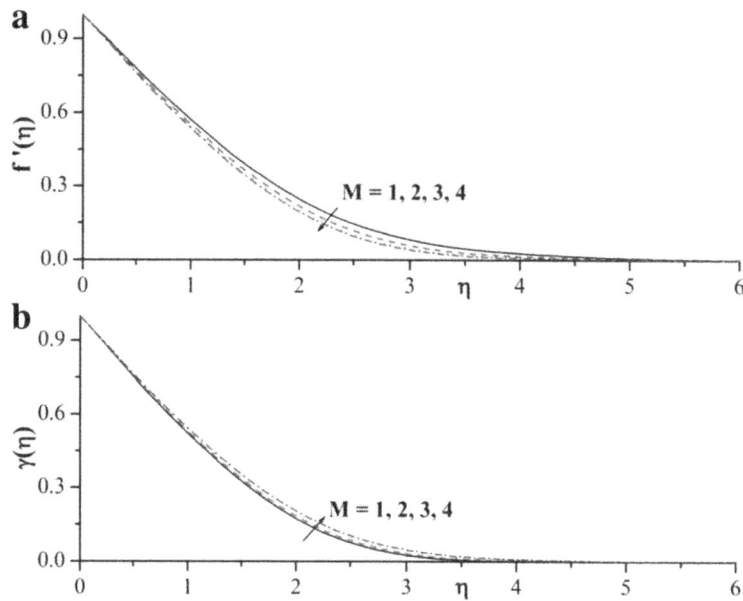

Fig. 8 a Velocity and **b** solute concentration profiles for variable values of *M*

wide range of nonlinear problems. In this method, we choose a finite value of $\eta \to \infty$ as η_6 in such a way that the boundary conditions are satisfied asymptotically. Table 1 depicts the validation of the current results by comparison with the existed literature for some special restricted cases (Khan and Pop (2010), Wang (1989), Gorla and Sidawi (1994). Further, the results are also compared with Khan and Pop (2010) and Goyal and Bhargava (2014) for numerous values of Nt and Nb which is tabulated in Tables 2 and 3. We notice that the comparison shows smart agreement for every value of Nt and Nb, which confirm that the current results are accurate. Numerical values of local Nusselt, Sherwood, and nanofluid Sherwood number are also depicted in Tables 4 and 5 for various

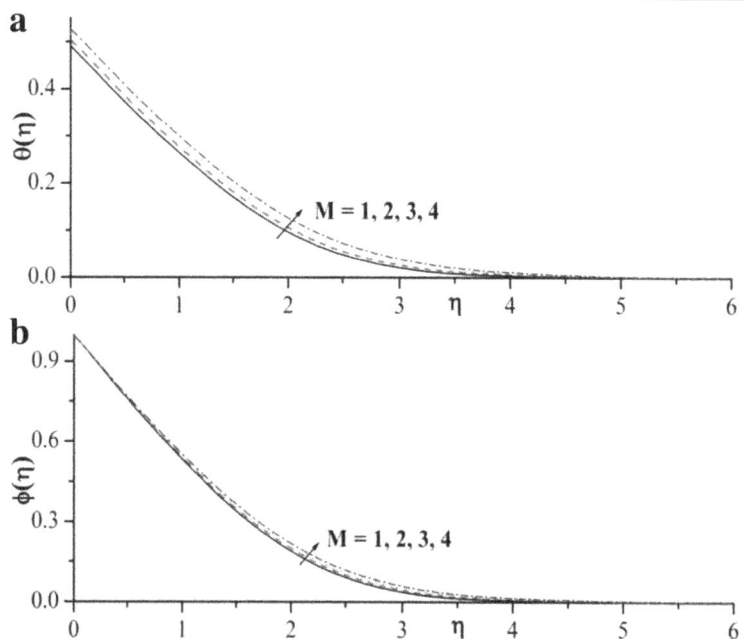

Fig. 9 a Temperature and **b** nanoparticle concentration profiles for variable values of *M*

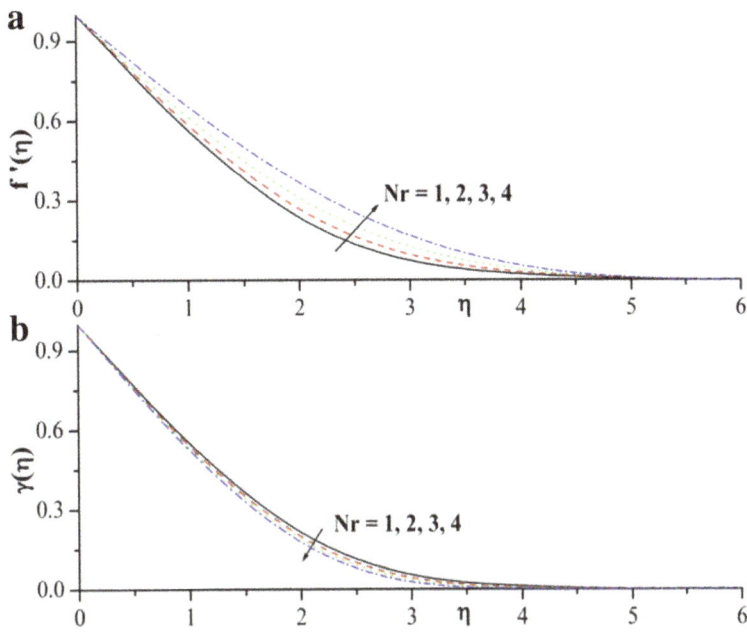

Fig. 10 a Velocity and **b** solute concentration profiles for variable values of Nr

concerned parameters. For numerical results, we considered the nondimensional parameter values as $Bi = 0.5$, $\lambda = 1$, $\alpha = 0.5$, $Ld = 0.5$, $Le = 5$, $Ln = 4$, $M = 1.5$, $Nb = 0.3$, $Nc = 0.1$, $Nd = 0.01$, $Nt = 0.3$, $Pr = 4$, $R = 0.5$, and $\theta_w = 1.2$. These values are kept as common in the entire study except the variations in respective figures and tables.

Results and discussion

The main aim of this section is to analyze the effects of various physical parameters like Biot number, magnetic parameter, Prandtl number, radiation parameter, thermophoresis parameter, Brownian motion, and viscoelastic parameter. Hence, Figs. 2, 3, 4, 5, 6, 7, 8, 9, 10,

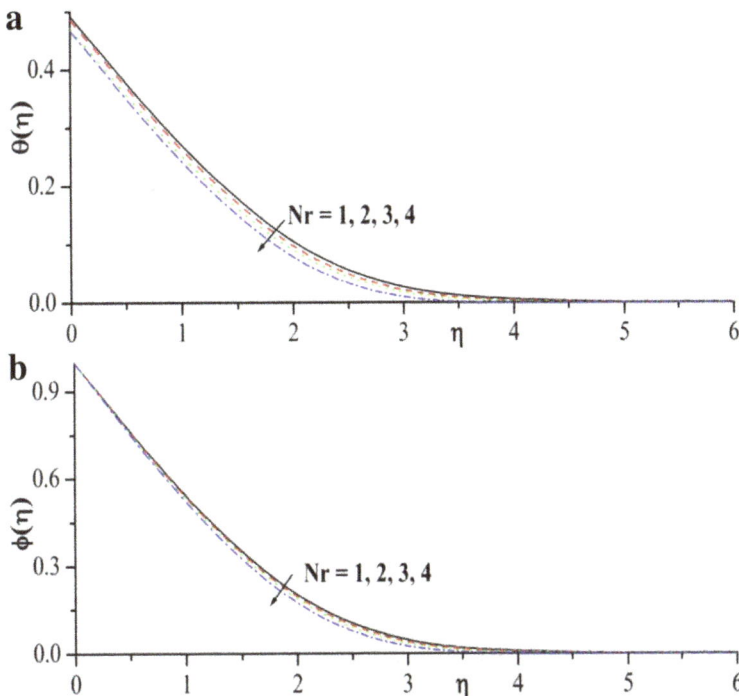

Fig. 11 a Temperature and **b** nanoparticle concentration profiles for variable values of Nr

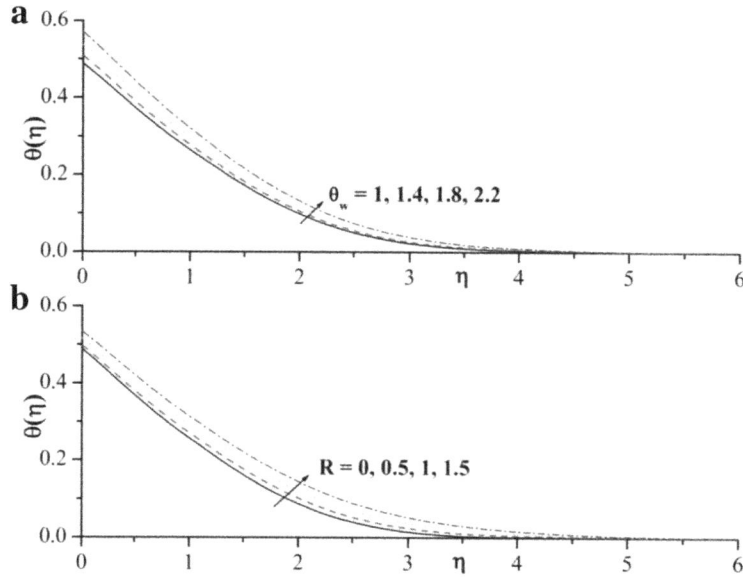

Fig. 12 a, b Temperature profiles for variable values of θ_w and R, respectively

11, 12, 13, 14, 15, 16, 17, and 18 have been plotted for such objective. Figure 2a describes the influences of double-diffusive buoyancy ratio parameter (Nc) on velocity and temperature profiles. It is observed that velocity profile is an increasing function of double-diffusive buoyancy ratio parameter; as a result, momentum boundary layer thickness increases. Figure 2b depicts that the solutal concentration reduced by increasing the buoyancy force parameter. Further, it is

observed that the corresponding boundary layer is also reduced. Figure 3a, b is plotted to visualize the effects of double-diffusive buoyancy ratio parameter on temperature and nanoparticle concentration profile. From these figures, we observed that increasing values of the double-diffusive buoyancy ratio parameter lead to decrease the temperature, nanoparticle concentration, and corresponding boundary layer thickness as shown in Fig. 3a, b.

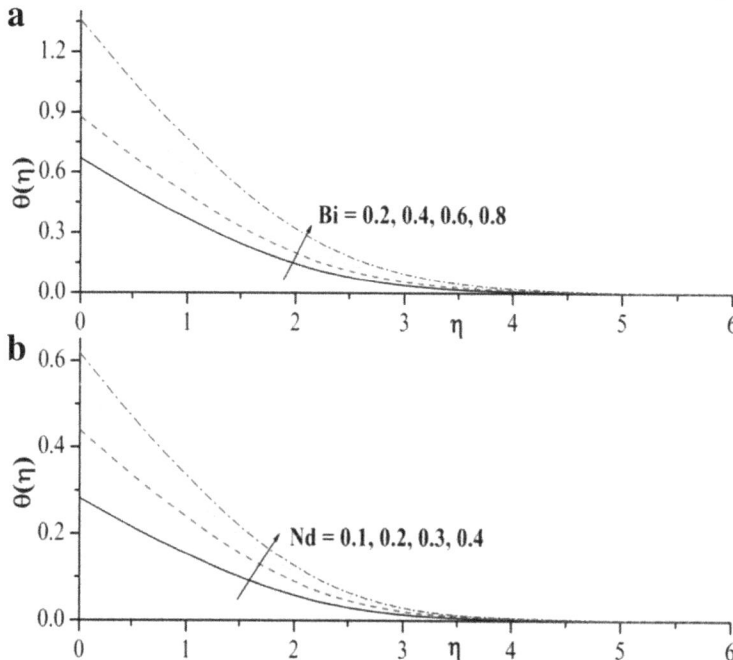

Fig. 13 a, b Temperature profiles for variable values of Bi and Nd, respectively

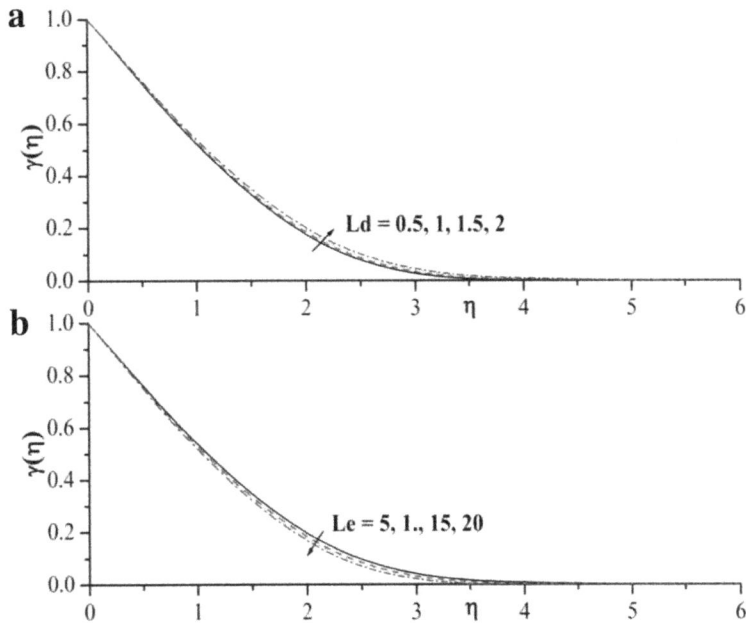

Fig. 14 a, b Solute concentration profiles for variable values of Ld and Le, respectively

Figure 4a, b shows the velocity and concentration profile for different values of mixed convection parameter (λ). It depicts that the velocity field and momentum boundary layer thickness increase by increasing mixed convection parameter as shown in Fig. 4a. Figure 4b illustrates that the solutal concentration profile and its boundary layer thickness are a decreasing function of mixed convection parameter. As viewed from Fig. 5a, b,

the temperature and nanoparticle concentration profiles decreased significantly with an increase in mixed convection parameter.

Figures 6a, b and 7a, b show the velocity, solutal concentration, temperature, and nanoparticle concentration profile for different values of viscoelastic parameter (α), respectively. From this plot, it is evident that increasing values of viscoelastic parameter oppose the motion of

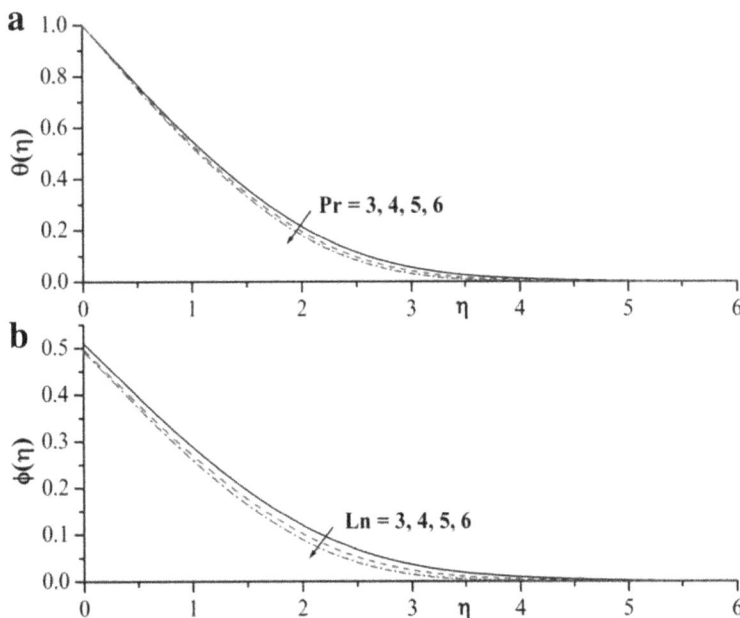

Fig. 15 a Temperature and **b** nanoparticle concentration profiles for variable values of Pr and Ln, respectively

a

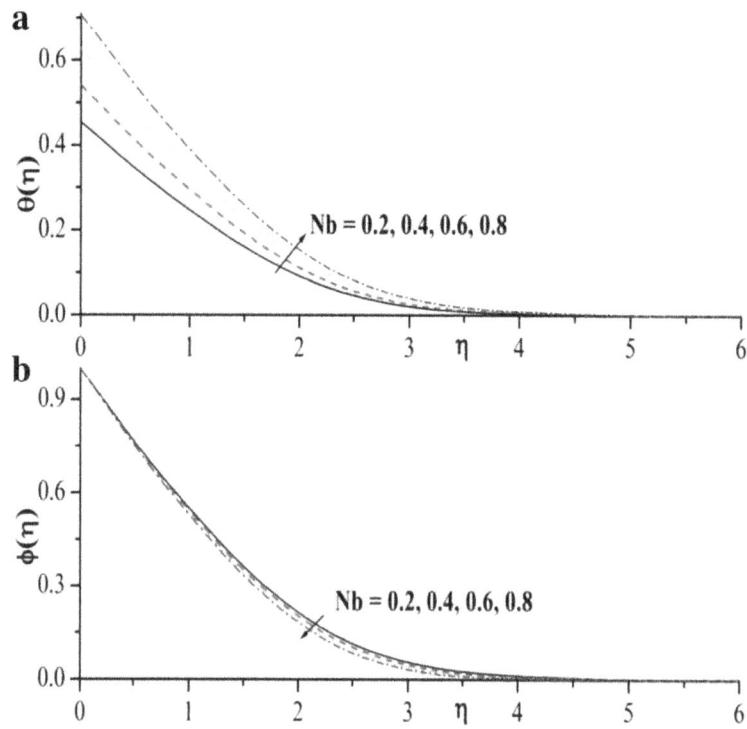

b

Fig. 16 a Temperature and **b** nanoparticle concentration profiles for variable values of Nb

a

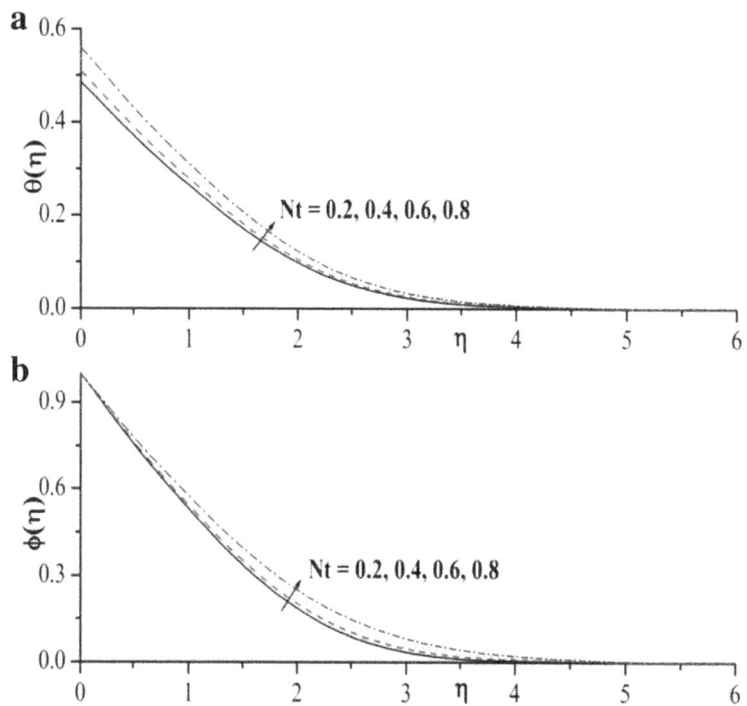

b

Fig. 17 a Temperature and **b** nanoparticle concentration profiles for variable values of Nt

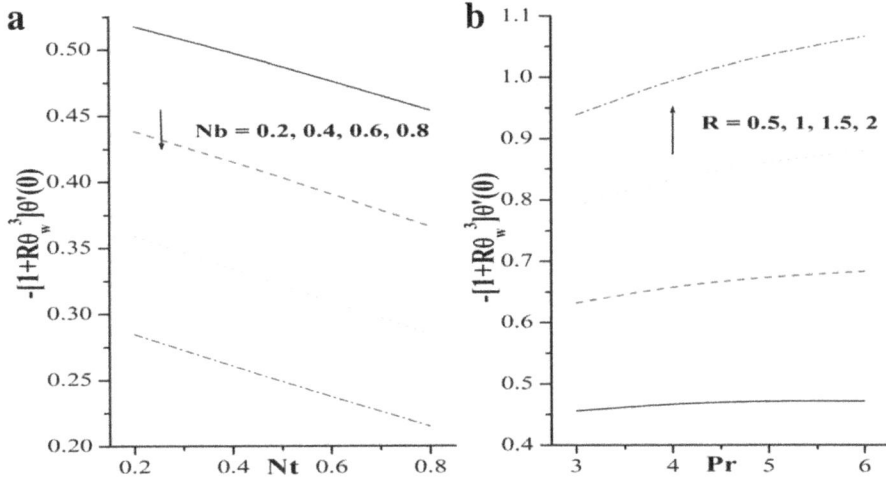

Fig. 18 a, b Influence of Nt and Pr with the various values of Nb and R on Nusselt number, respectively

the liquid close to the stretching sheet and assist the motion of the liquid faraway from the stretching sheet. Increasing values of viscoelastic parameter enables the liquid to flow at a faster rate, due to which there is decline in the heat transfer. This is responsible for the increase in momentum boundary layer, whereas the thermal, solute concentration, and nanoparticle concentration boundary layers reduce when the viscoelastic effects intensify.

Figure 8a, b, respectively, shows the effect of magnetic parameter (M) on the velocity and solute concentration profile. In general, the application of transverse magnetic field will result a restrictive type of force (Lorenz's force) similar to drag force which tends to resist the fluid flow and thus reducing its velocity. It is clear that, as the magnetic parameter increases, it reduced the velocity profile and enhanced the solute concentration profile.

Figure 9a, b reveals the temperature and nanoparticle concentration profile for different values of magnetic parameter. It is observed from the above figures that, for increasing the values of M, the temperature and nanoparticle concentration distributions increase and also the corresponding boundary layer thickness.

The effect of nanofluid buoyancy ratio parameter (Nr) on velocity and solute concentration profile are depicted in Fig. 10a, b. It is evident from this figure that the velocity profile increases and solute concentration profile decreases for increasing the values Nr. The temperature and nanoparticle concentration profiles for different values of nanofluid buoyancy ratio parameter are presented in Fig. 11a, b, respectively. Thermal and nanoparticle boundary layer thickness decrease by increasing the nanofluid buoyancy ratio parameter.

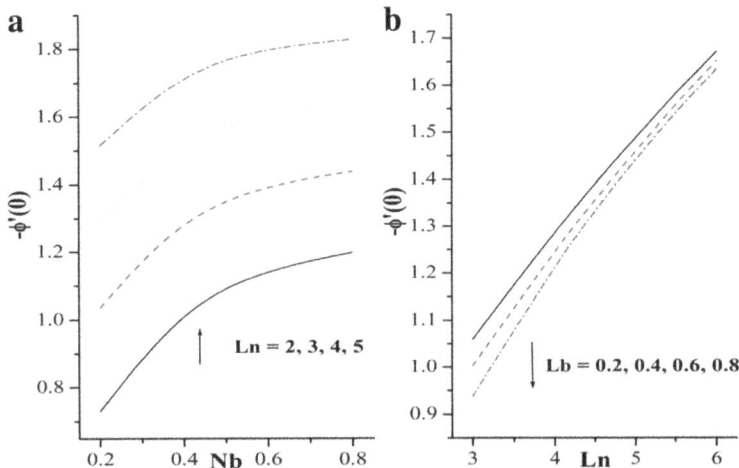

Fig. 19 a, b Influence of Nb and Ln with the various values of Ln and Ld on Sherwood number, respectively

Table 6 The numerical values of skin friction coefficient with different physical parameters with linear and nonlinear radiation

Bi	R	Pr	α	Nd	Ln	Le	Ld	Nb	Nt	λ	M	Nc	Nusselt number	
													Linear radiation	Nonlinear radiation
0.2													0.14372	0.26757
0.4													0.22566	0.41719
0.6													0.27708	0.50867
	0												0.25574	0.25574
	0.5												0.25407	0.46802
	1												0.24600	0.28380
		3											0.24744	0.45607
		4											0.25407	0.46802
		5											0.25664	0.47224
			1										0.27821	0.47469
			2										0.28283	0.48317
			3										0.28565	0.48839
				0.1									0.16878	0.30838
				0.2									0.05850	0.11343
				0.3									−0.07540	−0.10123
					3								0.26013	0.47983
					4								0.25407	0.46802
					5								0.24975	0.45967
						5							0.25407	0.46802
						10							0.24709	0.45463
						15							0.24131	0.44357
							0.5						0.25407	0.46802
							1						0.25425	0.46828
							1.5						0.25443	0.46854
								0.2					0.27452	0.50788
								0.4					0.23288	0.42742
								0.6					0.18971	0.34719
									0.2				0.27936	0.47871
									0.4				0.26948	0.45699
									0.6				0.25883	0.43399
										0			0.27034	0.45927
										2			0.27770	0.47455
										4			0.28247	0.48425
											1		0.27811	0.47471
											2		0.27120	0.46183
											3		0.26518	0.45060
												1	0.27720	0.47309
												2	0.27975	0.47793
												3	0.28196	0.48215

Figure 12a, b demonstrates the effect of temperature ratio and radiation parameter on temperature profile, respectively. Here, we observed that the temperature profile is an increasing function of temperature ratio and radiation parameter; as a result, thermal boundary layer thickness also increases. This is due to the fact that enhancement in the radiation parameter implies a decrease in the Rosseland radiation absorptive. Hence, the divergence of radiative heat flux q_r increases as absorption coefficient decreases. Therefore, the rate of radiative heat transferred to the fluid increases and consequently the fluid temperature and simultaneously the velocity of the fluid also increases.

Figure 13a, b describes the influences of Biot number and modified Dufour parameter on temperature profiles, respectively. The influence of Biot number (Bi) on the temperature profile is shown in Fig. 13a. The stronger convection leads to the maximum surface temperatures which appreciably enhance the temperature and the thermal boundary layer thickness. From Fig. 13b, it is noticed that the thermal boundary layer thickness increases by increasing the modified Dufour parameter (Nd). The variation of dimensionless solute concentration with Dufour Lewis number and regular Lewis number is illustrated Fig. 14a, b. An increase Dufour Lewis number enhances the solute concentration in the boundary layer thickness. But in case of regular Lewis number, it reduces the solute concentration in the boundary layer thickness as shown in Fig. 14b.

The effect of Prandtl number and nanofluid Lewis number on temperature and nanoparticle concentration profiles are exhibited in Fig. 15a, b, respectively. From Fig. 15a, we observed that an increase in the Prandtl number is seen to decrease the fluid temperature above the sheet. Physically it means that, the thermal boundary layer becomes thinner for the larger Prandtl number. The Prandtl number signifies the ratio of momentum diffusivity to thermal diffusivity. Hence, the Prandtl number can be used to increase the rate of cooling in conducting flows. From another plot, it is evident that increasing values of Lewis number reduces nanoparticle concentration profile; physically, it means that Lewis number lessens the mass diffusivity which in turn lessens the penetration depth of the concentration boundary layer as shown in Fig. 15b.

Figure 16a, b portraits the consequences of Brownian motion parameter on temperature and nanoparticle concentration profile. The Brownian motion parameter (Nb) will increase the random motion of the fluid particles and boundary layer thickness conjointly, which ends up in an additional heat to provide. Therefore, temperature profile will increase. However, nanoparticle concentration profiles show an opposite behavior because increasing Brownian motion parameter enhances the nanoparticle volume fraction transfer rate. This can be shown in Fig. 16b. It is interesting that the variation of the Brownian motion parameter does not show a significant influence on the concentration and temperature profiles, but its effect on the velocity profiles is obvious in the vicinity of the wall.

The development of the thermophoresis parameter (Nt) on temperature and nanoparticle concentration profiles is inspecting in Fig. 17a, b. Raising the values of thermophoresis parameter enhance both temperature $\theta(\eta)$ and concentration $\phi(\eta)$ profiles. Physically, thermophoretic parameter increases the density of the thermal boundary layer. As a result, temperature rises with the improvement in thermophoresis. Further, the boundary layer thickness is higher for larger values of thermophoresis parameter. It is a mechanism which little particle area unit force off from the new surface to a chilly one. As a result, it maximizes the temperature and nanoparticle concentration of the fluid. The combined effects of Brownian motion and thermophoresis parameters on the reduced Sherwood number are shown in Fig. 18a. An increase in the values of Brownian motion and thermophoresis parameters results a decrease in Sherwood number, but Sherwood number increases by increasing values of the radiation parameter and Prandtl number. This can be shown in Figs. 18b and 19a, b.

Conclusions

The present work analyzes the double diffusion boundary layer flow of a viscoelastic nanofluid over a stretching sheet including the nonlinear thermal radiation, mixed convection, and magnetic effects. The radiative heat flux term in the energy equality is presented by means of the nonlinear Rosseland diffusion approximation. It should also be concluded that in contrast to the linear Rosseland diffusion approximation, when use is made of the nonlinear one, the problem is also governed by the newly temperature ratio parameter θ_w. The results presented indicate quite clearly that θ_w, which is an indicator of the small/large temperature difference between the surface and the ambient fluid, has a relevant effect on heat transfer characteristics and temperature distributions within the flow region generated by an isothermal sheet stretched. Also, the effects of various standards of emerging parameters are discussed for velocity, temperature, nanoparticle concentration, and solute concentration. The key results of the present analysis can be listed as below.

- Magnetic field reduces the velocity the profile and enhances the temperature, solute, and nanoparticle concentration profiles.
- An increasing value of the regular buoyancy ratio, nanofluid buoyancy ratio, viscoelastic parameter, and mixed convection parameter is to increase the momentum boundary layer thickness and to decrease the thermal, solutal, and nanoparticle boundary layer thickness.
- Brownian motion parameter that has an opposite effect on temperature and nanoparticle concentration profiles but similar effect on temperature and nanoparticle concentration profiles is observed in case of thermophoresis parameter.

- Increasing values of temperature ratio parameter θ_w extinguishes the rate of heat transfer $|\theta'(0)|$ for fixed Pr and R.
- The temperature ratio parameter θ_w and the thermal radiation parameter R have the same effect. From a qualitative point of view, temperature increases within creasing θ_w and R. However, thermal boundary layer thickness decreases when the Prandtl number increases.
- In coolant factor, nonlinear thermal radiation is a superior copier to linear thermal radiation (Table 6).
- The temperature profile as well as thermal boundary layer thickness increases with an increase in both Biot and modified Dufour parameter.
- Solutal concentration profile increases for higher values of Dufour Lewis number whereas it decreases with increase in values of regular Lewis number.

Authors' contributions
GK has involved in conception and design of the problem. NGR has involved in analysis and interpretation of data. MS has involved in drafting the manuscript or revising it critically for important intellectual content. BJG has given final approval of the version to be published. All authors read and approved the final manuscript.

Competing interests
The authors declare that they have no competing interests.

Author details
[1]Department of Studies and Research in Mathematics, Kuvempu University Shankaraghatta, Shimoga, Karnataka 577 451, India. [2]Faculty of Engineering, Department of Engineering Mathematics, Christ University, Mysore Road, Bengaluru 560074, India.

References
Beg, O. A., Uddin, M. J., Rashidi, M. M., & Kavyani, N. (2014). Double-diffusive radiative magnetic mixed convective slip flow with Biot and Richardson number effects. *Journal of Engineering Thermophysics, 23*(2), 79–97.
Buongiorno, J. (2006). Convective transport in nanofluids. *ASME Journal Heat Transfer, 128*, 240–250.
Choi, S. U. S. (1995). Enhancing thermal conductivity of fluids with nanoparticles *Proceedings of the 1995 ASME International Mechanical Engineering Congress and Exposition, San Francisco, USA ASME FED 231/MD, 66*, 99–105.
Crane, L. J. (1970). Flow past a stretching plate. *ZAMP, 21*, 645–655.
Eastman, J. A., Choi, S. U. S., Li, S., Yu, W., & Thompson, L. J. (2001). Anomalously increased effective thermal conductivity of ethylene glycol-based nanofluids containing copper nanoparticles. *Applied Physics Letters, 78*, 718–720.
Gaikwad, S. N., Malashetty, M. S., & Rama Prasad, K. (2007). An analytical study of linear and non-linear double diffusive convection with Soret and Dufour effects in couple stress fluid. *International Journal of Non-Linear Mechanics, 42*, 903–913.
Ganesh Kumar, K., Rudraswamy, N. G., Gireesha, B. J., & Krishnamurthy, M. R. (2017). Influence of nonlinear thermal radiation and viscous dissipation on three-dimensional flow of Jeffrey nano fluid over a stretching sheet in the presence of joule heating. *Nonlinear Engineering*, https://doi.org/10.1515/nleng-2017-0014.
Gorla, R. S. R., & Sidawi, I. (1994). Free convection on a vertical stretching surface with suction and blowing. *Applied Scientific Research, 52*, 247–257.
Goyal, M., & Bhargava, R. (2014). Finite element solution of double-diffusive boundary layer flow of viscoelastic nanofluids over a stretching sheet. *Computational Mathematics and Mathematical Physics, 54*(5), 848–863.
Haq, R. U., Nadeem, S., Khan, Z. H., & Okedayo, T. G. (2014). Convective heat transfer and MHD effects on Casson nanofluid flow over a shrinking sheet. *Central European Journal of Physics, 12*(12), 862–871.

Hayat, T., Farooq, M., & Alsaedi, A. (2014). Melting heat transfer in the stagnation-point flow of Maxwell fluid with double-diffusive convection. *International Journal of Numerical Methods for Heat & Fluid Flow, 24*(3), 760–774.
Hsiao, K. L. (2017b). Micropolar nanofluid flow with MHD and viscous dissipation effects towards a stretching sheet with multimedia feature. *International Journal of Heat and Mass Transfer, 112*, 983–990.
Hsiao, K.-L. (2014). Nanofluid flow with multimedia physical features for conjugate mixed convection and radiation. *Computers & Fluids, 104*(20), 1–8.
Hsiao, K.-L. (2016). Stagnation electrical MHD nanofluid mixed convection with slip boundary on a stretching sheet. *Applied Thermal Engineering, 98*(5), 850–861.
Hsiao, K.-L. (2017a). To promote radiation electrical MHD activation energy thermal extrusion manufacturing system efficiency by using Carreau-Nanofluid with parameters control method. *Energy, 130*(1), 486–499.
Hsiao, K.-L. (2017b). Combined electrical MHD heat transfer thermal extrusion system using Maxwell fluid with radiative and viscous dissipation effects. *Applied Thermal Engineering, 112*, 1281–1288. doi:10.1016/j.applthermaleng.2016.08.208.
Kang, H. U., Kim, S. H., & OhE, J. M. (2006). Particles of thermal conductivity of nanofluid using experimental effect i n particle volume. *Experimental Heat Transfer, 19*(3), 181–191.
Khan, W. A., & Aziz, A. (2011). Double-diffusive natural convective boundary layer flow in porous medium saturated with a nanofluid over a vertical plate: prescribed surface heat, solute and nanoparticle fluxes. *International Journal of Thermal Sciences, 726*, 2154–2160.
Khan, W. A., & Gorla, R. S. R. (2011). Heat and mass transfer in non-Newtonian nanofluids over a non-isothermal stretching wall. *Proceedings of the Institution of Mechanical Engineers, Part N: Journal of Nanoengineering and Nanosystems, 225*(4), 155–163.
Khan, W. A., & Pop, I. (2010). Boundary layer flow of a nanofluid past a stretching sheet. *International Journal of Heat and Mass Transfer, 53*, 2477–2483.
Kuznetsov, A. V., & Nield, D. A. (2010). Natural convective boundary-layer flow of a nanofluid past a vertical plate. *International Journal of Thermal Sciences, 49*, 243–247.
Nadeema, S., Haq, R. U., & Khan, Z. H. (2014). Numerical study of MHD boundary layer flow of a Maxwell fluid past a stretching sheet in the presence of nanoparticles. *Journal of the Taiwan Institute of Chemical Engineers, 45*(1), 121–112.
Nield, D. A., & Kuznetsov, A. V. (2011). The Cheng–Minkowycz problem for the double-diffusive natural convective boundary layer flow in a porous medium saturated by a nanofluid. *International Journal of Heat and Mass Transfer, 54*, 374–378.
Rudraswamy, N. G., Gireesha, B. J., & Chamkha, A. J. (2015). Effects of magnetic field and chemical reaction on stagnation-point flow and heat transfer of a nanofluid over an inclined stretching sheet. *Journal of Nanofluids, 4*(2), 239–246.
Rudraswamy, N. G., Ganesh Kumar, K., Gireesha, B. J., & Gorla, R. S. R. (2016). Soret and Dufour effects in three-dimensional flow of Jeffery nanofluid in the presence of nonlinear thermal radiation. *Journal of Nanoengineering and Nanomanufacturing, 6*(4), 278–287.
Rudraswamy, N. G., Ganesh Kumar, K., Gireesha, B. J., & Gorla, R. S. R. (2017). Combined effect of Joule heating and viscous dissipation on MHD three dimensional flow of a Jeffrey nanofluid. *Journal of Nanofluids, 6*(2), 300–310.
Rudyak, V. Y., Belkin, A. A., & Tomilina, E. A. (2010). On the thermal conductivity of nnanofluids. *Technical Physics Letters, 36*(7), 660–662.
Sakiadis, B. C. (1961a). Boundary-layer behavior on continuous solid surfaces: I. Boundary layer equations for two-dimensional and axisymmetric flows. *AICHE, 7*, 26–28.
Sakiadis, B. C. (1961b). Boundary-layer behavior on continuous solid surfaces: II. The boundary layer on a continuous flat surface. *AICHE, 7*, 221–225.
Sharma, A. K., Dubey, G. K., & Varshney, N. K. (2012). Effect of kuvshinski fluid on double-diffusive convection-radiation interaction on unsteady MHD flow over a vertical moving porous plate with heat generation and soret effects. *Advances in Applied Science Research, 3*(3), 1784–1794.
Shehzad, S. A., Hayat, T., & Alsaedi, A. (2015). MHD flow of Jeffrey nanofluid with convective boundary conditions. *Journal of the Brazilian Society of Mechanical Sciences and Engineering, 37*, 873–883.
Wang, C. Y. (1989). Free convection on a vertical stretching surface. *ZAMM-Journal of Applied Mathematics and Mechanics/Zeitschrift für Angewandte Mathematik und Mechanik, 69*, 418–420.
Wang, X., Xu, X., & Choi, S. U. S. (1999). Thermal conductivity of nanoparticle fluid mixture. *Journal of Thermophysics and Heat Transfer, 13*, 474–480.

13

Experimental study of strain fields during shearing of medium and high-strength steel sheet

E. Gustafsson[1][*] (ID), L. Karlsson[1] and M. Oldenburg[2]

Abstract

Background: There is a shortage of experimentally determined strains during sheet metal shearing. These kinds of data are a requisite to validate shearing models and to simulate the shearing process.

Methods: In this work, strain fields were continuously measured during shearing of a medium and a high strength steel sheet, using digital image correlation. Preliminary studies based on finite element simulations, suggested that the effective surface strains are a good approximation of the bulk strains below the surface. The experiments were performed in a symmetric set-up with large stiffness and stable tool clearances, using various combinations of tool clearance and clamping configuration. Due to large deformations, strains were measured from images captured in a series of steps from shearing start to final fracture. Both the Cauchy and Hencky strain measures were considered, but the difference between these were found negligible with the number of increments used (about 20 to 50). Force-displacement curves were also determined for the various experimental conditions.

Results: The measured strain fields displayed a thin band of large strain between the tool edges. Shearing with two clamps resulted in a symmetric strain band whereas there was an extended area with large strains around the tool at the unclamped side when shearing with one clamp. Furthermore, one or two cracks were visible on most of the samples close to the tool edges well before final fracture.

Conclusions: The fracture strain was larger for the medium strength material compared with the high-strength material and increased with increasing clearance.

Keywords: Sheet metal, Experiment, Shearing, Force, Strain, Strain field

Background

Shearing is a widely used process in the sheet metal industry and has been extensively studied to increase the understanding of the shearing mechanism, but also to improve the industrial process with respect to parameters such as tool force, tool wear and sheared surface appearance. Tool forces were measured for various materials already by Izod (1906), who used a symmetric set-up with double tools to obtain a large stiffness in the lateral direction and a stable clearance. A refined set-up was devised by Gustafsson et al. (2014) which allowed the measurement of the lateral force component (the component that strives

to separate the tools and increase the clearance), in addition to the force in the shearing direction, with a minimal reduction of lateral stiffness. The lateral force was previously measured by Yamasaki and Ozaki (1991) but with friction losses in sliding guides and a low stiffness in the set-up that results in tool clearance changes during shearing. The same concept was improved by Kopp et al. (2016) for increased stiffness and with compensation of friction losses through calibrations. Still, the stiffness is lower than in symmetric double shearing and there is a significant risk of misinterpreted forces due to changed conditions in the sliding guides.

Closely related to shearing is blanking, a process which has an inherent clearance stability due to its rotational symmetric set-up. The appearance of the cut surfaces in the blanking process was studied by Crane (1927), who

*Correspondence: egu@du.se
[1]Dalarna University, SE-791 88 Falun, Sweden
Full list of author information is available at the end of the article

characterised the surface appearance in terms of various zones. Crane (1927) also conducted post-shearing tensile tests to relate the surface appearance with the deformation and fracture characteristics in the tensile testing. The properties of the sheared sheet is also influenced by the properties of the plastically deformed material in a region beneath the cut sheet surface. Micro-hardness testing of sheared samples was used by Weaver and Weinmann (1985) to study the deformation hardening in this region, and Dalloz et al. (2009) showed that a heavily deformed sub-surface region contains crack initiating voids.

The strains which arise in the sheet metal during shearing influence the properties of the sheared sheet. Furthermore, experimentally measured strains are a requisite for validation of newly developed shearing models, especially for initiation of fracture. Digital image correlation (DIC) is an established and frequently used technique for non-contact strain measurement. The fundamentals of DIC is covered by, for example, Sutton et al. (2009). This technique was used to study the local deformation during planar blanking by Stegeman et al. (1999), and by Tarigopula et al. (2008) to measure large plastic deformation during shear testing of dual phase steel. A method in which the metal flow line tilting angle was measured and converted to local shear strain, was used by Wu et al. (2012) to characterise the strain distribution within mechanically sheared edges of dual phase steels.

In this study the DIC-technique was used to evaluate the strain distribution on sheared sheets of a medium and a high strength steel, using various clearances and clamp configurations. The strain fields are presented at particular stages of tool penetration corresponding to a fixed tool displacement, crack initiation and final fracture. In addition to the strain fields, the force-displacement curves were determined for the various experimental conditions. Furthermore, a generic finite element simulation was done to compare a plain strain assumption with an actual three dimensional state at the surface, and hence, evaluate if surface strains can be used to approximate the bulk strains.

Finite element simulations

Finite element (FE) simulations were applied to compare the strains on the sheet surface and the interior of the sheet. If there is a good agreement between these strain fields, then a strain field measured on the sheet surface can be used also as a measure of the strains within the sheet. This would increase the relevance of the surface strain measurements since most of the shearing is of bulk material.

Model

FE analyses of shearing, modelled in both two dimensional (2D) plane strain and three dimensions (3D), were performed with a commercial general-purpose finite element software. Geometry and boundary conditions for the 2D plane strain model are described in detail by Gustafsson et al. (2014). The same conditions were applied to the 3D model. Both models were coarsely meshed except in the area where large deformations were expected and a denser mesh were required to resolve gradients in the state variables. Four-noded and fully integrated plane strain elements were used in the 2D model and eight-noded fully integrated selective reduced solid elements were used in the 3D model. Adaptive remeshing was used for the 2D model. Contacts were modelled with surface to surface formulations and both static and dynamic friction coefficients were 0.1.

Elastic tools and an isotropic elastic-plastic sheet material were used in both models. Poisson's ratio was 0.3 and Young's modulus was 210 GPa for all materials. The yield stress for the sheet material was described with an exponential hardening law according to Hollomon (1945), $\sigma_Y = K \bar{\varepsilon}_p^n$, where $\bar{\varepsilon}_p$ is the effective plastic strain and K and n are material specific parameters.

A constant tool acceleration of 100 mm s^{-2} was used in all simulations.

Strain comparison

Effective strains from FE simulations at the surface of a 3D simulation, the interior of a 3D simulation, and at plane strain are shown in Fig. 1. Due to element distortion in the 3D model, where no adaptive remeshing was applied, the strain field comparison was done at moderate tool displacement. The similar appearance of the strain fields in Fig. 1 suggest that the effective strain on the surface and inside the material are comparable. Plane strain was earlier shown to be a good approximation in terms of shearing forces by Gustafsson et al. (2014) for parallel tools and also for angled tools by Gustafsson et al. (2016a). This observation implies that strains on the sheet surface are representative for the interior strains and that plane strain simulations are once again a good approximation.

Measures of strain

Rate of deformation is the only strain rate measure that is energetically conjugate with the Cauchy stress. In a Cartesian coordinate system, the rate of deformation tensor is

$$D_{ij} = \frac{1}{2}\left(v_{i,j} + v_{j,i}\right), \tag{1}$$

where $v_{i,j}$ is the velocity gradient tensor written

$$v_{i,j}^{2D} = \begin{bmatrix} \frac{\partial v_1}{\partial x_1} & \frac{\partial v_1}{\partial x_2} \\ \frac{\partial v_2}{\partial x_1} & \frac{\partial v_2}{\partial x_2} \end{bmatrix} = \begin{bmatrix} \frac{\partial v_x}{\partial x} & \frac{\partial v_x}{\partial y} \\ \frac{\partial v_y}{\partial x} & \frac{\partial v_y}{\partial y} \end{bmatrix}, \tag{2}$$

for two dimensions. True (logarithmic) strain is related to rate of deformation as

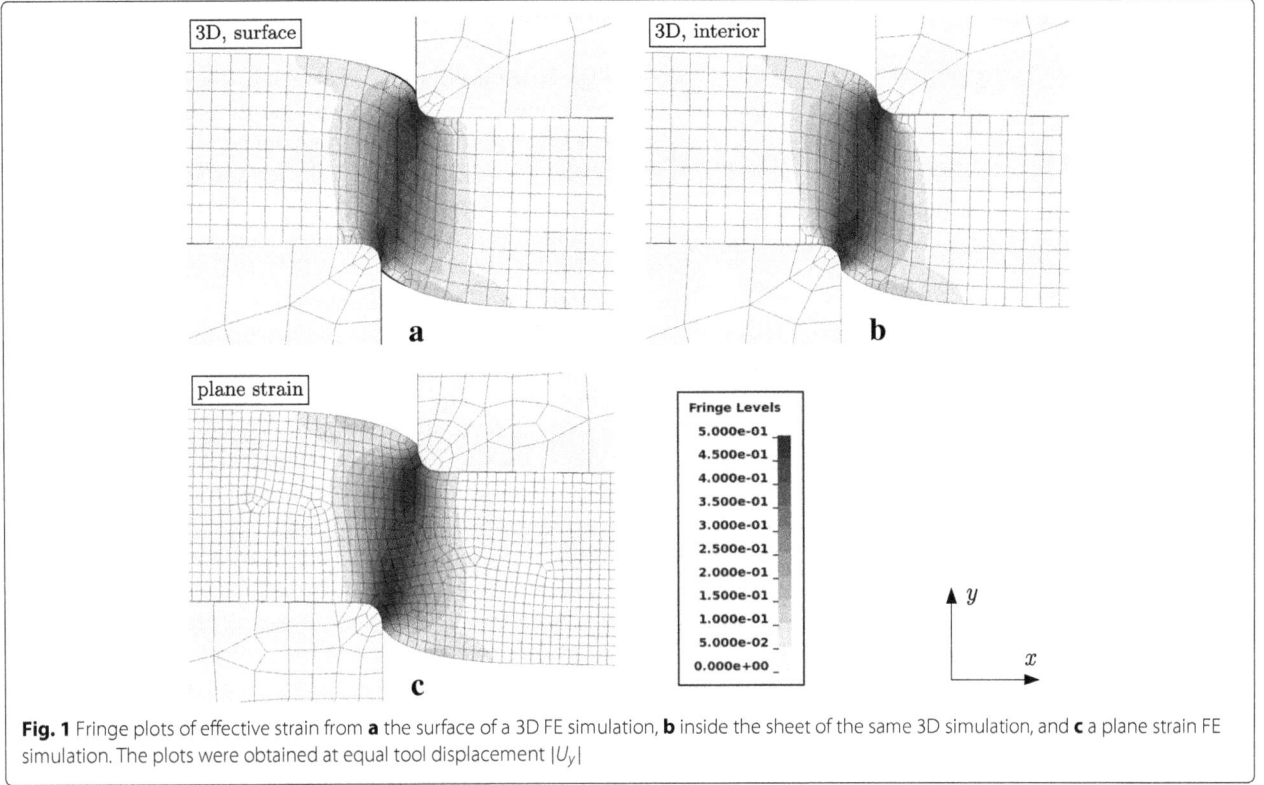

Fig. 1 Fringe plots of effective strain from **a** the surface of a 3D FE simulation, **b** inside the sheet of the same 3D simulation, and **c** a plane strain FE simulation. The plots were obtained at equal tool displacement $|U_y|$

$$\dot{\varepsilon}_{ij}^{\text{True}} = D_{ij}. \tag{3}$$

If rigid body rotations are present, the strain tensor may be more easily interpreted if expressed in the original orientation. In the original orientation the strain is

$$\dot{\varepsilon}_{ij}^{\prime\,\text{True}} = R_{mi}D_{mn}R_{nj}, \tag{4}$$

where R_{ij} is the rotation tensor. The rotation does not alter the strain levels and thus the effective strain,

$$\dot{\varepsilon}_{\text{eff}} = \sqrt{\frac{2}{3}\dot{\varepsilon}_{ij}\dot{\varepsilon}_{ij}}, \tag{5}$$

is unaffected.

When the displacements originate from a range of images taken of the deformation process, the continuous rate of deformation is unknown and the best approximation is a mean rate of deformation between two images. In that case it is more convenient to describe deformation in terms of a displacement gradient tensor $u_{i,j}$, for each interval between two images, instead of a mean velocity gradient tensor. In two dimensions

$$u_{i,j}^{\text{2D}} = \begin{bmatrix} \frac{\partial u_1}{\partial x_1} & \frac{\partial u_1}{\partial x_2} \\ \frac{\partial u_2}{\partial x_1} & \frac{\partial u_2}{\partial x_2} \end{bmatrix} = \begin{bmatrix} \frac{\partial u_x}{\partial x} & \frac{\partial u_x}{\partial y} \\ \frac{\partial u_y}{\partial x} & \frac{\partial u_y}{\partial y} \end{bmatrix}. \tag{6}$$

The Cauchy strain tensor for one increment is then

$$\Delta \varepsilon_{ij}^{\text{Cauchy}} = \frac{1}{2}\left(u_{i,j} + u_{j,i} \right), \tag{7}$$

and, if the increments are small enough, the sum over a number of increments is a good approximation of the true strain. The logarithmic Hencky strain for each increment may result in an even better approximation, but due to an unknown deformation path inside the increment, it is still an approximation. Hencky strain and the rotation tensor is obtained through a polar decomposition of the deformation gradient,

$$F_{ij} = \delta_{ij} + u_{i,j}, \tag{8}$$

where δ_{ij} is the Kronecker delta, into a rotation part and a stretch part $F_{ij} = R_{ik}U_{kj}$. The decomposition is accomplished by finding the eigenvalues $\lambda_{(n)}^2$ and normalised eigenvectors $N_i^{(n)}$ to the eigenvalue problem

$$\left(C_{ij} - \lambda^2 \delta_{ij} \right) N_i = 0, \tag{9}$$

where

$$C_{ij} = F_{ki}F_{kj} \tag{10}$$

is the right Cauchy-Green deformation tensor. Then, $\lambda_{(n)}$ are the principal stretches that act parallel to the $N_i^{(n)}$ directions. The number of eigenvalues and principal stretches n_D depends on the number of dimensions in the calculations; if the problem is two dimensional then $n_D = 2$. Further, the right stretch tensor is

$$U_{ij} = \sum_{n=1}^{n_D} \lambda_{(n)} N_i^{(n)} N_j^{(n)} \qquad (11)$$

and

$$R_{ij} = F_{ik} U_{kj}^{-1}, \qquad (12)$$

where U_{ij}^{-1} is the inverse of the right stretch tensor. The inverse of a tensor, in this case U_{ij}, can be calculated as

$$U_{ij}^{-1} = \frac{1}{2\epsilon_{rst} U_{r1} U_{s2} U_{t3}} \epsilon_{jmn} \epsilon_{ipq} U_{mp} U_{nq}, \qquad (13)$$

where ϵ_{ijk} is the Levi-Civita permutation called alternating tensor. The Hencky strain for the increment is therefore

$$\Delta\varepsilon_{ij}^{\text{Hencky}} = \sum_{n=1}^{n_D} \log \lambda_{(n)} N_i^{(n)} N_j^{(n)}, \qquad (14)$$

or rotated to the original configuration

$$\Delta\varepsilon_{ij}^{\prime\,\text{Hencky}} = R_{mi} \varepsilon_{mn} R_{nj}. \qquad (15)$$

Finally, the incremental effective strain is

$$\Delta\bar{\varepsilon} = \sqrt{\frac{2}{3} \Delta\varepsilon_{ij} \Delta\varepsilon_{ij}}, \qquad (16)$$

where $\Delta\varepsilon_{ij}$ is either the Cauchy or the Hencky strain tensor. These increments are summed to obtain accumulated effective strain.

Methods

Strains on the sheet xy-surface (labelled image area in Fig. 2) were obtained after a number of steps that can be summarised in: sample preparation, shearing of speckled sheet samples under high speed photography, digital image correlation, strain calculations and finally creating images to show the strain levels. The steps are explained further in the following subsections.

Materials and sample preparation

Two materials were used in the study: a medium strength (MS) construction steel, and a high strength (HS) wear plate steel. The mechanical properties are shown in Table 1 and the stress-strain curves in Fig. 3.

Samples with outer dimensions of $140 \times 150\,\text{mm}$ and a centred square, $100 \times 100\,\text{mm}$, hole were laser cut and machined from the sheet (Fig. 2). The outer xy-surface of the samples was prepared with high contrast random speckles at a characteristic size of a few image pixels. Abrasive blasting was applied to the surface and the rugged surface was sprayed with dull black paint. The surface was then ground with a fine abrasive paper to remove the paint on the most protruding parts of the steel surface to create a high contrast speckle pattern as shown in Fig. 4. Use of the traditional fine spray of dull black paint over a dull white coat was not applicable in this experiment due to limited adhesion and large shear strains in the sheet.

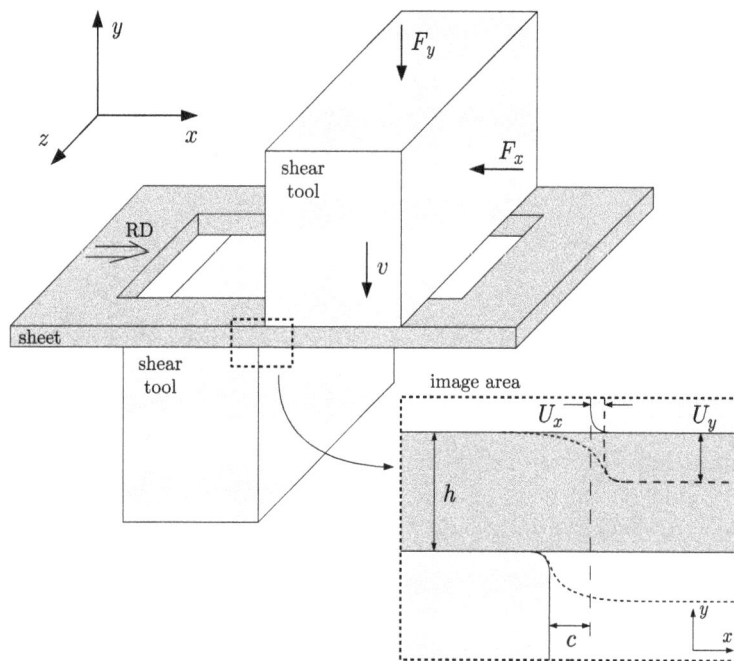

Fig. 2 Schematic representation of the rectangular sheet sample, with square hole, and tools together with definition of the coordinate system used. Reaction forces on the moving tool as result of the velocity v are F_x and F_y. Definitions of sheet thickness h, tool clearance c, and tool displacements U_x and U_y are shown in the magnified area. A speckle pattern was applied to the sheet xy-surface in the indicated image area. RD indicates the rolling direction

Table 1 Mechanical properties, evaluated from uniaxial tensile tests, of sheet metal grades used in the study. The thickness range of the sheared samples is also shown

Material strength	Yield strength, Rp02 [MPa]	Tensile strength, Rm [MPa]	Elongation, A80 [%]	Thickness [mm]
Medium	450	520	25	5.97–6.03
High	1080	1260	7	6.11–6.15

Shearing experiments

The rectangular sheet samples were sheared in a set-up with two parallel tools as schematically shown in Fig. 2. The speckle pattern on the sheet surface was imaged by a camera during the entire shearing process. Accurate measurements of tool displacements and shearing forces, and also stable tool clearances, were provided by the symmetric set-up illustrated in Fig. 5. Details on the tool displacement measurements are covered by Gustafsson et al. (2016a). The initial development of the symmetric experimental set-up, including details of the force measurements, are described by Gustafsson et al. (2014).

The various combinations of clearances and clamping configurations which were used for shearing the sheets are shown in Table 2. Two experiments were performed for each indicated combination in Table 2.

High-speed photography and image analysis

A camera equipped with a lens with 200 mm focal length, was used to obtain an image of the speckle pattern within an image area of 15×12 mm (see also Fig. 2). Images with a resolution of 1280×1024 pixels where captured at 1 kHz with about 200 µs exposure time. At this resolution and sample rate, the camera allowed approximately 1.6 s long image sequences which were sufficient for the shearing

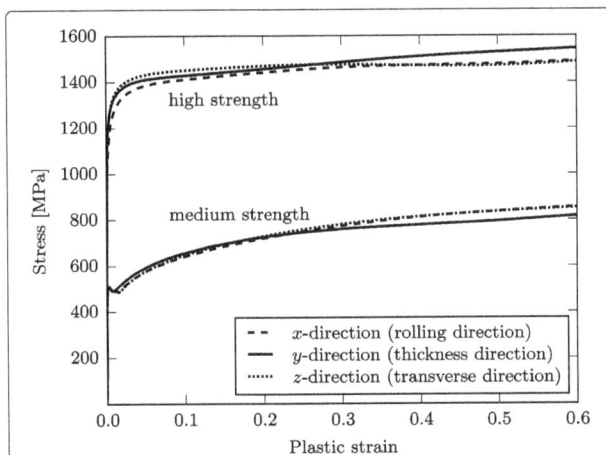

Fig. 3 Flow stress curves from uniaxial compression tests by Gustafsson et al. (2016b) in three directions for the two material grades. The x-, y-, and z-directions are defined in Fig. 2

process. A total of about 1600 images were hence captured for each shearing experiment (combination of material grade, clearance and clamping configuration). From each set of 1600 images, about 20 to 50 images were selected for further analysis as described in the next paragraph. The images were selected at regular intervals corresponding to a tool displacement of about 10 pixels (~0.1 mm), so the number of images for analysis depended on the tool displacement before fracture. Most of the selected images were from the last part of the 1600 images since the tool velocity increased from shearing start to final fracture. The maximum camera speed was only needed for the final part of the shearing. Due to the distinct final fracture, and the following rapid tool displacement, the image sequences was easily manually synchronised in time to the recorded tool displacements and forces.

Digital image correlation (DIC) was applied to the images that were selected as described above. In this technique, two images are compared and analysed using image analysis algorithms to compute parameters such as the displacements of a solid body. For a detailed account of DIC, the reader is referred to a textbook on this topic, such as Sutton et al. (2009). In short, points in the first image are located in the second image by correlation of a surrounding rectangular area of eligible size called interrogation window and returned as a displacement field. The number of vectors in the displacement field equals the number of selected points for analysis. Here, the analysed points were positioned in an equally spaced rectangular matrix, eight pixels apart (approximately 0.1 mm on the sheet sample). The imaged object should be prepared with a random speckle pattern of appropriate speckle size so that each interrogation window contain a unique image.

Because of the large deformations, the images were analysed consecutively (1+2, 2+3, 3+4, ...) which resulted in incremental displacement fields. The application if the DIC technique was performed with the open source JPIV software (JPIV r24, available under terms of GNU GPL, http://www.jpiv.vennemann-online.de).

The resulting displacement fields were used to calculate strain fields according to the following subsection.

Evaluation of strain

The displacement fields were used to calculate the effective strain (Eq. (16)) through either the Cauchy strain tensor (Eq. (7)) or the Hencky strain tensor (Eq. (14)). The strain tensor was calculated for each increment and the effective strain for each increment was added to the previous accumulated strain in material coordinates (Lagrange). The pixel size in the images of the strain fields was about 0.1 mm. In these images, strains were shown as mean values over 3×3 pixels which means that the effective length of measurement was about 0.3 mm.

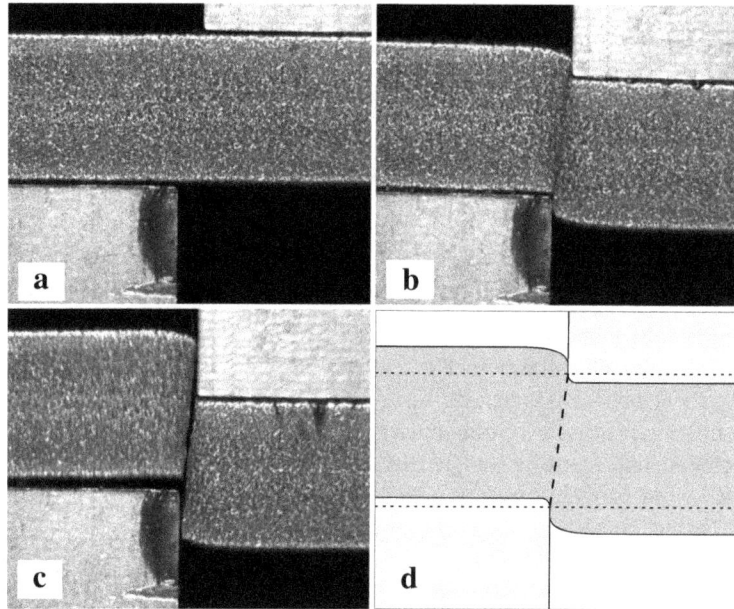

Fig. 4 Typical speckle pattern on the sheet strip *xy*-surface: **a** before shearing start, **b** just before final fracture, **c** final fracture immediately after **b**. The images were captured by the high-speed camera when shearing the high strength material at 0.15*h* clearance and with two clamps. Note that the dark area on the lower tool is spilled grease and no tool defect. Panel **d** shows a schematic figure of the shearing. Strains were plotted along the thick dashed line and the largest strain found between the two thinner horizontal dashed lines was tabulated

A comparison between accumulated effective strain fields obtained from the Cauchy strain tensor and the Hencky strain tensor is presented in the results section. Since there was no apparent difference between the strain fields, only the more commonly used Cauchy strain is presented later in the paper. The accumulated strain fields are presented at various stages of the shearing to allow comparison of various parameters. The strain field images are an efficient tool to visualise the appearance of the strain distribution, but the actual strain value is often difficult to read from the fringe plots. Therefore, the strains on a straight line between the lower and the upper tool edges as well as the maximum strain in the shear area were evaluated. The line was positioned with its ends at the transition between the curved and the vertical tool contours as schematically shown in Fig. 4d. The maximum strains were restricted to the area between the two horizontal lines shown in Fig. 4d.

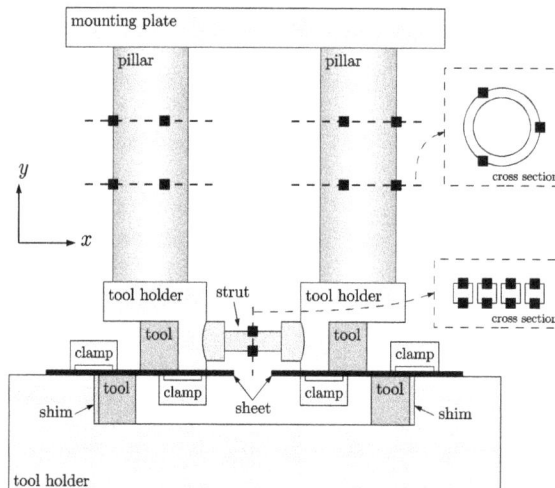

Fig. 5 Schematic front view of the experimental set-up showing strain gauge positions as *black squares* (Gustafsson et al. 2014). Sheet strips are clamped on both sides

Table 2 Combinations of clearances, material grades and clamping configurations used for shearing the sheet samples. The tool clearance is expressed as a fraction of the sheet thickness h

| Clearance | Medium strength | | High strength | |
	1 clamp	2 clamps	1 clamp	2 clamps
0.05h	x	x		
0.15h	x	x	x	x
0.25h	x	x	x	x

Results

Shearing experiments

Forces, F_x and F_y, versus tool displacement, $|U_y|$, from the shearing experiments on the medium strength and high strength materials are shown in Fig. 6. The force-displacement curves are presented to indicate the range of forces and displacements for which strains were evaluated (see following subsections for these results). Among

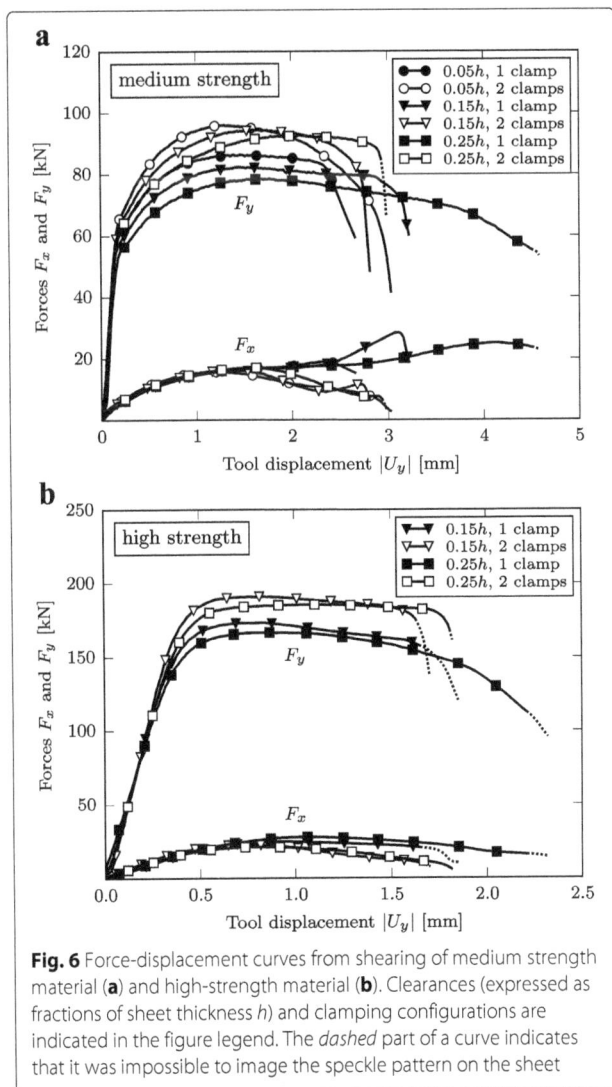

Fig. 6 Force-displacement curves from shearing of medium strength material (**a**) and high-strength material (**b**). Clearances (expressed as fractions of sheet thickness h) and clamping configurations are indicated in the figure legend. The *dashed* part of a curve indicates that it was impossible to image the speckle pattern on the sheet

the two symmetric sides of the set-up and the two experiments performed for each configuration, the difference in force was less than 1% and $|U_y|$ varied a few percent at the point of fracture. These variations were smaller than the expected accuracy in the strain measurements, and therefore, strains were only calculated from one series of images for each combination of shearing parameters and material. Strains were evaluated from shearing start to short before fracture, that is, during practically the whole shearing process. For some samples, however, the shearing velocity was to large to allow sharp image capturing and strain evaluation. In those cases it was impossible to record images during the later stages of the shearing, as indicated by the dotted force-displacement curves in Fig. 6.

Comparison of strain measures

There was no obvious visual difference between the Cauchy strain and the Hencky strain, as exemplified in Fig. 7. If, however, the difference $\bar{\varepsilon}^{\text{Cauchy}} - \bar{\varepsilon}^{\text{Hencky}}$ was calculated, then a difference between the strain measures of up to about 0.2 was revealed as shown in Fig. 7e and f. This is still a small difference compared with the strain levels: the strain difference 0.2 corresponds to a relative difference of about 7%. The distribution of the differences seemed random. All other material grades, clearances and clamp configurations displayed differences in the same range as shown in Fig. 7e, f.

Strain field measurements

Accumulated strains over the shearing cycle from start to short before fracture, as indicated by the end of the solid lines in Fig. 6, are shown for the medium and high-strength materials in Figs. 8 and 9, respectively. Most of the strain fields showed a thin band of large strain across the sheet. The strain band across the sheet was most straight for the 0.15h clearances; the curvature increased for both smaller and larger clearances. With two clamps the strain band was mostly symmetric. Shearing with one clamp resulted in a large strain concentration at the lower tool and a wider strain band as compared with shearing with two clamps. The extra width of the strain band appeared on the unclamped side of a line between the tool edges. In general, the strains before final fracture were larger in the medium strength material compared with the high strength material. Most samples had one or two visible cracks at the tool edges well before final fracture. At 0.05h clearance and with two clamps, shown in Fig. 8b, the cracks propagated so that they circumvent each other with large deformation in the material between the cracks before final fracture as result.

Strains on a straight line from the lower to the upper tool edge, as defined in Fig. 4d, were evaluated and plotted in Fig. 10. The line is indicated in the sheared samples by the dashed lines in Figs. 8 and 9. While the strain levels were

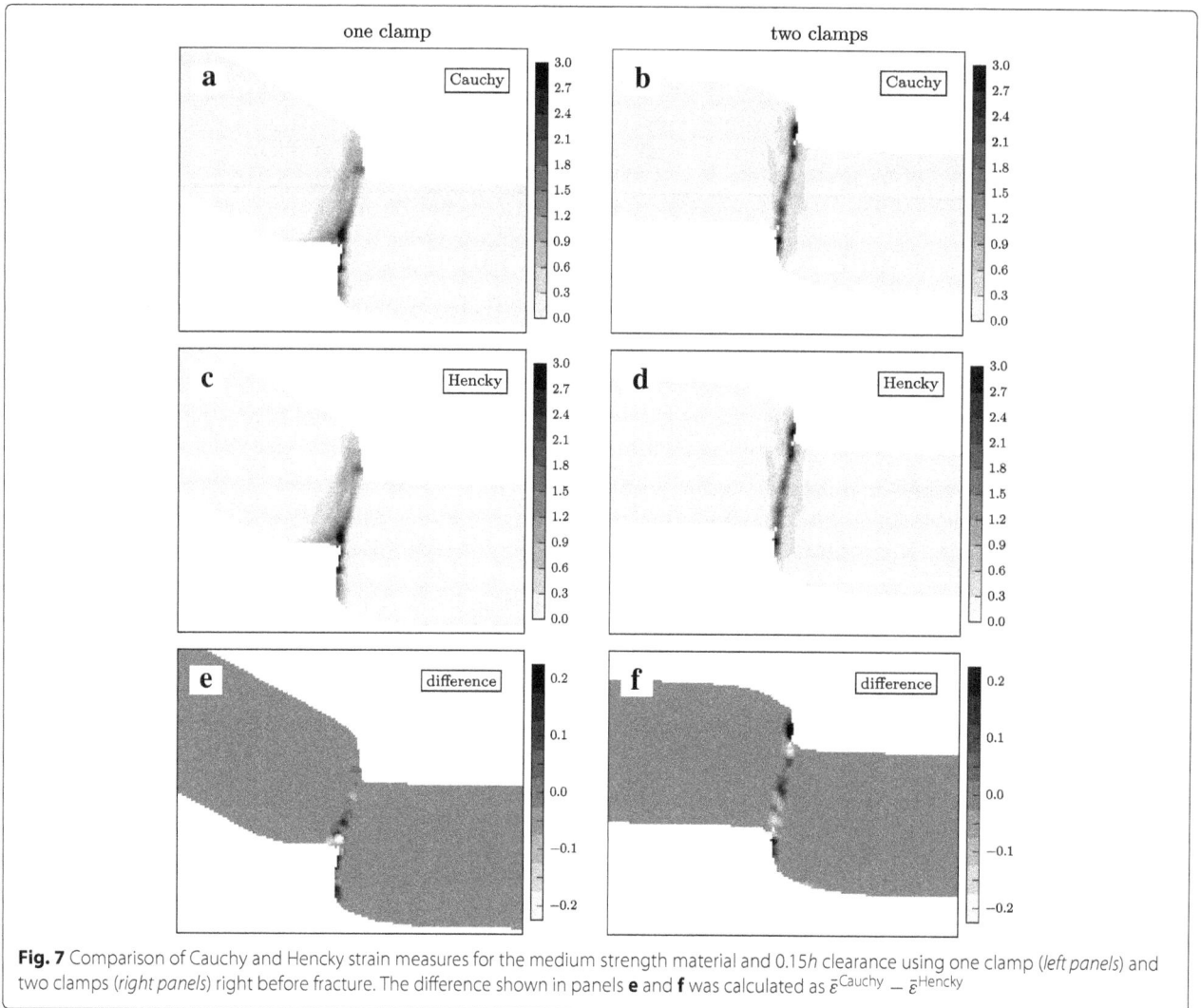

Fig. 7 Comparison of Cauchy and Hencky strain measures for the medium strength material and 0.15h clearance using one clamp (*left panels*) and two clamps (*right panels*) right before fracture. The difference shown in panels **e** and **f** was calculated as $\bar{\varepsilon}^{\text{Cauchy}} - \bar{\varepsilon}^{\text{Hencky}}$

easy to interpret, the line gave little information of the strain distribution. In case of one clamp (Fig. 10a, c), the strain concentration at the lower tool was captured, but the line then continues mostly outside of the band with elevated strain levels. With two clamps (Fig. 10b, d) the line mostly coincides with the high strain band, especially for the 0.05h and 0.15h clearances. At 0.25h clearance the strain band was more s-shaped and a peak on the strain curves was shown at each of the multiple intersections between the line and the strain band. Part of the curves, in the beginning or end of the line, may be missing if the material had a visible crack underneath as, for example, the beginning of the curve for 0.15h clearance in Fig. 10c.

The medium strength material sheared at 0.05h clearance with two clamps showed large strains in an area about halfway between the lower and upper tool edges (Fig. 10b). This was due to the observed behaviour where cracks propagate from each tool so that they wrapped around each other as seen in Fig. 8b. The remaining

material between the cracks is sheared to a much larger tool displacement $|U_y|$ and thus larger strains compared with the earlier fractured material at the sheet xz-surfaces. Also compared with samples sheared at larger clearance, where the cracks are more prone to meet in the middle, the strain at final fracture is much larger (compare Fig. 8b, d, and f).

The maximum strains, $\bar{\varepsilon}_{\text{final}}$, were determined and are shown in Table 3 together with corresponding tool displacement $|U_y|_{\text{final}}$. The positions of the maximum strains are indicated by white circles in Figs. 8 and 9. For all experimental conditions in Figs. 8 and 9, except Figs. 8e and 9c (0.25h clearance and one clamp for both material strength levels), the position of the maximum strain fell, approximately, on the line between the tool edges along where strains were evaluated.

When there was a visible crack before the final fracture, also the strains at crack initiation are of interest to compare. Therefore, the maximum strains at the first

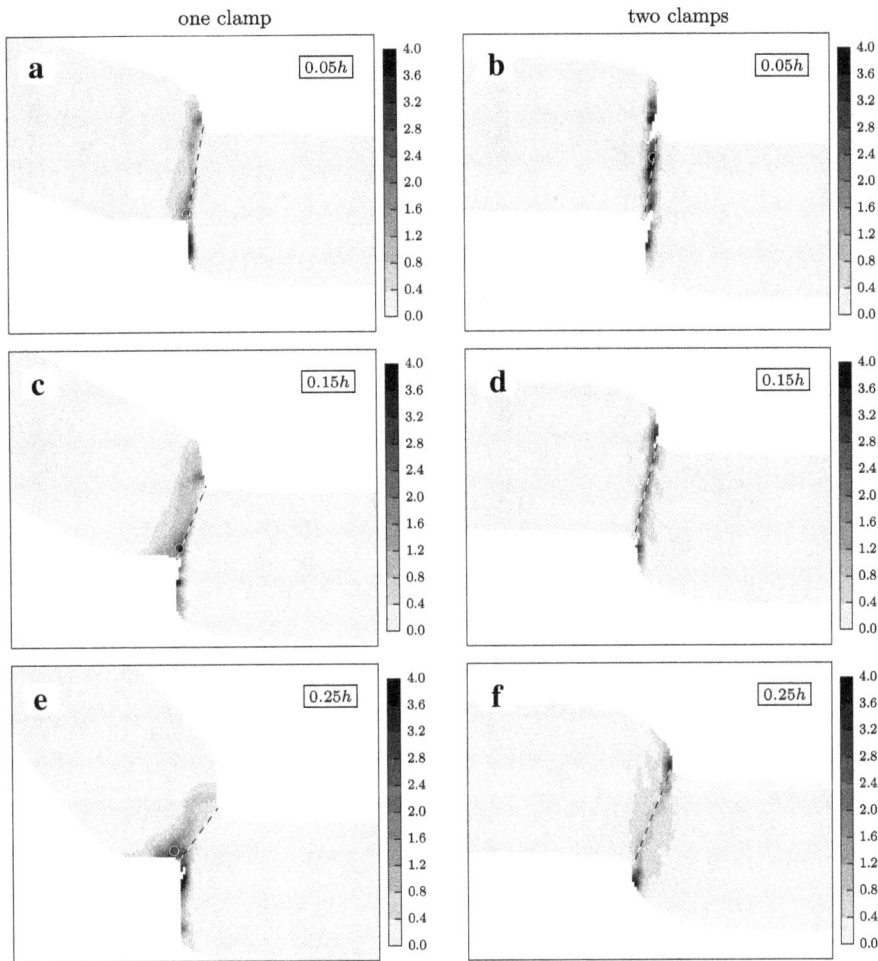

Fig. 8 Effective Cauchy strain fields (Eqs. 5 and 7) short before final fracture from shearing of the medium strength material using one clamp (*left panels*) and two clamps (*right panels*). Clearances, expressed as fractions of sheet thickness h, are indicated in each panel. Strains along the *dashed black and white line* are plotted in Fig. 10. The positions of the maximum strains in the shear bands are indicated by *white circles*, and this maximum strain is shown in Table 3

appearance of a crack on the imaged sheet surface (crack initiation) and the corresponding applied tool displacements are shown in Table 3 as $\bar{\varepsilon}_{\text{init}}$ and $|U_y|_{\text{init}}$, respectively. Strain fields and line plots for these tool displacements are shown in Figs. 11, 12 and 13, in the same way as earlier shown at final fracture. Here, the maximum strain is located close to the drawn line for all combinations of materials and shearing parameters, as shown by the dashed lines and white circles in Figs. 11 and 12. When the high strength material was sheared with two clamps and 0.25h clearance the lapse from crack initiation to fracture was very short. Initiated cracks were hard to distinguish from the preceding narrow shear bands, which made the exact moment of crack initiation, and thus also $\bar{\varepsilon}_{\text{init}}$ in Table 3, somewhat hard to determine. Overall, the fracture strains were larger for the medium strength material compared with the high strength material, and

increased with increased tool clearance. With two clamps, however, the maximum strain at final fracture, $\bar{\varepsilon}_{\text{final}}$, decreased with increased tool clearance.

With exception of $\bar{\varepsilon}_{\text{final}}$ for the medium strength material, 0.05h with two clamps, and 0.25h with one clamp, the maximum strains presented in Table 3 are at such positions on the sheets where they are comparable to the strain at the crack tip and thus they are approximations of the fracture strain. The fracture strains are therefore larger at the final fracture compared with at crack initiation.

Figures 8, 9, 11, and 12 show strain at various tool displacements for all combinations of shearing parameters. A comparison of the strains at a given tool displacement, same for all combinations of shearing parameters, can also be of interest. The sample with the least tool displacement at fracture was the high strength material at 0.15h

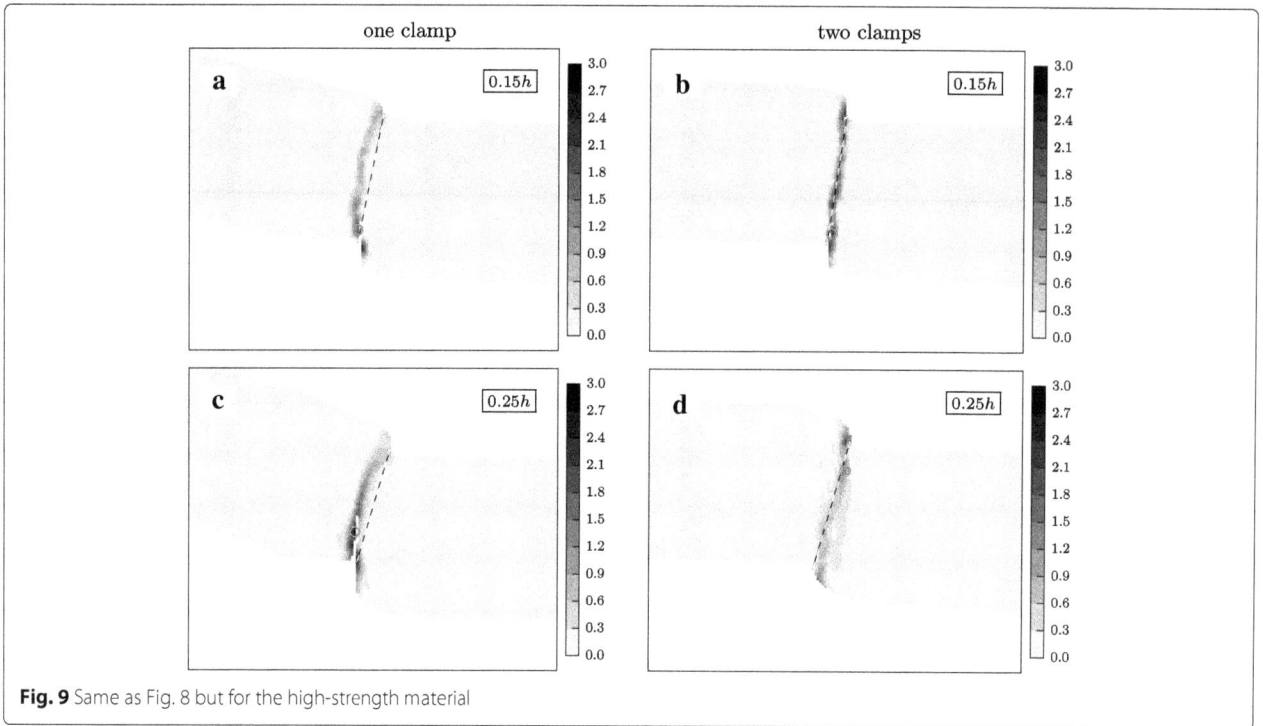

Fig. 9 Same as Fig. 8 but for the high-strength material

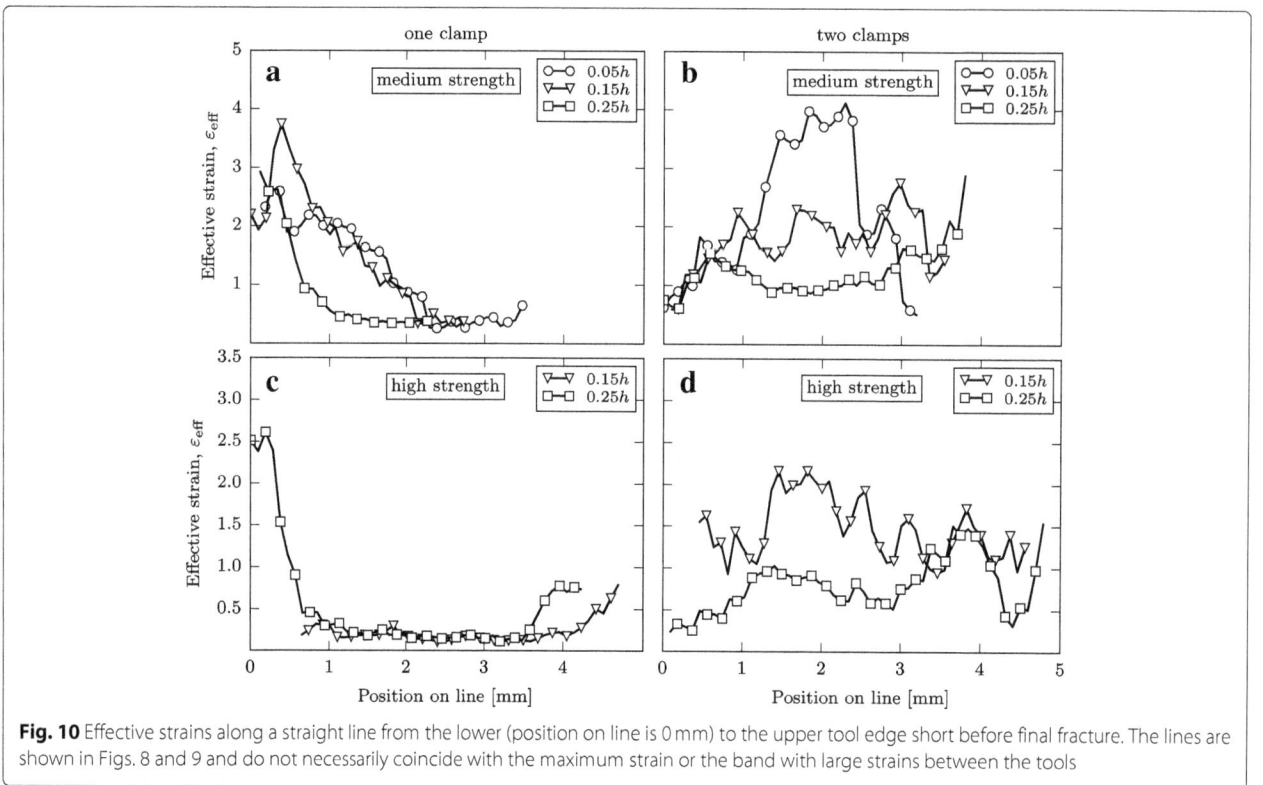

Fig. 10 Effective strains along a straight line from the lower (position on line is 0 mm) to the upper tool edge short before final fracture. The lines are shown in Figs. 8 and 9 and do not necessarily coincide with the maximum strain or the band with large strains between the tools

Table 3 Maximum strains at crack initiation $\bar{\varepsilon}_{init}$ and at final fracture $\bar{\varepsilon}_{final}$, and corresponding tool displacements, $|U_y|_{init}$ [mm] and $|U_y|_{final}$ [mm], respectively

| | Medium strength | | | | | | High strength | | | |
| | 1 clamp | | | 2 clamps | | | 1 clamp | | 2 clamps | |
	0.05h	0.15h	0.25h	0.05h	0.15h	0.25h	0.15h	0.25h	0.15h	0.25h
$\bar{\varepsilon}_{init}$	1.7	1.7	3.0	1.1	1.2	1.4	0.7	0.6	0.6	1.3
$|U_y|_{init}$	2.0	2.0	4.0	1.8	2.1	2.8	1.2	1.6	1.1	1.8
$\bar{\varepsilon}_{final}$	2.6	3.7	3.2	4.1	2.8	1.8	2.1	2.9	2.4	1.5
$|U_y|_{final}$	2.6	3.2	4.6	3.0	2.8	3.0	1.8	2.3	1.7	1.8

clearance at a tool displacement of 1.6 mm. Therefore, strains at a tool displacement of 1.6 mm, for all sheared combinations for the medium and high strength materials, are shown in Figs. 14 and 15, respectively. Since the medium strength material fractured at a tool displacement much larger than 1.6 mm, the strain fields for this material was also evaluated at a tool displacement of 2.6 mm and are shown in Fig. 16. In all cases, the strains at a given tool displacement were larger at small clearances than at large clearances. At 1.6 mm tool displacement the strains were considerably smaller in the medium strength material compared with the high-strength material due

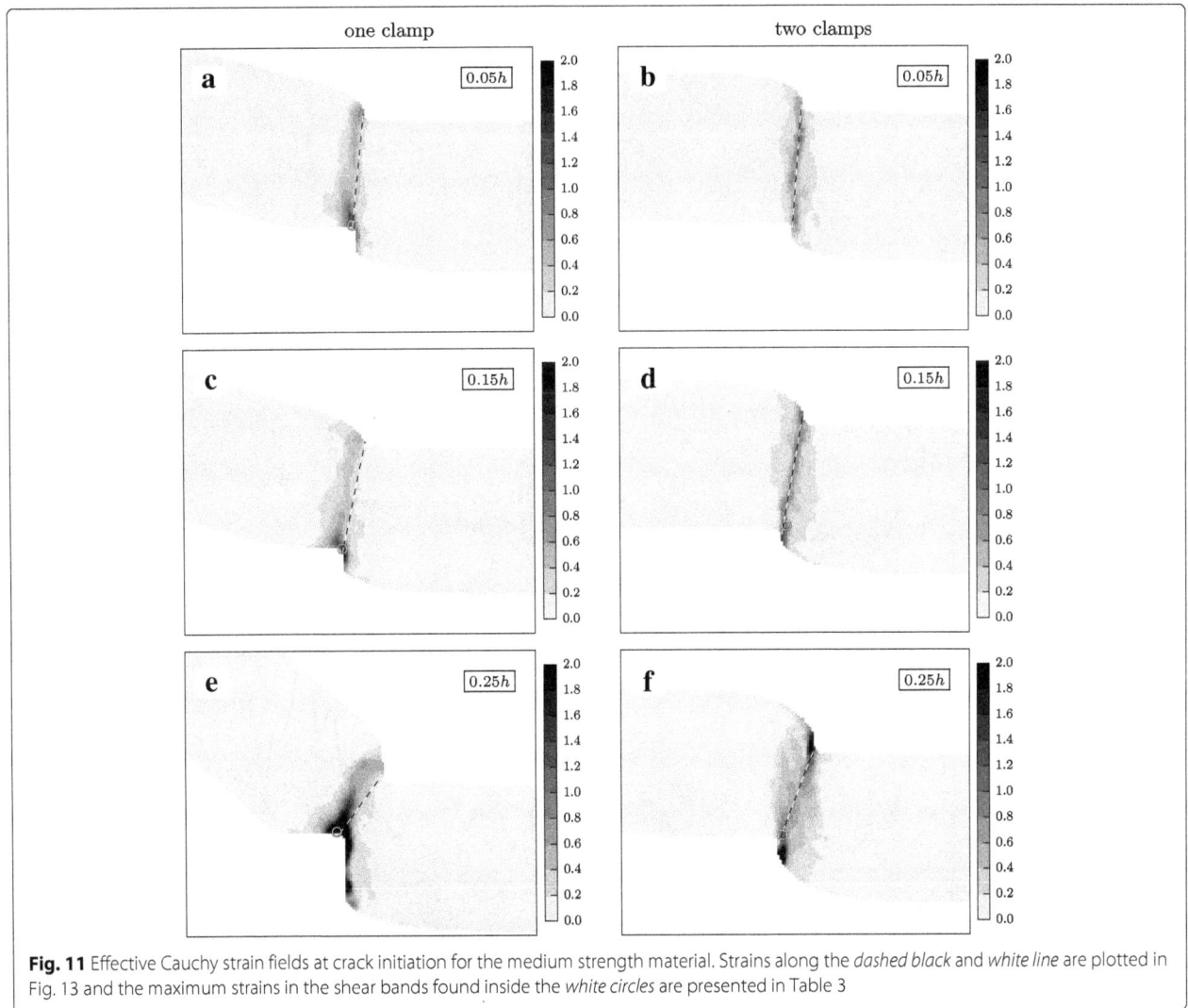

Fig. 11 Effective Cauchy strain fields at crack initiation for the medium strength material. Strains along the *dashed black* and *white line* are plotted in Fig. 13 and the maximum strains in the shear bands found inside the *white circles* are presented in Table 3

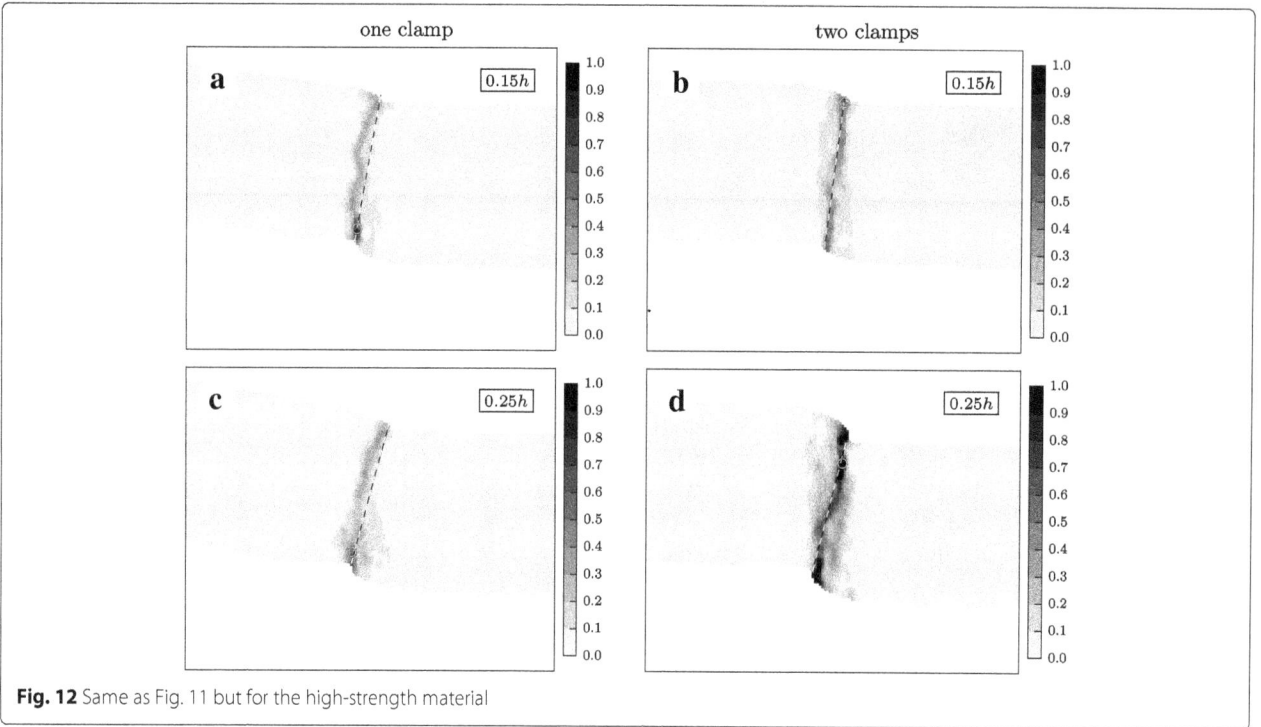

Fig. 12 Same as Fig. 11 but for the high-strength material

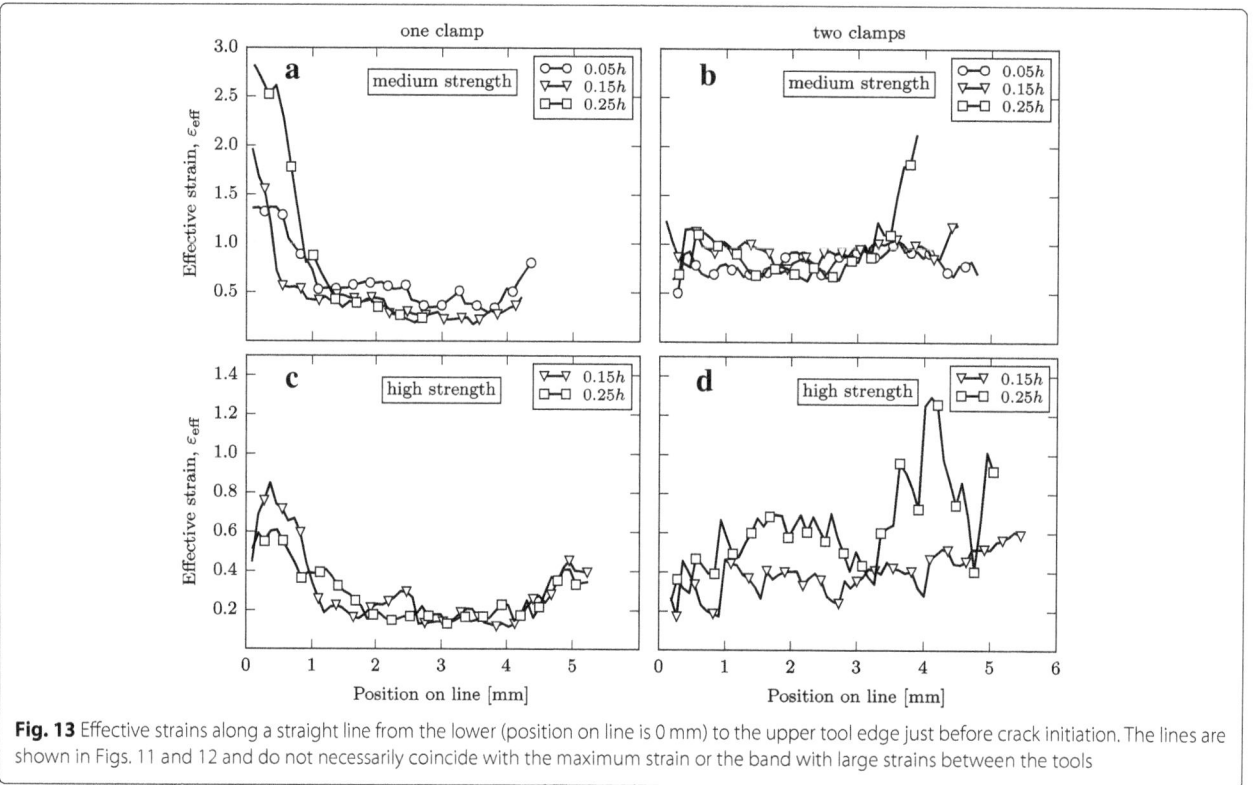

Fig. 13 Effective strains along a straight line from the lower (position on line is 0 mm) to the upper tool edge just before crack initiation. The lines are shown in Figs. 11 and 12 and do not necessarily coincide with the maximum strain or the band with large strains between the tools

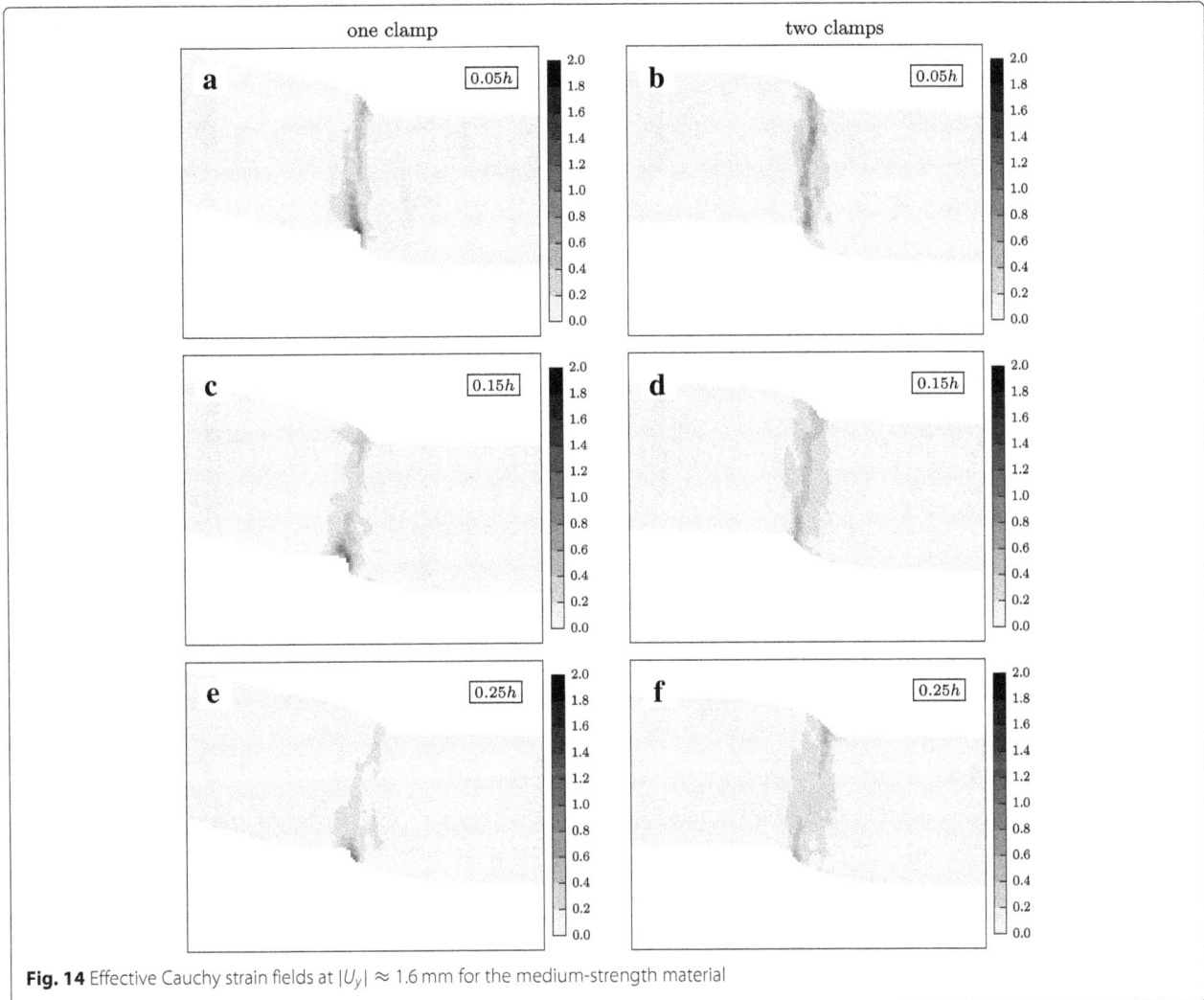

Fig. 14 Effective Cauchy strain fields at $|U_y| \approx 1.6$ mm for the medium-strength material

to a much wider strain band (larger deformed volume) in the medium-strength material. The medium-strength material is also more ductile and withstands larger strains before fracture.

Discussion

Comparison of effective strains from plane strain simulation, with strains on the strip surface in 3D simulations suggest that plane strain simulations are sufficient to describe the strain distribution in sheet metal shearing. The slightly larger strain in the middle of the sheared area seen in the plane strain plot (Fig. 1c) compared with the 3D strain plot (Fig. 1a and b) can be the result of inconsistent element sizes. The plane strain elements have half the size of the 3D elements and, thus, better resolve the large strain gradients.

The sheet samples were unintentionally delivered with bevelled edges (0.2–0.3 mm), which are seen as black borders between the speckled sheet surface and the tools in Fig. 4a and b; in panel c the black border is thicker since unloading of elastic energy caused the fractured sheet to lift from the tool. Edge effects influence the result of the digital image correlation technique since some part of the analysed area extends outside the speckled surface. This bevel on the edges causes additional wasted information and can also locally affect the strains in the speckled surface since there is no contact between the tools and the bevelled edges. The purpose of the hole in the samples was to decrease the forces to similar levels as previously obtained during shearing in the experimental set-up (Gustafsson et al. 2016b; Gustafsson et al. 2016a) with a maintained image area and visibility of the imaged surface.

There was a good agreement between the Cauchy and Hencky strain tensors in all calculations. Furthermore, the final accumulated strains were relatively insensitive to variations in the number of strain increments. This number was chosen as a compromise between few increments with larger image distortion, and many increments

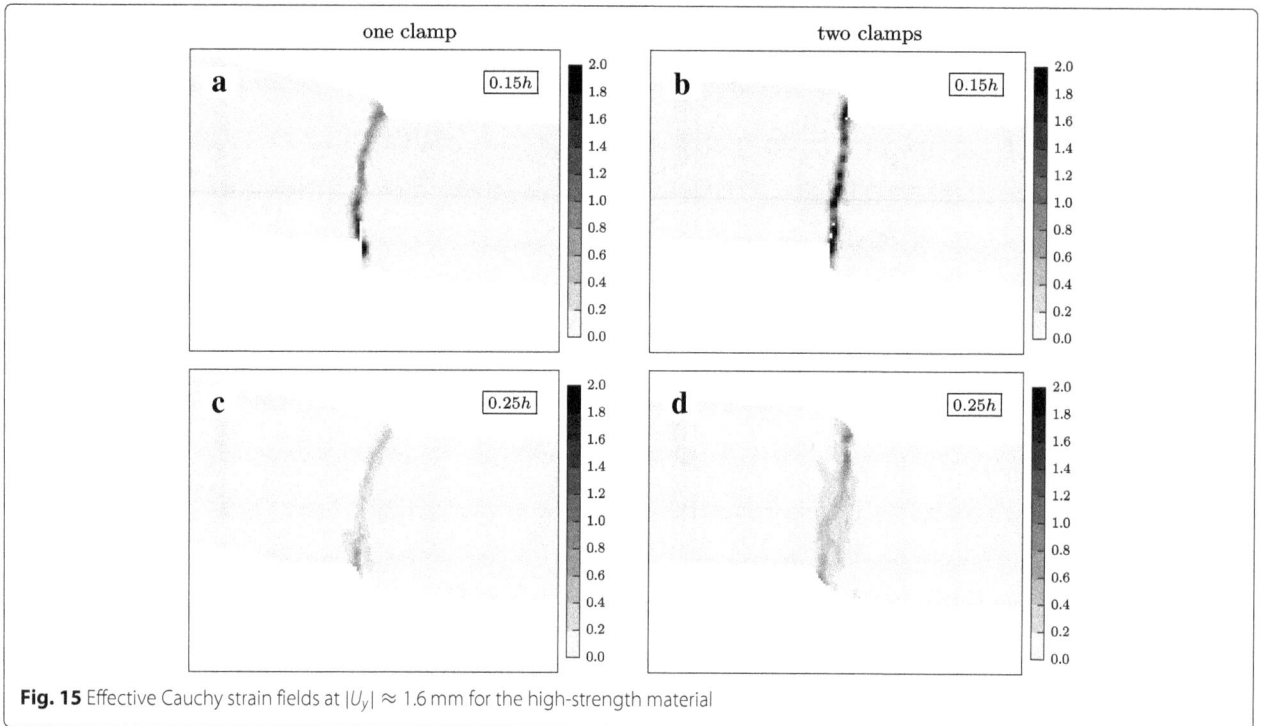

Fig. 15 Effective Cauchy strain fields at $|U_y| \approx 1.6\,\text{mm}$ for the high-strength material

with more accumulated noise. The maximum number of increments was limited by the image capturing speed; the minimum number of increments was limited by the image correlation algorithm. With too few increments, there were unresolved areas around the narrow bands of largest strain due to high image distortion. The agreement between the Cauchy and Hencky strains indicate that the strain increments used were small enough for the accumulated strain to be close to the true strain.

In general, $\bar{\varepsilon}_\text{final}$ was larger than $\bar{\varepsilon}_\text{init}$ which means that the first part of the crack (close to the tool) is formed at a smaller strain than the strain needed for the last part of the crack propagation before final fracture. This can be the result of a varying stress state along the shear band, but, most probably, it is an effect of a limited resolution in the measurements. According to Dalloz et al. (2009), the depth of a heavily deformed zone is 0.2 mm when shearing a 1.5 mm thick sheet. That zone is probably larger in this study on 6 mm thick sheet, but still not much larger than the effective measurement length of 0.3 mm. Therefore, both $\bar{\varepsilon}_\text{init}$ and $\bar{\varepsilon}_\text{final}$, but possibly $\bar{\varepsilon}_\text{init}$ to a larger extent, are probably underestimated in the measurements. Due to the nature of fracture, as an extremely local phenomenon, the resolution problem is inevitable, at least with the used technology. With knowledge of the resolution, however, the results are still useful.

Strains at the sheet edges that are in contact with the tools are in some extent influenced by relative movement in the contact region, since the interrogation windows at the sheet edges extends outside the sheet. The effects are largest at the vertical contact that is created during shearing, since the relative movement is largest there.

Conclusions

The digital image correlation (DIC) technique was used for reliable measurements of strains during sheet metal shearing. FE simulations show that strains on the sheet strip surface are comparable with strains inside the material and strains from a plane strain approximation. That observation extends the benefit of surface strain measurements as they are also relevant for the bulk material. The presented strains are useful for future validation of shearing models, and, coupled with a material model, to establish a fracture stress and to calibrate a fracture criteria. Some conclusions based directly on the strain measurements are:

- Most of the deformation in the sheet during shearing was concentrated to a "band" between the tools. The strain band was most straight at $0.15h$ clearance. The band was, however, slightly thinner at $0.05h$ clearance (medium strength material) especially when shearing with one clamp.
- With two clamps the strain band was almost symmetric, whereas with one clamp the strain band

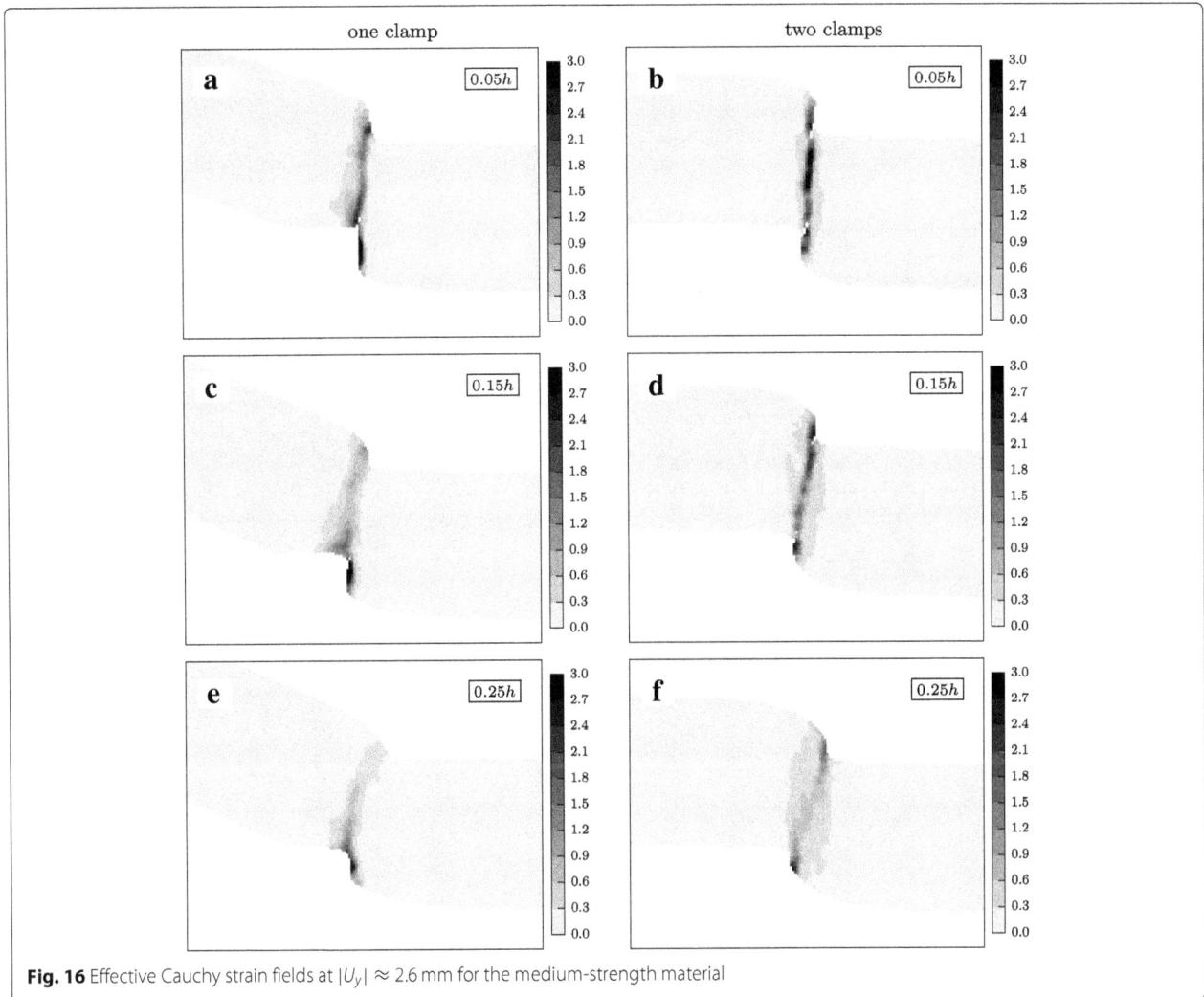

Fig. 16 Effective Cauchy strain fields at $|U_y| \approx 2.6$ mm for the medium-strength material

was shifted towards the free strip end, and a large strain concentration was formed around the tool on the free side.

- Strains were larger in the high strength material compared with the medium strength material at the same tool displacement due to a wider strain band (larger deformed volume) in the medium strength material.

- The fracture strain was larger for the medium strength material compared with the high strength material and increased with increasing clearance.

Acknowledgements
This work was performed at Dalarna University within the Swedish Steel Industry Graduate School, with financial support from Högskolan Dalarna (Dalarna University), Jernkontoret (Swedish Steel Producers' Association), Länsstyrelsen i Gävleborg (County Administrative Board of Gävleborg), Region Dalarna (Regional Development Council of Dalarna), Region Gävleborg (Regional Development Council of Gävleborg), Sandvikens kommun (Municipality of Sandviken) and SSAB. Carl-Axel Norman is acknowledged for building the experimental set-up and preparing the material. SSAB provided the shearing tools and sheet metal samples.

Authors' contributions
EG conducted the experiments, collected the data, and wrote the major part of the paper. MO and LK are EG's PhD supervisors, who guided and supported this work and contributed with their expertise and advices. All authors read and approved the final manuscript.

Competing interests
The authors declare that they have no competing interests.

Author details
[1] Dalarna University, SE-791 88 Falun, Sweden. [2] Luleå University of Technology, SE-971 87 Luleå, Sweden.

References
Crane, EV (1927). What happens in shearing metal. *Machinery, 30*, 225–230.
Dalloz, A, Besson, J, Gourgues-Lorenzon, AF, Sturel, T, Pineau, A (2009). Effect of shear cutting on ductility of a dual phase steel. *Engineering Fracture Mechanics, 76*(10), 1411–1424.
Gustafsson, E, Oldenburg, M, Jansson, A (2014). Design and validation of a sheet metal shearing experimental procedure. *Journal of Materials Processing Technology, 214*(11), 2468–2477.

Gustafsson, E, Karlsson, L, Oldenburg, M (2016a). Experimental study of forces and energies during shearing of steel sheet with angled tools. *International Journal of Mechanical and Materials Engineering, 11*(1), 10. doi:10.1186/s40712-016-0063-1.

Gustafsson, E, Oldenburg, M, Jansson, A (2016b). Experimental study on the effects of clearance and clamping in steel sheet metal shearing. *Journal of Materials Processing Technology, 229*, 172–180.

Hollomon, JH (1945). Tensile deformation. *Transactions of the American institute of mining and metallurgical engineers, 162*, 268–290.

Izod, EG (1906). Behaviour of materials of construction under pure shear. *Proceedings of The Institution of Mechanical Engineers, 70*, 5–55.

Kopp, T, Stahl, J, Demmel, P, Tröber, P, Golle, R, Hoffmann, H, Volk, W (2016). Experimental investigation of the lateral forces during shear cutting with an open cutting line. *Journal of Materials Processing Technology, 238*, 49–54.

Stegeman, YW, Goijaerts, AM, Brokken, D, Brekelmans, WAM, Govaert, LE, Baaijens, FPT (1999). An experimental and numerical study of a planar blanking process. *Journal of Materials Processing Technology, 87*, 266–276.

Sutton, MA, Orteu, JJ, Schreier, H (2009). *Image correlation for shape, motion and deformation measurements*: Springer. ISBN:0387787461.

Tarigopula, V, Hopperstad, OS, Langseth, M, Clausen, AH, Hild, F, Lademo, OG, Eriksson, M (2008). A study of large plastic deformations in dual phase steel using digital image correlation and FE analysis. *Experimental Mechanics, 48*(2), 181–196.

Weaver, HP, & Weinmann, KJ (1985). A study of edge characteristics of sheared and bent steel plate. *Journal of Applied Metalworking, 3*(4), 381–390.

Wu, X, Bahmanpour, H, Schmid, K (2012). Characterization of mechanically sheared edges of dual phase steels. *Journal of Materials Processing Technology, 212*, 1209–1224.

Yamasaki, S, & Ozaki, T (1991). Shearing of inclined sheet metals: Effect of inclination angle. *JSME international journal Ser 3, Vibration, control engineering, engineering for industry, 34*(4), 533–539.

14

Parametric optimization of MRR and surface roughness in wire electro discharge machining (WEDM) of D2 steel using Taguchi-based utility approach

M. Manjaiah[1*], Rudolph F. Laubscher[1], Anil Kumar[2] and S. Basavarajappa[3]

Abstract

Background: This paper reports the effect of process parameters on material removal rate (MRR) and surface roughness (Ra) in wire electro discharge machining of AISI D2 steel.

Findings: The wire electro discharge machining characteristics of AISI D2 steel have been investigated using an orthogonal array of design. The pulse on time and servo voltage are the most significant parameter affecting MRR and surface roughness during WEDM process. The simultaneous performance characteristics MRR and surface roughness was optimized by Taguchi based utility approach. The machined surface hardness is higher than the bulk material hardness due to the repetitive quenching effect and contained various oxides in the surface recast layer.

Methods: The experiments were performed by different cutting conditions of pulse on time (T_{on}), pulse off time (T_{off}), servo voltage (SV) and wire feed (WF) by keeping work piece thickness constant. Taguchi L_{27} orthogonal array of experimental design is employed to conduct the experiments. Multi-objective optimization was performed using Taguchi based utility approach to optimize MRR and Ra.

Results: Analysis of means and variance on to signal to noise ratio was performed for determining the optimal parameters. It reveals that the combination of Ton3, Toff1, SV1, WF2 parameter levels is beneficial for maximizing the MRR and minimizing the Ra simultaneously. The results indicated that the pulse on time is the most significant parameter affects the MRR and Ra.

Conclusions: The melted droplets, solidified debris around the craters, cracks and blow holes were observed on the machined surface for a higher pulse on time and lower servo voltage. Recast layer thickness increased with an increase in pulse on time duration. The machined surface hardness of D2 steel is increased due to the repetitive quenching effect and formation oxides on the machined surface.

Keywords: WEDM, D2 steel, MRR, Surface roughness, Utility approach, Recast layer and XRD

Background

Introduction

Tool steel, D2, is an important material for die-making industries. There is remarkable demand for hardened steel in cold-forming molds, press tool, and die-making industries due to their excellent wear resistance, high compressive strength, greater dimensional stability, and high hardness (Novotny & M 2001). However, the high carbon and chromium contents of these alloys which increase the mechanical strength and hardness of the material due to the formation of ultra hard and abrasive carbides result in high cutting temperature, larger stresses causing greater tool wear, and cutting tool failure in conventional machining of D2 steel. Leading to poor machinability and surface integrity, hence, it is considered as difficult material to cut (Jomaa et al. 2011). Nevertheless, industry, extremely the press tool sector, dies, molds are currently under pressure to compete on price and lead times. Manufacturers have therefore unavoidably needed to adopt enhanced technologies to achieve better surface quality, dimensional accuracy

* Correspondence: manjaiahgalpuji@gmail.com
[1]Department of Mechanical Engineering Science, University of Johannesburg, Kingsway Campus, Johannesburg 2006, South Africa
Full list of author information is available at the end of the article

and lower cost of production (Coldwell et al. 2003). This resulting development of new processes for machining hard-to-cut materials is required. Therefore, wire electro discharge machining (WEDM) is a thermo-electric process to machine hard to cut material of electrically conductive and gained popularity in machining complex shapes with desired accuracy and dimensions (Dhobe et al. 2013). Several researchers (Beri et al. 2010; Dhobe et al. 2014) have attempted to improve the performance characteristics such as material removal rate (MRR), surface roughness, and surface and subsurface properties. But with full potential utilization, this process still not completely solved due to its complexity, stochastic nature, and number of process variables involved in this process. In order to optimize the process parameters during WEDM of D3 steel Lodhi and Agarwal (2014) performed L_9 orthogonal array of experiments. Single objective optimization was performed by the Taguchi-based signal to noise (S/N) ratio approach. The discharge current and pulse on time is the most significant factors affecting the surface roughness. Singh and Pradhan (2014) investigated the effect of process parameters such as pulse on time, pulse off time, servo voltage, and wire feed on MRR and surface roughness during machining of AISI D2 steel using brass wire. The second order regression model has been developed to correlate the input parameters with the output responses. The authors concluded that pulse on time and pulse off time are significantly affecting the MRR and pulse on time and servo voltage is significantly affecting the surface roughness. Ugrasen et al. (2014) made a comparison of machining performances between the multiple regression model and group method data handling technique in WEDM AISI 402 steel using molybdenum wire. It was reported that the group method data handles the technique for prediction than multiple regression analysis. The EDM process parameters were optimized using a gray relation approach during machining of M2 steel. The electrode rotational speed majorly affects the MRR, electrode wear rate, and overcut followed by voltage and spark time (Purohit et al. 2015). Lin et al. (Lin et al. 2006) analyzed the material removal and surface roughness on the process parameters during WEDM of SKD11 steel. The process parameters were optimized using back propagation neural network (BPNN)—genetic algorithm (GA) method, and the predicted accuracy BPNN has close relation with the significance of optimum results. The optimal results can be achieved by nonlinear optimization of GA. Lin et al. (Lin et al. 2006) studied the effect of process parameters in EDM on the machining characteristics of SKH 57 high-speed steel by the L_{18} orthogonal array of Taguchi design. It is reported that the MRR increased with peak current and pulse on time duration as it increased up to 100 µs after it falls down. Surface roughness increased with peak current but dropped as the pulse duration increased. From the analysis of variance

(ANVOA) machining polarity, peak current affects the MRR and electrode wear, but only peak current significantly affected the surface roughness. Recently, increased in the production of micro/miniature parts based on the replication technologies are hot embossing, microinjection molding, and bulk forming technology. The forming tool yields sufficient tool life and integrated microstructural characteristics and high load bearing capacities, which due to the requirements, the hardened steels are used. Mechanical properties of these materials are limited to structuring technologies which are in use. Due to this, WEDM offers an alternative technology to obtain structural and dimensional accuracies in the miniature die- and mold-making applications (Uhlmann et al. 2005). Since, WEDM technologies apply to micromanufacturing technological applications (Löwe & Ehrfeld 1999). This is especially due to thermal material removal mechanism, allowing an almost all stress/force-free machining independently from the mechanical properties of machined material. In combination with the geometry and accuracy, the WEDM can process materials like hardened steel, silicon, cemented carbide, die steels, and electrically conductive ceramics with submicron precision. Hence, WEDM can be economically used to machine microparts, microdie, and mold especially in single and batch production mode (Dario et al. 1999; Receveur et al. 2007). In this paper, wire electro discharge machining characteristics of D2 steel were studied and determined the optimum process parameters for maximizing MRR and minimizing surface roughness using utility approach.

Experimental methodology
Work piece material and electrode
Experiments were performed on AISI D2 steel as work piece material; the general chemical composition is given in Table 1. The electrode material is zinc-coated brass wire of 0.25-mm diameter. The experiments were conducted on Electronica ECOCUT Wire EDM under power pulse mode (discharge current 12A) using de-ionized water as dielectric fluid with pressure of 12 kg/mm^2.

Experimental details
In the current research work, four process parameters such as pulse on time (T_{on}), pulse off time (T_{off}), servo voltage (SV), and wire feed (WF) were selected for conducting the experiments. The range of experiments was determined from the previous experiments by the author. Each process parameter was investigated at three levels to study the nonlinearity effect of parameters. The

Table 1 Chemical composition of AISI D2 steel

Composition	Wt (%)	Composition	Wt (%)
C	1.51	Ni	0.43
Cr	11.87	Mo	0.67

Table 2 Process parameters and their levels

Process parameters	Level 1	Level 2	Level 3
Pulse on time (µs)	110	120	130
Pulse off time (µs)	30	36	42
Servo voltage (V)	20	40	60
Wire feed (m/min)	2	4	6

selected process parameters and their levels are given in Table 2. Each trial of experiments was conducted three times as per the L_{27} orthogonal array of experiments, and their mean response are listed in Table 3.

Measurement

The MRR and surface roughness were selected as output responses. The tool wear is not important in WEDM, once the wire passes through the work piece; it is scrap, not reusable, and hence not considered in the present

study. The initial weights of work were weighed using an electronic balance (0.0001 g accuracy). The wire electrode and work were connected to negative and positive terminals of power supply, respectively. Towards the end of each trial, the work was removed and weighed on a digital weighing machine. The machining time was determined using stopwatch. The MRR is computed as:

$$MRR = \frac{WRW}{\rho * t} \tag{1}$$

where WRW is the work piece removal weight, ρ is the density of the work piece, t is the machining time.

The surface roughness of the machined samples were measured by using a Hommel surface profilometer (HOMMEL-ETAMIC T8000). The center line average surface roughness of the specimen were measured with 0.8-mm cutoff length. The roughness values were measured

Table 3 Experimental plan with mean responses, corresponding S/N ratios and multi-response S/N ratio values

Trial no.	T_{on} (µs)	T_{off} (µs)	SV (V)	WF (m/min)	MRR (mm³/min)	Ra (µm)	η_1	η_2	η
1	110	30	20	2	2.6325	1.31	8.4074	−2.3454	3.0309
2	110	30	40	4	1.701	1.5	4.6141	−3.5218	0.5461
3	110	30	60	6	0.7695	1.4	−2.2758	−2.9225	−2.5991
4	110	36	20	4	2.592	1.31	8.2727	−2.3454	2.9636
5	110	36	40	6	1.531	1.41	3.6995	−2.9843	0.3575
6	110	36	60	2	0.6885	1.28	−3.2419	−2.1442	−2.6930
7	110	42	20	6	1.863	1.36	5.4043	−2.6707	1.3667
8	110	42	40	2	1.2555	1.43	1.9763	−3.1067	−0.5651
9	110	42	60	4	0.6075	1.33	−4.3291	−2.4770	−3.4030
10	120	30	20	4	6.561	2.2	16.3394	−6.8484	4.7454
11	120	30	40	6	4.1148	1.85	12.2870	−5.3434	3.4717
12	120	30	60	2	1.9035	1.52	5.5911	−3.6368	0.9770
13	120	36	20	6	5.913	2.3	15.4362	−7.2345	4.1008
14	120	36	40	2	3.8475	1.7	11.7036	−4.6089	3.5472
15	120	36	60	4	1.782	1.54	5.0182	−3.7504	0.63387
16	120	42	20	2	5.2245	2.3	14.3609	−7.2345	3.5631
17	120	42	40	4	3.1995	1.97	10.1016	−5.8893	2.1061
18	120	42	60	6	1.377	1.59	2.7787	−4.0279	−0.6246
19	130	30	20	6	8.505	2.85	18.5935	−9.0969	4.7482
20	130	30	40	2	8.0595	2.46	18.1262	−7.8187	5.15373
21	130	30	60	4	3.7665	1.9	11.5188	−5.5750	2.9718
22	130	36	20	2	9.396	3.05	19.4589	−9.686	4.8864
23	130	36	40	4	6.966	2.7	16.8597	−8.6272	4.1161
24	130	36	60	6	3.1995	2.1	10.1016	−6.4443	1.8286
25	130	42	20	4	8.343	2.85	18.4264	−9.0969	4.6647
26	130	42	40	6	5.7105	2.6	15.1335	−8.2994	3.4170
27	130	42	60	2	2.7135	1.97	8.6706	−5.8893	1.3906

at five different locations of the work piece across the machined surface in the transverse direction of cutting, and the average of five roughness values was taken as an arithmetic surface roughness (Ra). The measured values of surface roughness and computed MRR are listed in Table 3.

The cross-sectioned machined surface morphologies were observed and measured using scanning electron microscopy (SEM, VEGA3 TESCAN). The microhardness of machined surface is measured using CLEMEX microhardness tester under a condition of 25-g load and 15-s dwell time. For each sample, average hardness value was taken from at least five test readings.

Results and discussion
Effect on MRR
The S/N ratios of the experimental responses are given in Table 3, and the average values of the MRR for each process parameter and the respective levels are plotted in Fig. 1. The main effect plot shows the effect of process parameters on the response of MRR. It is observed from the figure that the MRR is increased linearly with increase in pulse on time. This is due to the increase in pulse on time duration that causes more discharge energy onto the work piece which leads to melting the most amount of material and evaporation. The increase in discharge energy and enhancement in pulse on time leads to faster in cutting speed which causes higher MRR. As the pulse of time increases from 30 to 42 μs which is of smaller range, the MRR decreases with a lesser amplitude of variations. This is because the number of discharges within the desired period of time becomes smaller due to the time between the two pulses increases (pulse off time) which leads to lower cutting speed. Also, there may be due to the reduction in melting rate and spark ignition ratio in the plasma channel which causes lower MRR. Similarly, with the increase in

servo voltage, decreased MRR was attained. This is due to the reduced spark intensity caused by the discharge gap. The increase in servo voltage leading to an increased discharge gap causes the reduction of the spark intensity. This reduces the intensity of spark impinging on work piece surface which helps less melting and evaporation. The increase in the wire feed rate is from 2 to 6 m/min; there is not much difference in MRR observed. This may be due to the spark generation from the surface of zinc-coated wire which is of similar manner for smallest deviation in the wire feed rate. It may increase for larger deviation of wire feed rate movement due to higher cutting speed and faster number spark generation from the new surface of the wire.

Effect of process parameters on Ra Figure 2 shows the effect of process parameters on Ra. The increase in surface roughness was observed with increased pulse on time. This is because of increased pulse on time which produces larger discharge energy between electrode and the work piece. It ceases to melt more amounts of material which helps to create a larger and deeper crater. This influences the increase in surface roughness (Manjaiah et al. 2015; Kuruvila and V 2011). In the increase in pulse on time from 110 to 130 μs, drastic increase surface roughness was observed as seen in Fig. 2. At higher pulse on time, the pulse energy in the plasma channel is more and time to impinges on machined surface is more. The large number of sparks and high-intensity spark creates deeper craters and more melted droplets surrounded by globule of debris compared to lower pulse on time as observed from the SEM micrographs in Fig. 3.

As the increase in pulse off time, there were no changes in the surface roughness during machining of D2 steel. This is due to the time between the pulse width which is inconsequentially increased and which is not

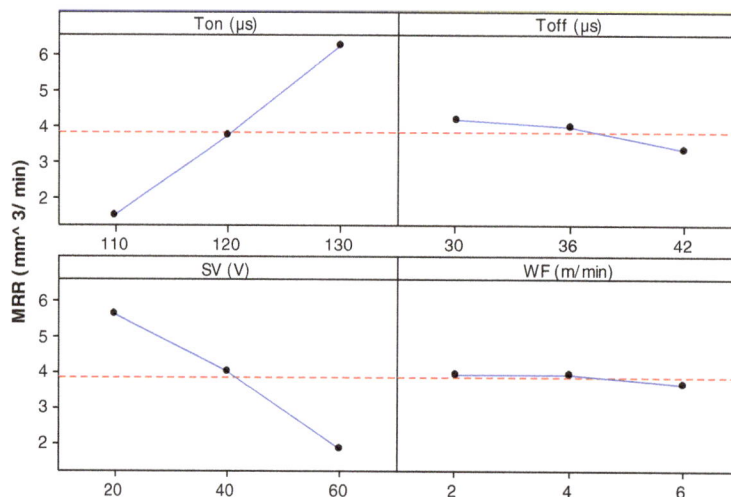

Fig. 1 Effect of process parameters on MRR of D2 steel

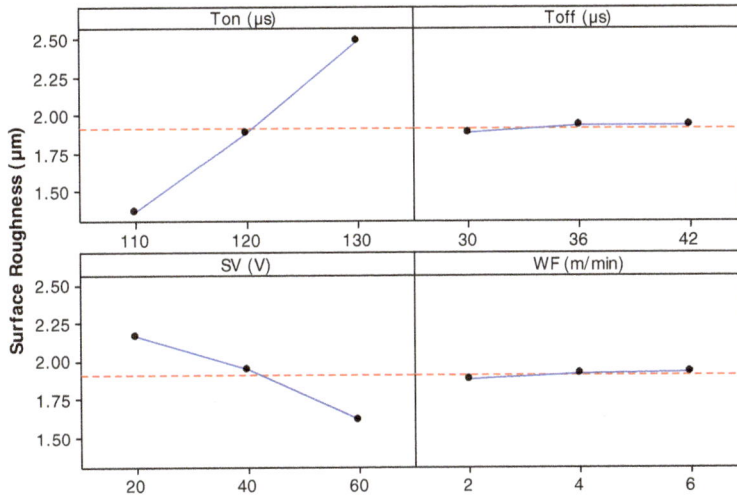

Fig. 2 Effect of process parameters on surface roughness of D2 steel

much affecting the ratio of pulse energy in the plasma channel between the electrode and work piece surface. It is observed that the surface roughness increases with increase of pulse on time and decreases with the increase of servo voltage. With increase in servo voltage, the average discharge gap gets widened resulting into better surface accuracy due to stable machining. During the increase in servo voltage, the discharge gap widens means the intensity of the spark impinges on the machined surface is low and causes for smaller craters leading to lower surface roughness (Shahali et al. 2012). At lower servo voltage, the discharge gap was reduced causes higher intensity of spark penetrates into the material, melts more amount of material, forms deeper crater on the machined surface, and causes worsening of the surface leads to higher surface roughness. Effect of

wire feed on surface roughness is similar to MRR. A minor change in the surface roughness with respect to wire feed is observed. This may be due to the no changes in cutting speed caused by the increase in wire feed which leads to melting and evaporation of material.

ANOM and ANOVA Analysis of means was performed to signal to noise ratio values to determining the optimum process parameter levels. In this paper, the Taguchi design with a utility approach concept (Manjaiah et al. 2014) is proposed for optimizing the multiple responses such as MRR and Ra. Here, MRR is to be maximized, and Ra is to be minimized. Hence, larger the better signal to noise ratio characteristic is selected, and for Ra, smaller the better signal to noise ratio is

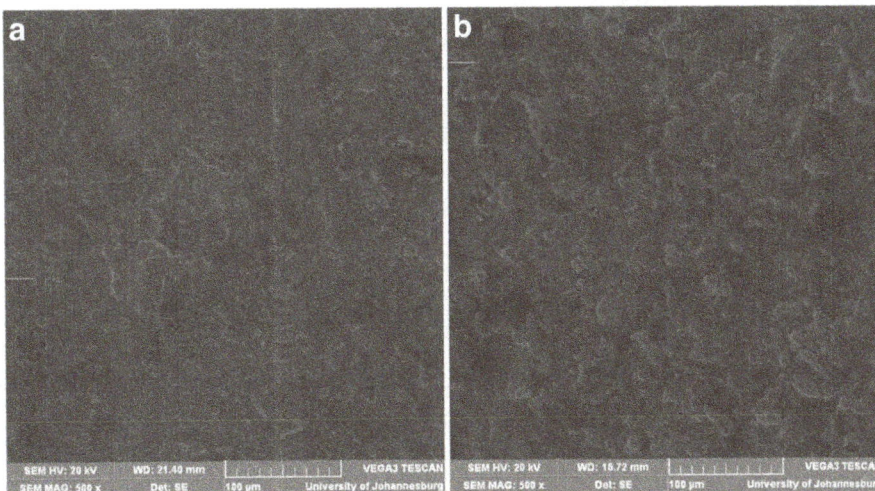

Fig. 3 SEM micrographs of D2 machined steel. **a** Pulse on time—110 μs. **b** Pulse on time—130 μs

selected. The S/N ratios associated with responses are depicted in the following Eqs. (2) and (3).

$$\eta_1 = -10 \log \left[\frac{1}{\mathrm{MRR}^2} \right] \qquad (2)$$

$$\eta_2 = -10 \log R_a{}^2 \qquad (3)$$

In the utility concept, the multi-response S/N ratio is given by Manjaiah et al. (Manjaiah et al. 2014; Chalisgaonkar & Kumar 2013) as follows:

$$\eta = w_1 \eta_1 + w_2 \eta_2 \qquad (4)$$

where w_1 and w_2 are the weighting factors associated with S/N ratio for each of the machining characteristics, MRR and Ra, respectively. In the present study, weighting factor of 0.5 for each of the machining response characteristics is considered, which gives equal priorities to both MRR and Ra for simultaneous optimization (Chalisgaonkar and Kumar 2013). The calculated values of the S/N ratio for each characteristic and the multi-response S/N ratio for each trial in the orthogonal array are listed in Table 3. Analysis of means (ANOM) is used to predict the optimum level of process parameters, and the results of optimum values are listed in Table 4. In the simultaneous optimization of MRR and Ra, the combination of process parameters is T_{on3}, T_{off1}, SV_1, and WF_2, which is beneficial for maximizing the MRR and minimizing the Ra simultaneous.

The relative significance of individual process parameters was investigated through the analysis of variance (ANOVA). Table 5 lists the summary of the ANOVA of multi-response S/N ratio values. It is found that the pulse on time has the highest contribution (63.86 %) followed by servo voltage (33 %). However, pulse off time and wire feed have least effect on optimizing the multi-objective response in WED-machining of D2 steel. It is revealed that the major influencing factor is pulse on time and servo voltage on both MRR and surface roughness. This means that the larger pulse on time leading greater discharge and intense spark removes lumps of material from the work piece surface leading to form a larger deeper crater causes greater MRR and surface roughness. The dominant parameters are duty cycle (pulse on time and pulse off time) and servo voltage (discharge gap). Causing a lower pulse off time and

Table 4 ANOM based on S/N ratio values

Parameter	Level 1	Level 2	Level 3	Optimum
T_{on}	−0.1106	2.5023	3.6864	3
T_{off}	2.5607	2.1935	1.324	1
SV	3.7856	2.4612	−0.1687	1
WF	2.1435	2.1494	1.7852	2

Table 5 Analysis of variance based on S/N ratio values

Source	Degrees of freedom	Sum of square	Mean square	F	P	% contribution
T_{on}	2	406.583	203.292	407.33	0.000	63.86
T_{off}	2	8.25	4.125	8.27	0.003	1.141
SV	2	210.603	105.301	210.99	0.000	33.00
WF	2	0.674	0.337	0.68	0.522	–
Error	18	8.984	0.499			3.14
Total	26	635.094	24.426			100

$S = 0.7065$, R-Sq = 98.6 %, R-Sq(adj) = 98.0 %

servo voltage, the larger number with high-intensity spark discharge leads to faster removal of material in a given time (Garg et al. 2012; Nourbakhsh et al. 2013).

Confirmation experiment The optimum process parameters (combination of collective optimization of MRR and surface roughness) have been determined by S/N ratio analysis using utility approach. The S/N ratio analysis of utility values was performed using MINITAB 16 statistical software. Taguchi approach for predicting the mean response characteristics and determination of confidence intervals for the predicted mean has been applied. Three trials of confirmation experiments for each response have been performed at optimal setting process parameters, and average values have been reported. The average values of confirmation experiments have been reported in Table 6. It is found that the prediction error is within the 95 % confidence interval (CI). For calculating the CI, the following equations has been used (Gaitonde et al. 2008):

$$\mathrm{CI} = \sqrt{F_{(1,\nu_e)} \, V_e \left(\frac{1}{n_{\mathrm{eff}}} + \frac{1}{n_{\mathrm{ver}}} \right)} \qquad (5)$$

where V_e is the degrees of freedom for error = 8, $F_{(1,\mathrm{Ve})}$ is the F value for 95 % CI = 4.4138, V_e is the variance of error = 0.499, $n_{\mathrm{eff}} = \frac{N}{1+\nu}$, N is the total trial number = 27, V is the degrees of freedom of p process parameters = 8, and η_{ver} is the validation test trial number = 3.

In the current study, the prediction error that is the difference between the η_{opt} and η_{obt} is 0.1405 dB, which is within the CI value of ±1.033 dB and, hence, justifies that the adequacy of additivity of the model.

Recast layer thickness Figure 4 shows the microstructure and the cross-sectioned machined surfaces. The recast layer was observed after the etching with Nital

Table 6 Confirmation experiments

Optimum values	Predicted S/N ratio	Experimental S/N ratio	Error	CI
T_{on3}, T_{off1}, SV_1, WF_2	16.1717 dB	16.0312 dB	0.1405 dB	1.033 dB

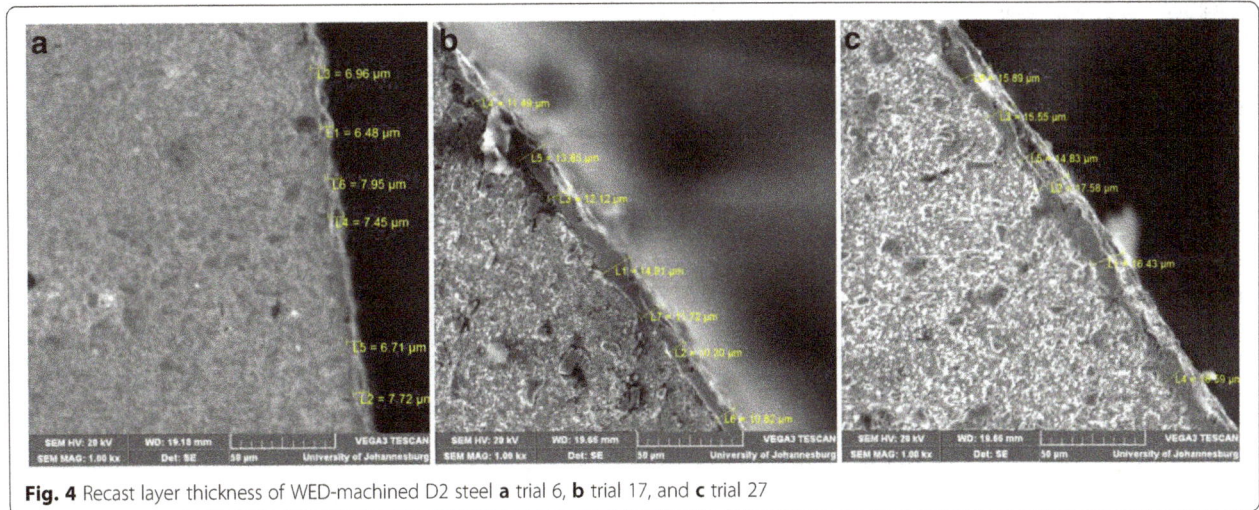

Fig. 4 Recast layer thickness of WED-machined D2 steel **a** trial 6, **b** trial 17, and **c** trial 27

2 %. The cross-sectioned surface shows deposition of a recast layer on the machined surface. This EDM characteristic shows the superimposition of craters due to metal evaporation during machining particles of different size melted and re-solidified on the surface which affects the surface properties of the component. The recast layer was measured for a varying pulse on time. It shows that the recast layer melting and deposition rate is increasing with an increase pulse on time duration. From Fig. 4a–c, the layer deposition is of the average of 7.2, 12.15, and 16.16 μm, respectively. Indeed, the damage of surface consists of recast layer zone and heat-affected zone which was found to be limited in its thickness and could not recognize easily. Hence, the depth of damaged zone measured from the extreme surface to the end of the recast zone. The recast layer zone increased with increase in pulse on time. This is due to the high discharge energy melting more material which is not flushed away from the surface caused by the less thermal conductivity of D2 steel at lower heat

transfer rate. The conclusion is made to reduce the recast layer zone by machining at lower pulse on time (110 μs) and higher servo voltage (60 V). It is found that wire feed does not have significant effect on the recast layer thickness. The outcome of the study is in agreement with the Hascalyk and Cayda (Ahmet Hasçalýk 2004) who pointed out the recast layer deposition related to EDM parameters.

Microhardness The microhardness of the machined surface measured for varying depth from the surface. It shows that the surface hardness increased due to repetitive heating and cooling. The thermal impact produced on the surface is accompanied by rapid quenching effect. The transient thermal waves produce a recast layer on the machined surface with heat-affected zone causes for increased hardness values. As seen from Fig. 5, the surface hardness increased from the bulk hardness because of over tempered martensite (Ahmet Hasçalýk 2004) produced by heating and cooling during machining. This

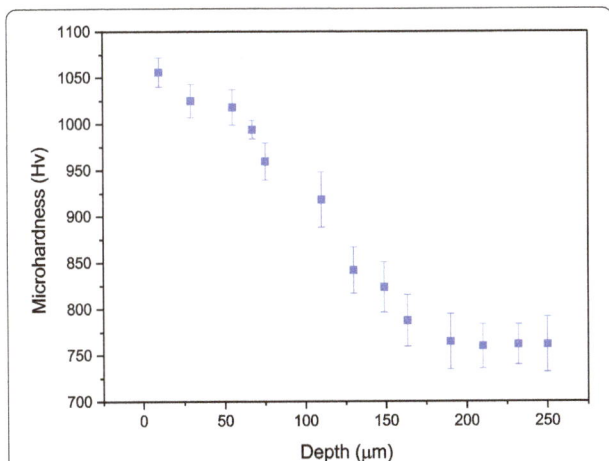

Fig. 5 Microhardness versus depth from the machined surface (trial 27)

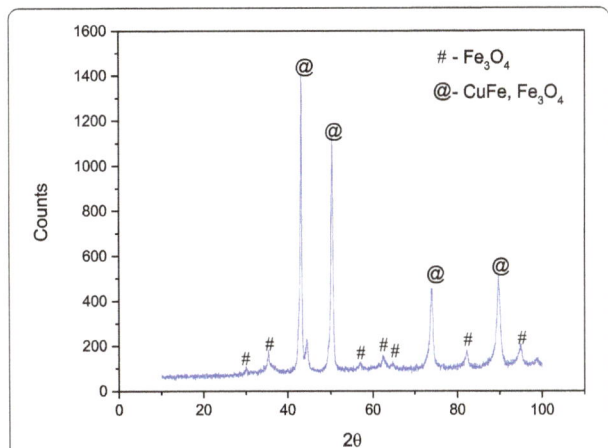

Fig. 6 XRD pattern of machined surface

rapid heating and cooling increases the carbon content in the recast layer. The surface hardness is also associated with primary and secondary hard particles and precipitates like oxides which are formed during machining.

XRD The X-ray diffractometry (XRD) analysis was done on an X-ray diffractometer system, model XPERT-PRO, PANalytical. The range of 2θ from 10° to 100° was used at a scan speed of 2°/min. The obtained XRD pattern of the machined surface is given in Fig. 6. It shows the presence of Fe_3O_4 and CuFe elements. This indicated the Cu is transferred from the wire surface and disassociated onto the material surface. Ferrous oxides are formed due to dissociation of dielectric fluid and react with Fe which leads to form Fe_3O_4 on the machined surface. This is responsible for the improvement of microhardness of the machined surface. In addition, the mechanical and physical properties such as toughness and wear resistance may also improve due to the increased hardness.

Conclusions

In this experimental study, the effect of WEDM process parameters such as pulse on time, pulse off time, servo voltage, and wire feed on machining characteristics was investigated and optimization was performed using utility approach. Summarizing the main feature of the results as follows conclusively.

1. The optimization of WED-machining parameters to determine the optimal combinations to maximize MRR and minimize surface roughness (Ra) was carried out using Taguchi-based utility approach during machining of D2 steel.
2. The pulse on time and servo voltage are the most significant parameters affecting MRR and Ra. This is because the increased pulse on time has higher electrode discharge energy, causes more melting and formation of deeper crater on the machined surface.
3. The recast layer thickness was increased with increase in pulse on time duration due to higher discharge energy caused more melting and re-deposition of molten metal. The wire feed and pulse off time do not have effect on it.
4. The specimen hardness near the outer zone can reach 1056 Hv. This hardening effect arises from the formation of oxides (Fe_3O_4) repetitive quenching effect.
5. XRD analysis shows that the machined surface gets oxidized due to reaction between the de-ionized water and work piece material. Evaporation of wire material caused for the presence of Cu on the machined surface.

Future scope

Further studies can be carried out on the effect of different wire material and wire diameter on machining characteristics of AISI D2 steel. Also, the effect of WEDM process parameters on the residual stresses can be evaluated.

Authors' contributions
MM and AK conducted experiments and collected the data. MM wrote the paper and presented discussion. RFL and SB are MM and AK's scientific supervisors, who guided and supported this work and contributed with theirs expertise and advices. All authors read and approved the final manuscript.

Competing interests
The authors declare that they have no competing intrests.

Author details
[1]Department of Mechanical Engineering Science, University of Johannesburg, Kingsway Campus, Johannesburg 2006, South Africa. [2]Department of Studies in Mechanical Engineering, University B.D.T. College of Engineering, Davangere 577 004, Karnataka, India. [3]Indian Institute of Information Technology, Dharwad 580029, Karnataka, India.

References
Ahmet Hasçalýk, U. Ç. (2004). Experimental study of wire electrical discharge machining of AISI D5 tool steel. *J Mater Process Technol, 148*(3), 362–367.
Beri, N., Maheshwari, S., Sharma, C., & Kumar, A. (2010). Technological advancement in electrical discharge machining with powder metallurgy processed electrodes: a review. *Mater Manuf Process, 25*(10), 1186–1197.
Chalisgaonkar, R., & Kumar, J. (2013). Optimization of WEDM process of pure titanium with multiple performance characteristics using Taguchi's DOE approach and utility concept. *Front Mech Eng, 8*(2), 201–214.
Coldwell, H., Woods, R., Paul, M., Koshy, P., Dewes, R., & Aspinwall, D. (2003). Rapid machining of hardened AISI H13 and D2 moulds, dies and press tools. *J Mater Process Technol, 135*(2-3), 301–311.
Dario, P., Carrozza, M. C., Croce, N., Montesi, M. C., & Cocco, M. (1999). Non-traditional technologies for microfabrication. *J Micromechanics Microengineering, 5*(2), 64–71.
Dhobe, M. M., Chopde, I. K., & Gogte, C. L. (2013). Investigations on surface characteristics of heat treated tool steel after wire electro-discharge machining. *Mater Manuf Process, 28*(10), 1143–1146.
Dhobe, M. M., Chopde, I. K., & Gogte, C. L. (2014). Optimization of wire electro discharge machining parameters for improving surface finish of cryo-treated tool steel using DOE.Mater. *Manuf Process, 29*, 1381–1386.
Galtonde, V. N., Karnik, S. R., & Davim, J. P. (2008). Multiperformance optimization in turning of free-machining steel using Taguchi method and utility concept. *J Mater Eng Perform, 18*(3), 231–236.
Garg, M. P., Jain, A., & Bhushan, G. (2012). Modelling and multi-objective optimization of process parameters of wire electrical discharge machining using non-dominated sorting genetic algorithm-II. *Proc Inst Mech Eng Part B J Eng Manuf, 226*(12), 1986–2001.
Jomaa, W., Fredj, N. B., Zaghbani, I., & Songmene, V. (2011). Non-conventional turning of hardened AISI D2 tool steel. *Int J Adv Mach Form Oper, 2*, 1–41.
Kuruvila, N., & V., R. H. (2011). Parametric influence and optimization of wire Edm of hot die steel. *Mach Sci Technol, 15*(1), 47–75.
Lin, Y.-C., Cheng, C.-H., Su, B.-L., & Hwang, L.-R. (2006). Machining characteristics and optimization of machining parameters of SKH 57 high-speed steel using electrical-discharge machining based on Taguchi method. *Mater Manuf Process, 21*(8), 922–929.
Lodhi, B. K., & Agarwal, S. (2014). Optimization of machining parameters in WEDM of AISI D3 steel using Taguchi technique. *Procedia CIRP, 14*, 194–199.
Löwe, H., & Ehrfeld, W. (1999). State-of-the-art in microreaction technology: concepts, manufacturing and applications. *Electrochim Acta, 44*, 3679–3689.
Manjaiah, M., Narendranath, S. S. B., & Gaitonde, V. N. (2014). Some investigations on wire electric discharge machining characteristics of titanium nickel shape memory alloy. *Trans Nonferrous Met Soc China, 24*(10), 3201–3209.
Manjaiah, M., Narendranath, S., & Basavarajappa, S. (2015). Wire electro discharge machining performance of TiNiCu shape memory alloy. *Silicon, 8*, 467–475.

Nourbakhsh, F., Rajurkar, K. P., & Cao, J. (2013). Wire electro-discharge machining of titanium alloy. *Procedia CIRP, 5*, 13–18.

Novotny, P.,M. (2001). Tool and Die Steels. Encyclopedia of Materials: Science and Technology, pp. 9384–9389. doi:10.1016/B0-08-043152-6/01697-1

Purohit, R., Rana, R. S., Dwivedi, R. K., Banoriya, D., & Singh, S. K. (2015). Optimization of electric discharge machining of M2 tool steel using grey relational analysis. *Mater Today Proc, 2*(4-5), 3378–3387.

Receveur, R. A. M., Lindemans, F. W., & De Rooij, N. F. (2007). Microsystem technologies for implantable applications. *J Micromechanics Microengineering, 17*(5), R50–R80.

Shahali, H., Yazdi, M. R. S., Mohammadi, A., & Iimanian, E. (2012). Optimization of surface roughness and thickness of white layer in wire electrical discharge machining of DIN 1.4542 stainless steel using micro-genetic algorithm and signal to noise ratio techniques. *Proc Inst Mech Eng Part B J Eng Manuf, 226*(5), 803–812.

Singh, V., & Pradhan, S. K. (2014). Optimization of WEDM parameters using Taguchi technique and response surface methodology in machining of AISI D2 steel. *Procedia Eng, 97*, 1597–1608.

Ugrasen, G., Ravindra, H. V., Prakash, G. V. N., & Keshavamurthy, R. (2014). Comparison of machining performances using multiple regression analysis and group method data handling technique in wire EDM of Stavax material. *Procedia Mater Sci, 5*, 2215–2223.

Uhlmann, E., Piltz, S., & Doll, U. (2005). Machining of micro/miniature dies and moulds by electrical discharge machining—recent development. *J Mater Process Technol, 167*(2-3), 488–493.

15

Strain and stress conditions at crack initiation during shearing of medium- and high-strength steel sheet

E. Gustafsson[1]* **iD**, S. Marth[2], L. Karlsson[1] and M. Oldenburg[2]

Abstract

Background: Strain and stress conditions in sheet metal shearing are of interest for calibration of various fracture criteria. Most fracture criteria are governed by effective strain and stress triaxiality.

Methods: This work is an attempt to extend previous measurements of strain fields in shearing of steel sheets with the stress state calculated from the measured displacement fields. Results are presented in terms of von Mises stress and stress triaxiality fields, and a comparison was made with finite element simulations. Also, an evaluation of the similarities of the stress conditions on the sheet surface and inside the bulk material was presented.

Results: Strains and von Mises stresses were similar to the surface and the bulk material, but the stress triaxiality was not comparable. There were large gradients in strain and stress around the curved tool profiles that made the result resolution dependent and comparisons of maximum strain and stress values difficult.

Conclusions: The stress state on the sheet surface calculated from displacement field measurements is useful for validation of a three-dimensional finite element model.

Keywords: Sheet metal, Experiment, Shearing, Strain, Stress, Crack initiation

Background

Shearing is a common process in the sheet metal industry. The constant development of new sheet materials with various shearing properties makes it desirable to have a model of the shearing process that can predict the appropriate shearing parameters. That model must consider the fracture in addition to the large plastic deformations. Numerous fracture criteria, based on strain and stress conditions, exist which can be used in finite element (FE) models of the shearing, for example, maximum effective strain, maximum shear stress, or combined stress and strain criteria (Cockcroft and Latham 1968; Johnson and Cook 1985). Such criteria need to be calibrated against experimentally measured strain and stress conditions. The ductile fracture is in general governed by effective strain and stress triaxiality as observed by McClintock (1968). Low triaxiality results in shear dimple rupture whereas high triaxiality results in void coalescence as modelled by Rice and Tracey (1969).

Strain and stress conditions can be calculated from full-field measurements of displacement as shown by Marth et al. (2016). In that work, incremental displacement fields were obtained from captured images during the experiments by the digital image correlation (DIC) technique, thorough reviewed by Hild and Roux (2006). With DIC, the displacements are measured on sub-pixel level, see for example Sjödahl (1994). Accuracy in DIC measurements are covered in detail by Sjödahl (1997).

Strains were measured during planar blanking by Stegeman et al. (1999) and in experiments with a symmetric shearing set-up by Gustafsson et al. (2016b). The purpose of the present work was to extend these experiments with stress calculations using the method described by Marth et al. (2016), to determine the strain and stress conditions at crack initiation. The calculated stresses were compared with stresses from FE simulations at the stage of crack initiation.

*Correspondence: egu@du.se
[1]Dalarna University, SE-791 88 Falun, Sweden
Full list of author information is available at the end of the article

Methods

Materials and yield stress model

The medium-strength (MS) construction steel SSAB Domex 420 MC, and the high-strength (HS) wear plate steel SSAB Hardox 400, with mechanical properties according to Table 1, were used in the study. The yield stress for the sheet materials was described with the exponential hardening law suggested by Hollomon (1945), $\sigma_Y = K\bar{\varepsilon}_p^n$, where $\bar{\varepsilon}_p$ is the effective plastic strain and K and n are material specific parameters. These parameters were fitted to compression test data provided by Gustafsson et al. (2016b) and are shown in Table 1 for the sheared material grades.

Shearing experiment

The shearing experiments were performed with the method and set-up developed by Gustafsson et al. (2014). This set-up uses symmetry to balance the force defined as F_x in Fig. 1. Therefore, no guides with friction losses are needed, and the forces, F_x and F_y, which are measured in the set-up, are accurate. The design also features large stiffness in the x-direction and thus a stable tool clearance c. Tool displacements, U_x and U_y, were measured with linear transducers as described by Gustafsson et al. (2016a). Tool clearance and clamping of the sheet samples were varied in the experiments: clearances $0.05h$, $0.15h$ and $0.25h$, where h is the sheet thickness used and the sheet was clamped on one or both sides as shown in Fig. 1.

Images of the approximately magnified area shown in Fig. 1 were captured during the shearing to register the in-plane deformation of the sheet. For this purpose, the xy-surface of the sheet samples was prepared with a random speckle pattern for subsequent digital image correlation. Further details on the experiments are described by Gustafsson et al. (2016b).

Image analysis and strain evaluation

The digital image correlation (DIC) technique, implemented with the commercial software ARAMIS, was used to calculate the deformation gradients on the surface. The evaluated area was partitioned into sub-areas, called facets, and then, these facets were traced

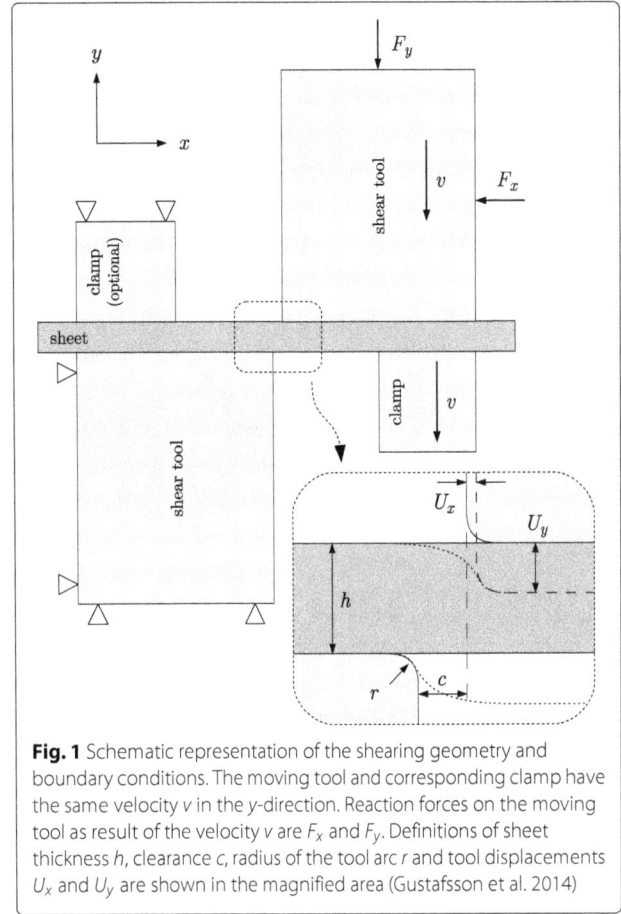

Fig. 1 Schematic representation of the shearing geometry and boundary conditions. The moving tool and corresponding clamp have the same velocity v in the y-direction. Reaction forces on the moving tool as result of the velocity v are F_x and F_y. Definitions of sheet thickness h, clearance c, radius of the tool arc r and tool displacements U_x and U_y are shown in the magnified area (Gustafsson et al. 2014)

through cross-correlation of the images captured during the shearing. The facet size was 64×64 pixels, and the facet step size was 8 pixels in each direction. The in-plane strain field (x- and y-components) was calculated from the measured deformations, and under assumption of plastic incompressibility, the missing strain component (z-direction) was calculated, as shown by Kajberg and Lindkvist (2004).

Evaluation of stress from measured strain

The strain tensor, obtained as described in previous subsection, was used to calculate the stress tensor by a radial return algorithm based on isotropic von Mises plasticity, as described by Marth et al. (2016), but without using a stepwise modelling of the hardening relation. Instead of this stepwise modelling, the plastic work hardening of the materials was modelled with the Hollomon hardening law using the material parameter presented in Table 1. From the stress tensor, the effective von Mises stress was calculated as

$$\bar{\sigma} = \sqrt{\frac{3}{2}s_{ij}s_{ij}}, \quad (1)$$

Table 1 Mechanical properties in terms of yield strength Rp02, tensile strength Rm and elongation A80, evaluated from uniaxial tensile tests of sheet metal grades used in the study

Material strength	Rp02 [MPa]	Rm [MPa]	A80 [%]	h [mm]	K [MPa]	n [—]
Medium	450	520	25	5.97–6.03	880	0.127
High	1080	1260	7	6.11–6.15	1550	0.0345

The range of sheet thickness h for the sheared samples is also shown. Finally, K and n are the hardening law parameters for the two material grades as fitted to compression test data

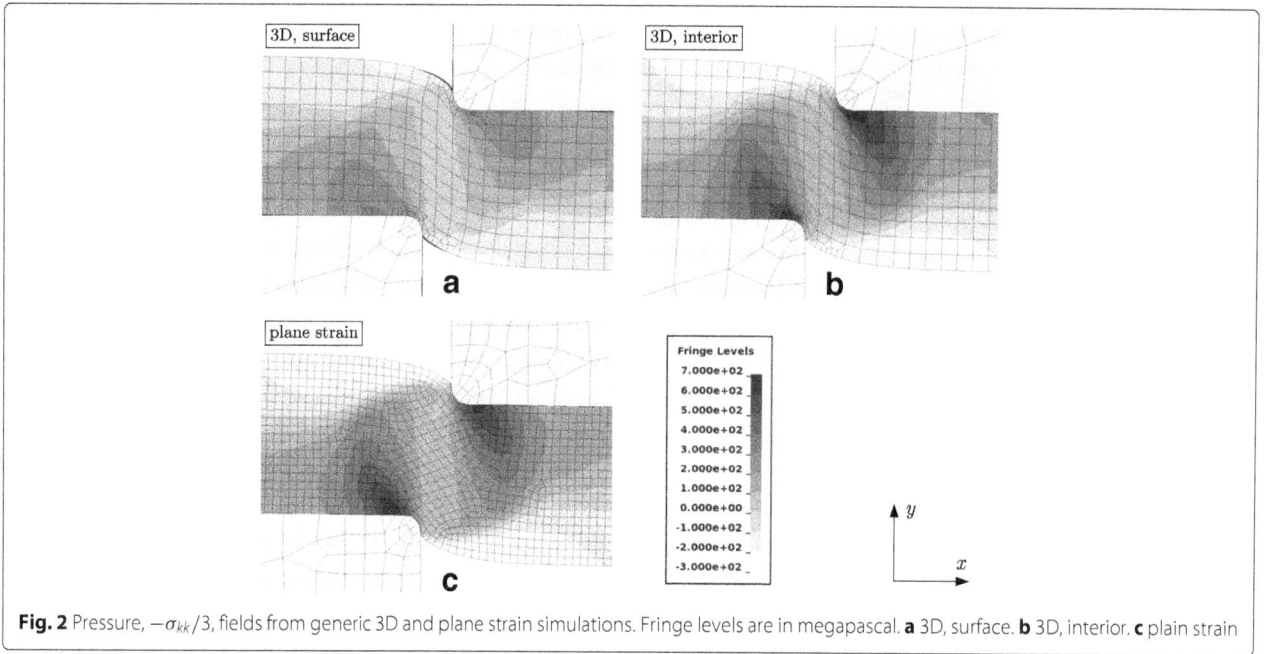

Fig. 2 Pressure, $-\sigma_{kk}/3$, fields from generic 3D and plane strain simulations. Fringe levels are in megapascal. **a** 3D, surface. **b** 3D, interior. **c** plain strain

where $s_{ij} = \sigma_{ij} - \sigma_{kk}\delta_{ij}/3$ is the deviatoric stress tensor. Taking the mean stress, $\sigma_m = \sigma_{kk}/3$, the stress triaxiality state can be calculated as

$$\eta = \frac{\sigma_m}{\bar{\sigma}}. \tag{2}$$

Since the strain values used in this method were obtained on the material surface, a plane stress approach was used to evaluate the surface stress conditions. A plane strain approach would otherwise be the intuitive choice to analyse the stress state inside the material, but that approach would violate the plastic incompressibility condition.

Finite element simulations

Plane strain FE analyses of the shearing were performed with a commercial general-purpose finite element software. Geometry and boundary conditions for the model

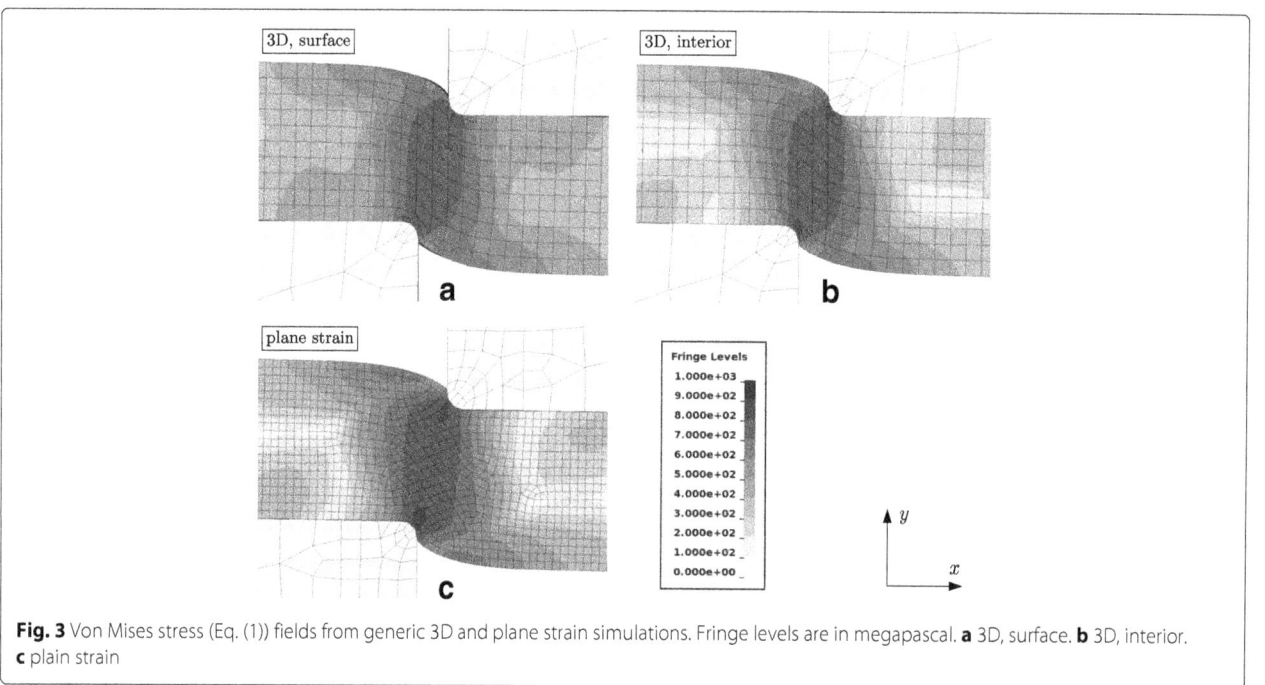

Fig. 3 Von Mises stress (Eq. (1)) fields from generic 3D and plane strain simulations. Fringe levels are in megapascal. **a** 3D, surface. **b** 3D, interior. **c** plain strain

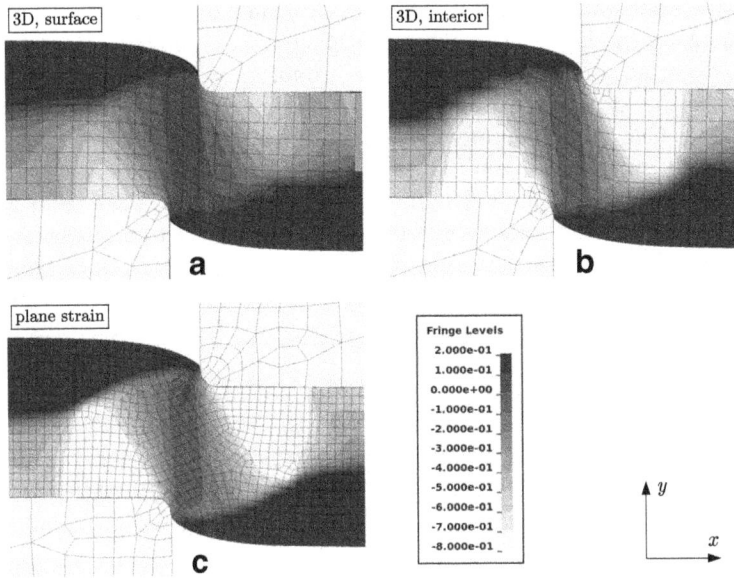

Fig. 4 Stress triaxiality (Eq. (2)) fields from generic 3D and plane strain simulations. **a** 3D, surface. **b** 3D, interior. **c** plain strain

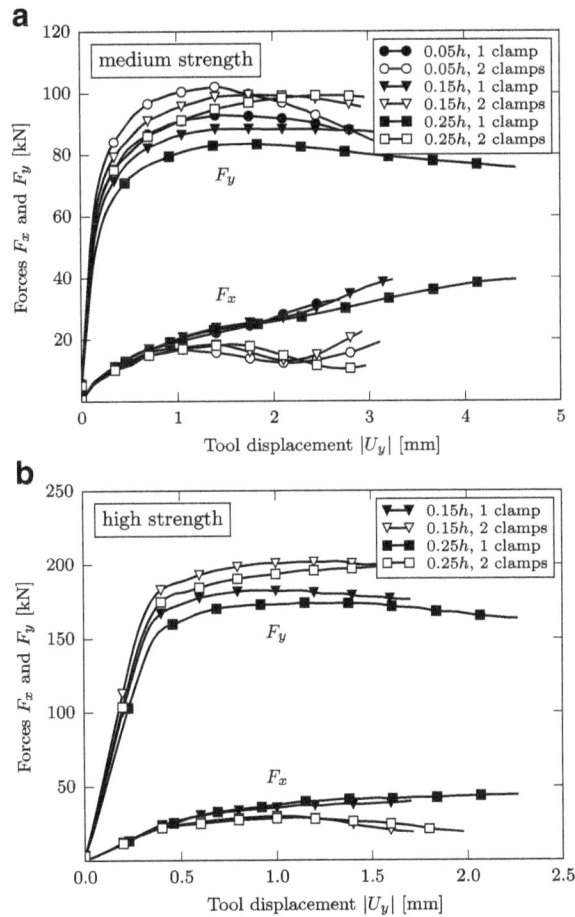

Fig. 5 FE simulated force-displacement curves when shearing the medium-strength material (**a**) and the high-strength material (**b**). Clearances (expressed as fractions of sheet thickness h) and clamping configurations are indicated in the figure legend

are described in detail by Gustafsson et al. (2014). The model was coarsely meshed except in the area where large deformations were expected and a denser mesh was required to resolve gradients in the state variables. The undeformed element size in the dense area was 0.1 mm. Four-noded and fully integrated plane strain elements were used for sheet and tools. Adaptive remeshing was applied to avoid large aspect ratios of the elements. Contacts were modelled with surface to surface formulations, and both static and dynamic friction coefficients were 0.1.

A model with elastic tools and an isotropic elastic-plastic sheet material, where Poisson's ratio was 0.3 and Young's modulus was 210 GPa, was used. A constant tool acceleration of 100 mm s^{-2} was used in all simulations. This is a rough approximation of the experimental conditions, but still satisfactory since no rate or temperature effects were modelled.

For a preliminary study of the stress state on the sheet surface and in the interior of the material, generic FE models were used in three dimensions (3D) and in plane strain. For this purpose, a slightly different geometry, material parameters and, most notably, no remeshing were applied in the 3D model. This 3D model was therefore only useful at relatively small tool displacements.

Comparison between experiments and simulations

Comparisons were made between results from FE simulations and results based on measurements, in terms of effective von Mises stress (Eq. (1)) and stress triaxiality (Eq. (2)). Furthermore, effective strain and tool forces versus tool displacement were presented for comparison with corresponding experimental data presented by Gustafsson et al. (2016b). An agreement between these simulations and experiments (effective strain and force vs. displacement) should be seen as a validation of the FE model. The effective plastic strain fields from the FE simulations

$$\bar{\varepsilon}_p = \int \sqrt{\frac{2}{3}D_{ij}^p D_{ij}^p} dt,\qquad(3)$$

where D_{ij}^p is the plastic component of the rate of deformation tensor, which can be compared with the effective strains measured by Gustafsson et al. (2016b), since the elastic part of the latter is negligible.

All presented strains and stresses are from the stage of crack initiation. The fracture was not modelled in the FE simulations, and a comparison with experiments is therefore irrelevant after crack initiation.

Results

First, results from generic FE simulations are presented to show how the stress state varies between the surface and the bulk material, and that plane strain is a satisfactory

approximation of the conditions in the bulk material ("Generic FE simulations of the stress state" section). Second, the FE model of the experimental shearing conditions is validated against previously published tool forces and strain fields ("Validation of the FE model at experimen- tal conditions" section). Third, a comparison between the stress state from FE simulations and from calculations based on the measured displacement fields is presented, in terms of effective von Mises stress ("Comparison of von Mises stress" section) and stress triaxiality ("Comparison of stress triaxiality" section). The stress calculated from measured displacement fields will hereafter be referred to as experimental stress. Finally, a summary of the comparison between simulated and experimentally based stress and strain conditions closes this section ("Summary of measured and simulated results" section).

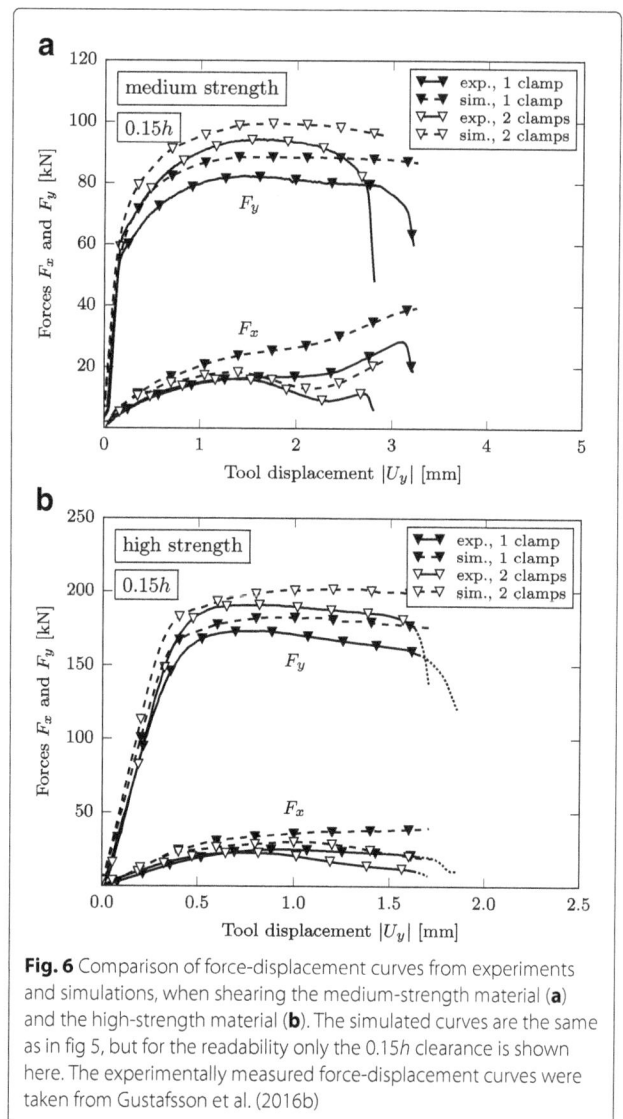

Fig. 6 Comparison of force-displacement curves from experiments and simulations, when shearing the medium-strength material (**a**) and the high-strength material (**b**). The simulated curves are the same as in fig 5, but for the readability only the 0.15h clearance is shown here. The experimentally measured force-displacement curves were taken from Gustafsson et al. (2016b)

Generic FE simulations of the stress state

Strains on the sheet surface and inside the sheet are almost the same during shearing, as showed by Gustafsson et al. 2016b. Here, the same 3D and plane strain FE models were used to investigate the stress state in terms of mean stress (pressure), von Mises stress and triaxiality. Results from the generic FE simulations are presented in Figs. 2, 3 and 4. Plane strain was a good approximation of the interior stress state according to the evaluated parameters (Figs. 2b–c, 3b–c and 4b–c), but the surface stress, in terms of pressure and triaxiality, was not comparable to the stress in the bulk material (Figs. 2a–b and 5a–b).

There was, however, a good agreement between the von Mises stress on the surface and the interior (Fig. 3a–b). This was expected since the effective strain was in good agreement between surface and the interior in a previous

comparison by Gustafsson et al. (2016b). The von Mises stress gradients were largest outside of the plastic zone. This was also expected since the gradient, after the yield locus is reached, is proportional to the hardening modulus which is much smaller than the Young's modulus.

In the area between the arcs of the tools, the pressure was lower on the surface compared to the interior and the triaxiality was larger. The triaxiality in the material between the tools was in general slightly positive on the surface and slightly negative in the bulk material at this state of the shearing.

Validation of the FE model at experimental conditions

Tool forces from the FE simulations, up to the tool displacement $|U_y|$ where final fracture occurred in the experiments, are presented in Fig. 5. All trends in the

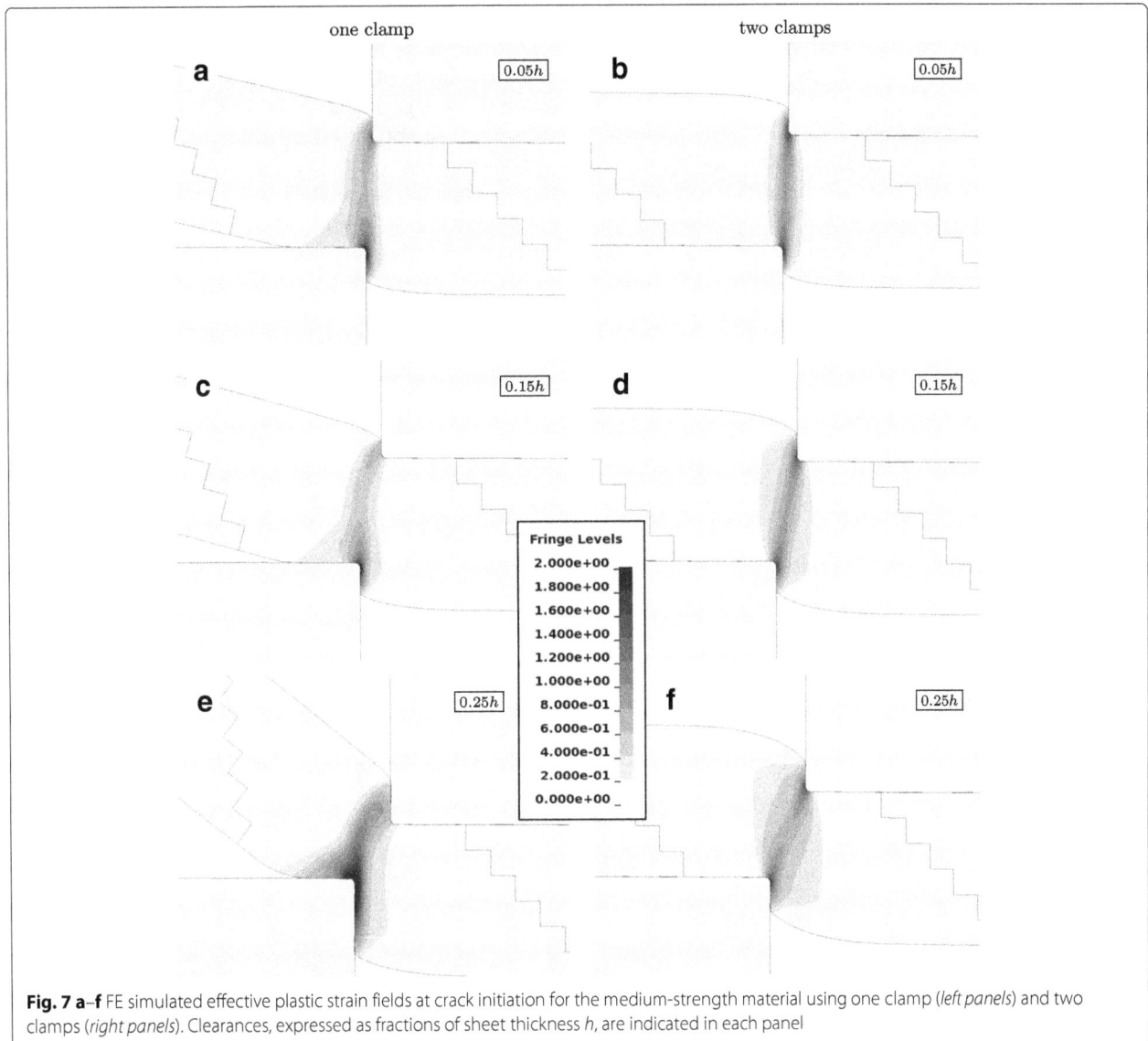

Fig. 7 a–f FE simulated effective plastic strain fields at crack initiation for the medium-strength material using one clamp (*left panels*) and two clamps (*right panels*). Clearances, expressed as fractions of sheet thickness *h*, are indicated in each panel

forces from variations of the tool clearance and clamping configurations, and the general shape of the force curves, were the same in the simulations shown here as in the experiments by Gustafsson et al. (2016b). There was, however, a slight overestimation of both force components F_x and F_y in the simulations, as shown in Fig. 6. In general, the overestimation increased along with the tool displacement, $|U_y|$, for both materials, but for the medium-strength material, the overestimation in F_y was almost constant. The drop in forces after crack initiation, seen in the experiments, was not captured by the simulations since crack formation was not accounted for in the model.

Effective plastic strains from FE simulations for the two material grades at the stage of crack initiation are shown in Figs. 7 and 8. There was a good agreement between the experiments presented by Gustafsson et al. (2016b) and the simulations shown here. The large gradients close to the arc of the tool were better resolved in the simulations, and a comparison of the largest strains was therefore not straightforward.

Comparison of von Mises stress
Von Mises stresses from FE simulations are presented in Figs. 9 and 11, and the corresponding experimental stresses are presented in Figs. 10 and 12. As for the strains, there was a good agreement between the von Mises stresses from FE simulations and experiments. The experimental results showed less details compared with the simulations, especially at the sheet edges. Missing information at the edges also resulted in a shape of the imaged area without a distinct indentation of the tools into the sheet (Figs. 10 and 12). For some experiments, white areas of missing data extends further into the sheet from the edge. This is most pronounced in Fig. 10e, where the deformations were largest.

Comparison of stress triaxiality
Stress triaxiality from FE simulations are shown in Figs. 13 and 15 and from experiments in Figs. 14 and 16. The calculated triaxialities contained much noise, especially in areas with small von Mises stresses due to the definition as the ratio of mean stress to von Mises stress. Therefore, the data in Figs. 14 and 16 was low-pass filtered to improve the readability. In general, the triaxiality was larger in the area close to the vertical surface of the tools next to the arc of the tool than in an area close to the arc itself. The areas of the largest triaxiality, along the horizontal sheet edge that is not in contact with the tool, were not subject to notable effective strains, and therefore, no cracks were observed there. Since the simulations (Figs. 13 and 15) show the stress state in the bulk material and the experiments (Figs. 14 and 16) show the stress state on the surface, these triaxiality fields were not comparable.

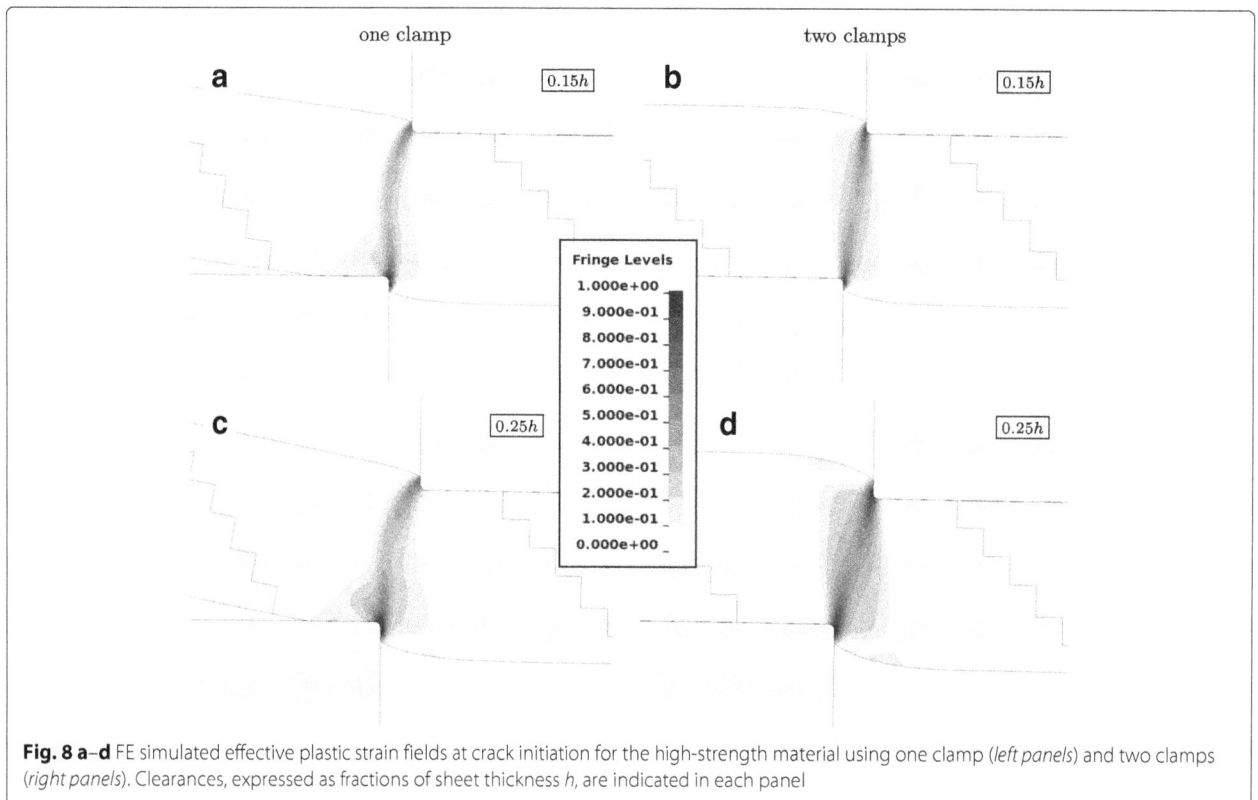

Fig. 8 a–d FE simulated effective plastic strain fields at crack initiation for the high-strength material using one clamp (*left panels*) and two clamps (*right panels*). Clearances, expressed as fractions of sheet thickness *h*, are indicated in each panel

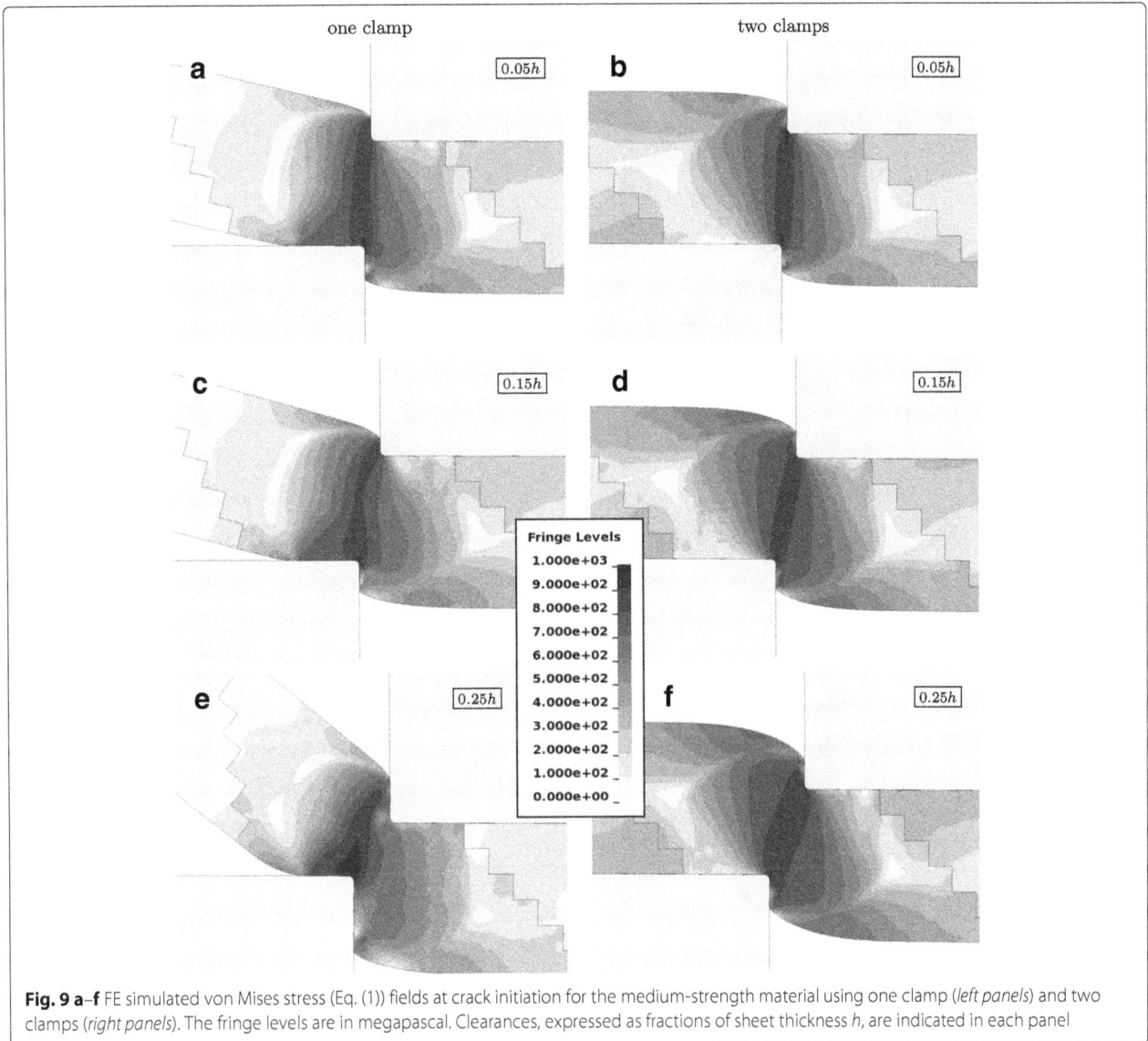

Fig. 9 a–f FE simulated von Mises stress (Eq. (1)) fields at crack initiation for the medium-strength material using one clamp (*left panels*) and two clamps (*right panels*). The fringe levels are in megapascal. Clearances, expressed as fractions of sheet thickness *h*, are indicated in each panel

Summary of measured and simulated results

In addition to the fringe plots of stress and strain presented (Figs. 7, 8, 9, 10, 11, 12, 13, 14, 15 and 16), a numeric summary of these parameters, at crack initiation, is provided in Table 2. The parameters were evaluated in three regions of the sheet in the vicinity of the arc-shaped 2D profile of the lower tool, as indicated in Fig. 17. At point I, where the curvature of the tool profile changes, the material is heavily loaded and large stresses and strains are therefore expected there. Large strain gradients from the simulated data were found close to the tool; hence, point II was chosen a short distance away from the curved tool profile. There were large stress triaxialities in the simulated data in region III, where the sheet is in contact with the vertical tool profile just

below point I. Moreover, cracks were sometimes found in this region.

The summary in Table 2 shows that strains at point I were larger in the simulations, $\bar{\varepsilon}_I^{sim}$, than in the experiments, $\bar{\varepsilon}_I^{exp}$. The simulations indicated a large strain gradient between points I and II, but such gradient could not be revealed from the experiments since points I and II could not be distinguished due to the experimental resolution. There was a reasonable agreement between the simulated and experimentally based von Mises stresses for the high-strength material; the agreement was less satisfactory for the medium-strength material. There was no clear trend of stress triaxiality for either material strength, tool clearance or clamping configuration, but there was still a large spread of measured values.

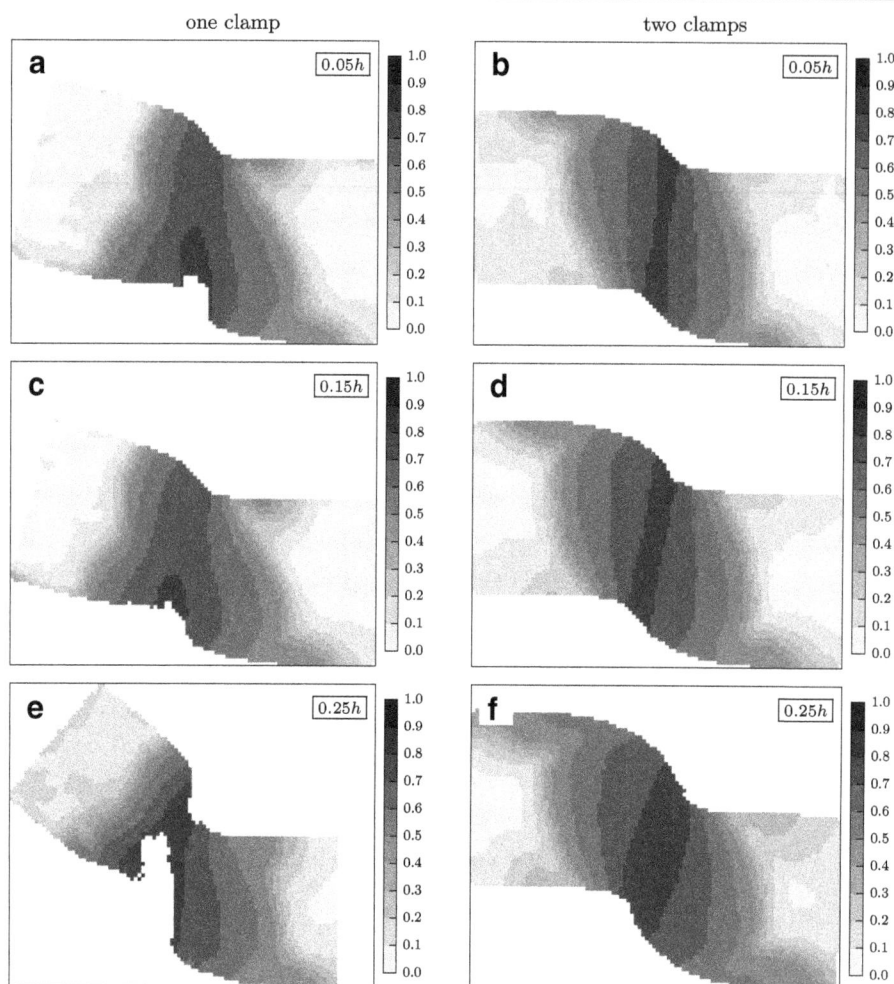

Fig. 10 a–f Experimentally based von Mises stress (Eq. (1)) fields at crack initiation for the medium-strength material using one clamp (*left panels*) and two clamps (*right panels*). The fringe levels are in gigapascal. Clearances, expressed as fractions of sheet thickness *h*, are indicated in each panel

Discussion

Tool forces were slightly overestimated by the applied FE model already from small tool displacements, as shown in Fig. 6. Possible reasons include an incorrect material yield strength and hardening and also the friction coefficient between sheet and tools. The materials were characterised by accurate compression tests, but only data from the thickness direction were used for the isotropic model and there was no perfect fit of the exponential hardening law to the compression test data. The friction coefficient was studied in the sensitivity analyses by Gustafsson et al. (2014) and had large effects on the tool forces. An increased friction coefficient resulted in largely increased F_x and increased F_y along with the tool displacement.

The large strain gradients in the sheet material around the curved tool profile made the comparison between measurements and simulations difficult, as the results were resolution dependent. These gradients, expressed as relative increase of strain from point II to point I (a distance of about 0.2 mm), varied from 26 % to almost 90 %, depending on shearing configuration and material strength. There was also a limitation in the measurement technique that prevented accurate measurements close to the edges as discussed by Gustafsson et al. (2016b). Also, due to the same phenomenon, measured strains after crack initiation would likely be overestimated since the image correlation algorithm and strain calculations interpret relative movement of the fractured parts as strain.

The comparison of the stress state between experiments and simulations is problematic since the calculations based on the measured strain give the stress state on the surface and the simulations in plane strain give the stress state in the bulk material. The obtained von Mises stress was still almost identical between the two methods, but the triaxiality differed in both distribution and levels. Possibly, the triaxiality from both methods are correct,

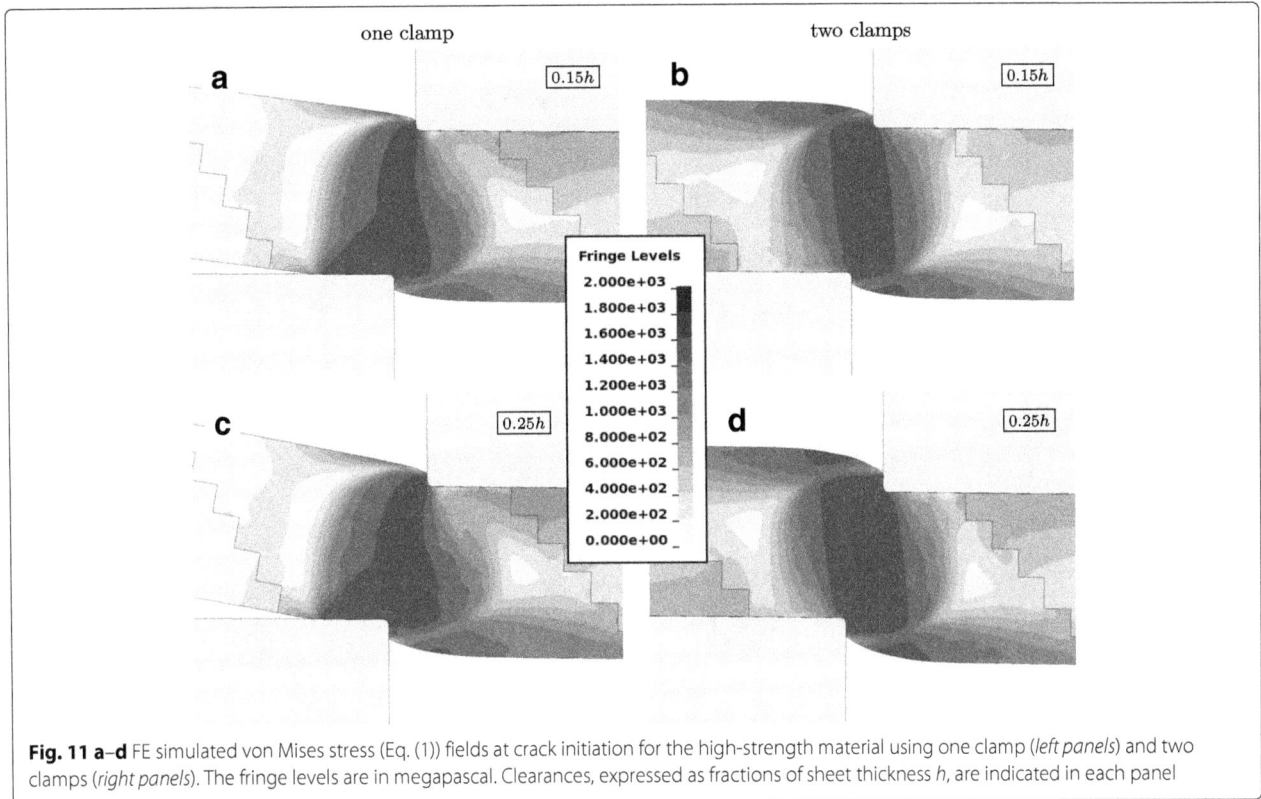

Fig. 11 a–d FE simulated von Mises stress (Eq. (1)) fields at crack initiation for the high-strength material using one clamp (*left panels*) and two clamps (*right panels*). The fringe levels are in megapascal. Clearances, expressed as fractions of sheet thickness h, are indicated in each panel

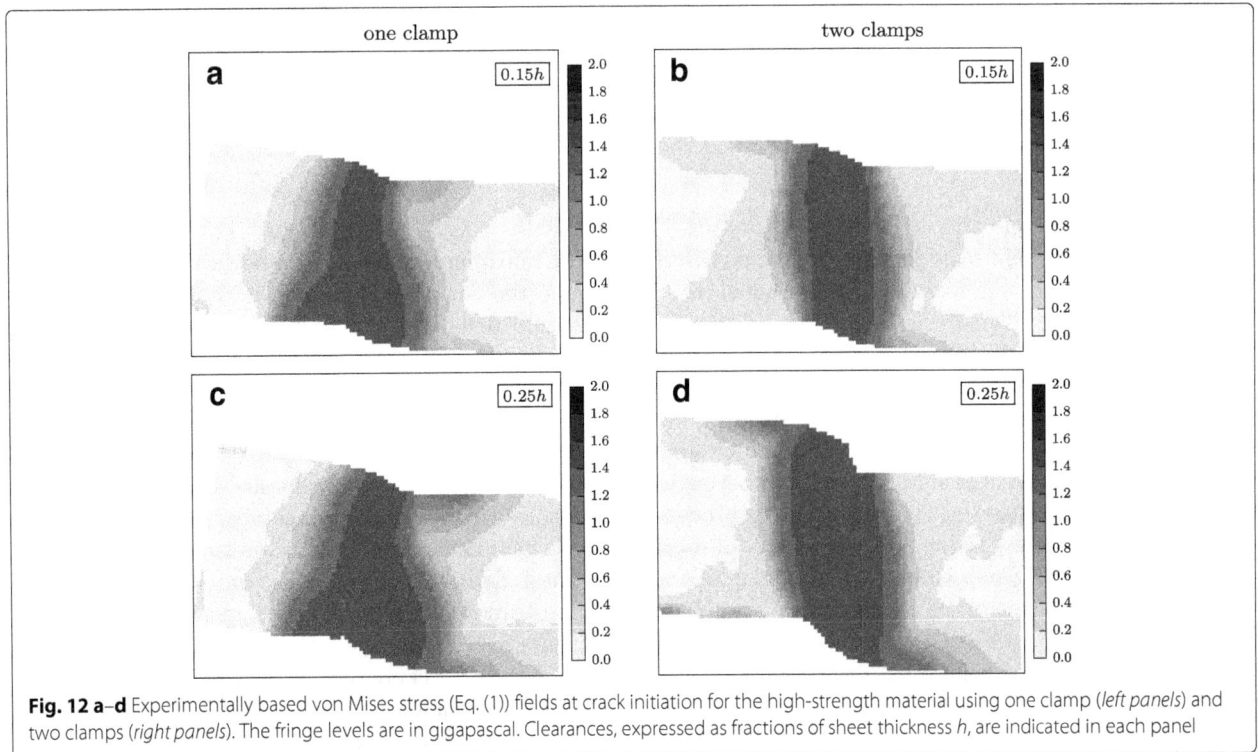

Fig. 12 a–d Experimentally based von Mises stress (Eq. (1)) fields at crack initiation for the high-strength material using one clamp (*left panels*) and two clamps (*right panels*). The fringe levels are in gigapascal. Clearances, expressed as fractions of sheet thickness h, are indicated in each panel

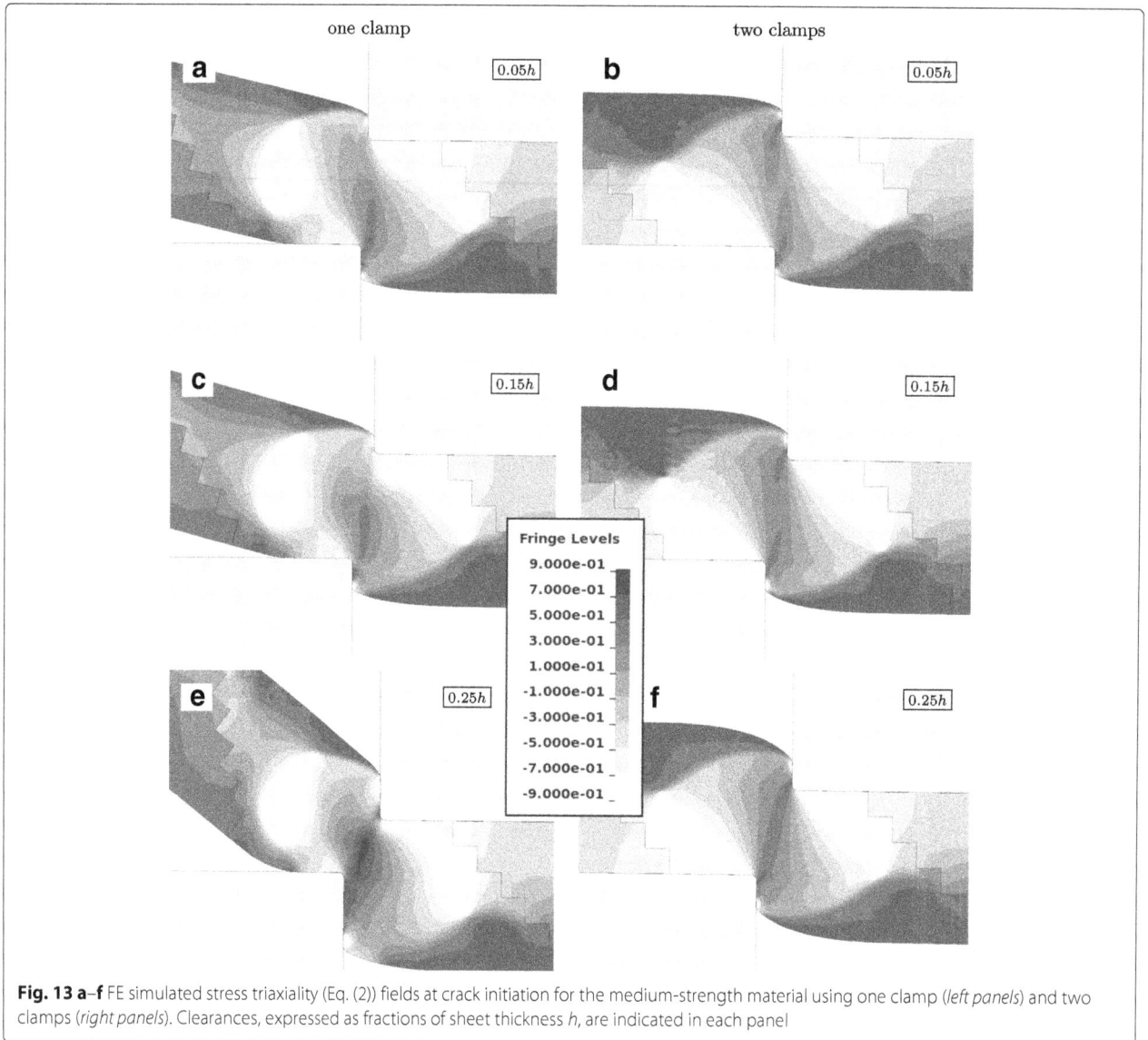

Fig. 13 a–f FE simulated stress triaxiality (Eq. (2)) fields at crack initiation for the medium-strength material using one clamp (*left panels*) and two clamps (*right panels*). Clearances, expressed as fractions of sheet thickness *h*, are indicated in each panel

simulations for the bulk material and calculations from measurements for the surface, although not equal. For a correct comparison, either the simulations need to be in three dimensions and the triaxiality taken on the surface or the calculations from measured strains need to be done with the same plain strain approximation. A surface comparison would maybe be the most intuitive choice since strains are measured on the surface. On the other hand, the surface is a small part of the total shearing process, and therefore, the stress in the bulk material is of more interest. Consequently, in a strictly experimental aspect, the calculations from measured displacements should use the plane strain approximation to provide bulk material stresses. The present experimental results are still useful for validation of a three-dimensional finite element simulation.

The maximum strain was found in the sheet material around the curved tool profile. Within the zone that is plastically deformed, the maximum triaxiality was found in the material in contact with the vertical surface of the tool below the curved tool profile. Therefore, a larger impact of triaxiality on the crack initiation will result in a larger burr on the sheared samples.

Although gradients in the studied parameters are better resolved in the simulations compared with calculations from experiments, the result is still largely dependent on the mesh density. The gradients, and also the maximum value of the effective strain, will increase with smaller elements in contact with the tools and decrease with larger elements. Therefore, if the limitation in the digital image correlation technique close to the edges of the sheet is ignored, the mesh density in the finite element

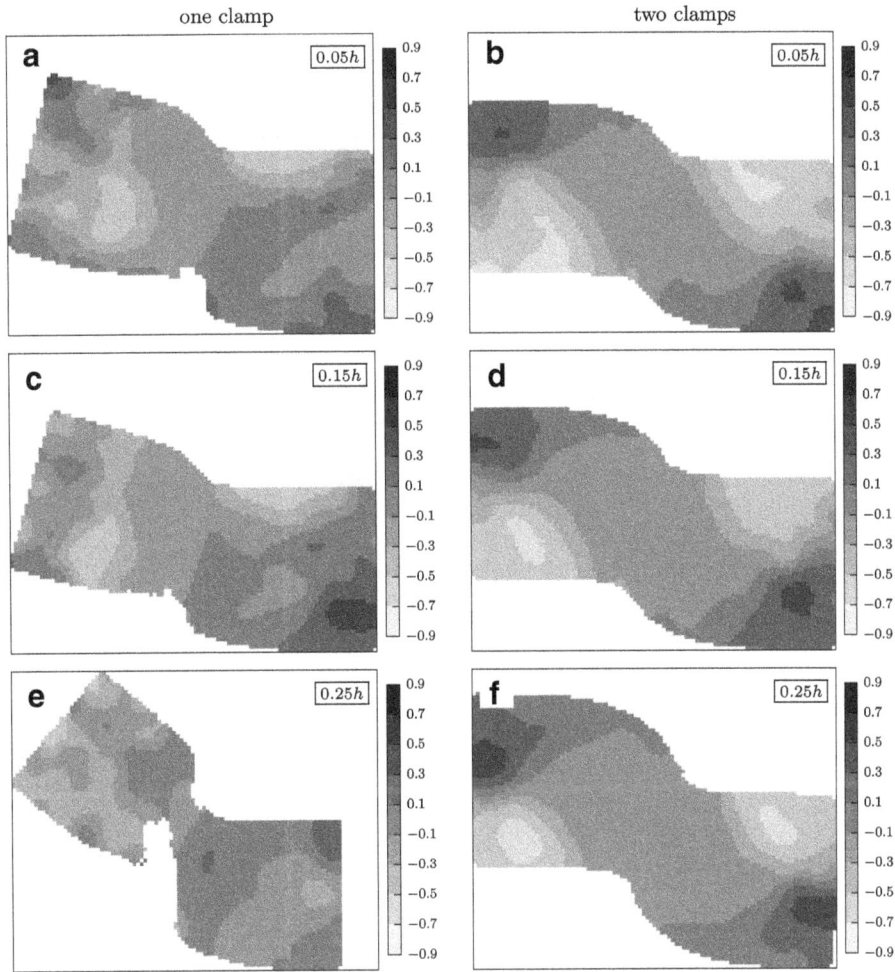

Fig. 14 a–f Experimentally based stress triaxiality (Eq. (2)) fields at crack initiation for the medium-strength material using one clamp (*left panels*) and two clamps (*right panels*). Clearances, expressed as fractions of sheet thickness h, are indicated in each panel

simulations can be matched to the resolution in the measurements to obtain comparable results. This matching was initially done in the simulations, but the experimental results required filtering that reduced the effective resolution and the limitation in the measurements close to the edges was substantial. The measurements at the edges were also negatively influenced by the sample preparation as discussed by Gustafsson et al. (2016b). In the present state, the experiments are best used if the data close to the edges are ignored and the remaining data are used for validation of a finite element model that later can provide the missing experimental data.

A possible weak point in the present study is the identification of the point of crack initiation, which is the point during the shearing where all comparisons are made. The point of crack initiation was taken when the crack became visible on the surface in the captured images. Possibly, micro-cracks existed earlier, and the cracks on the surface were maybe formed earlier or later than in the bulk material. A visible crack in a heavily sheared surface is quite subjective, and possibly, the tool displacement at crack initiation, that was also used in the FE simulations, was misinterpreted by a few tenth of a millimetre. Still, this would unlikely change the stress triaxiality to the magnitude that the observed spread would vanish or a clear trend become visible.

Conclusions

The stress and strain conditions at crack initiation in shearing of two steel sheet grades with various clearance and clamping were studied by calculations based on measured displacement fields and by finite element simulations. The finite element model was first validated against experimentally measured tool forces and strain fields on the sheet surface. Conclusions from the study were as follows:

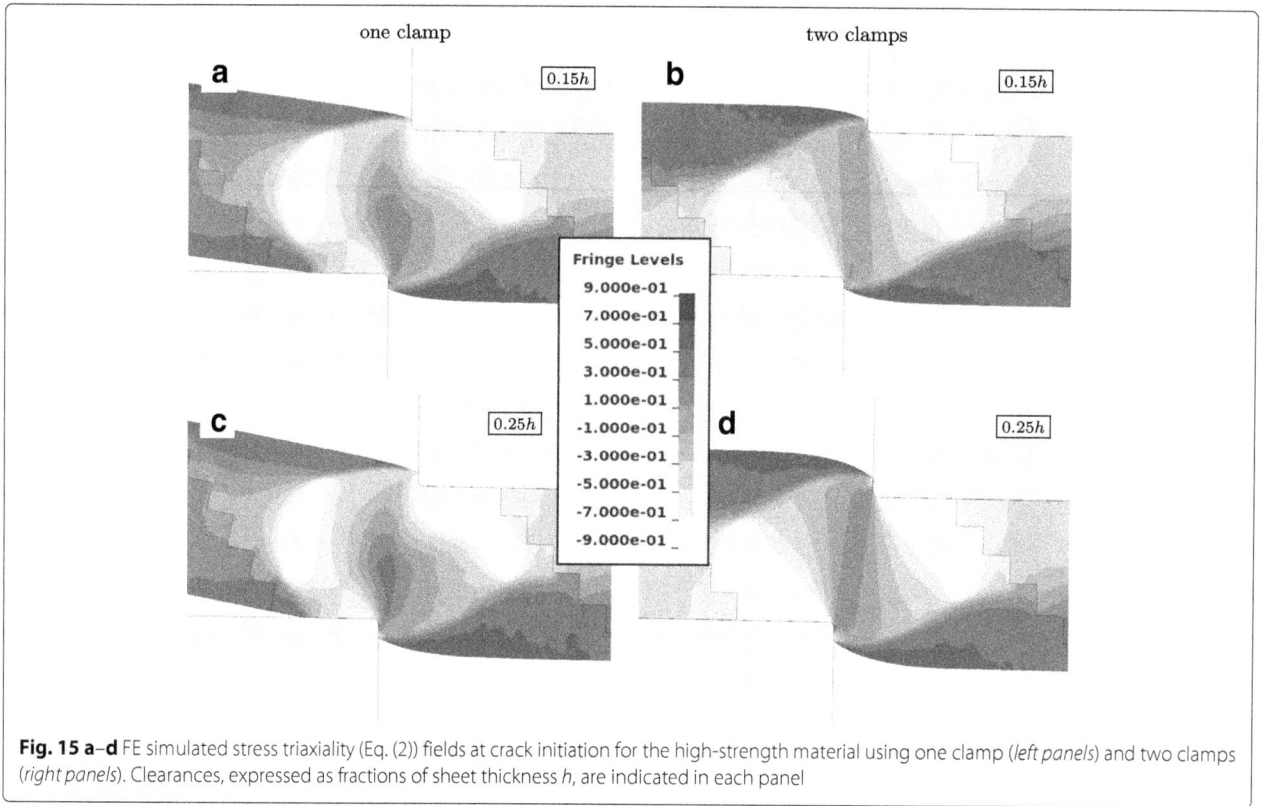

Fig. 15 a–d FE simulated stress triaxiality (Eq. (2)) fields at crack initiation for the high-strength material using one clamp (*left panels*) and two clamps (*right panels*). Clearances, expressed as fractions of sheet thickness *h*, are indicated in each panel

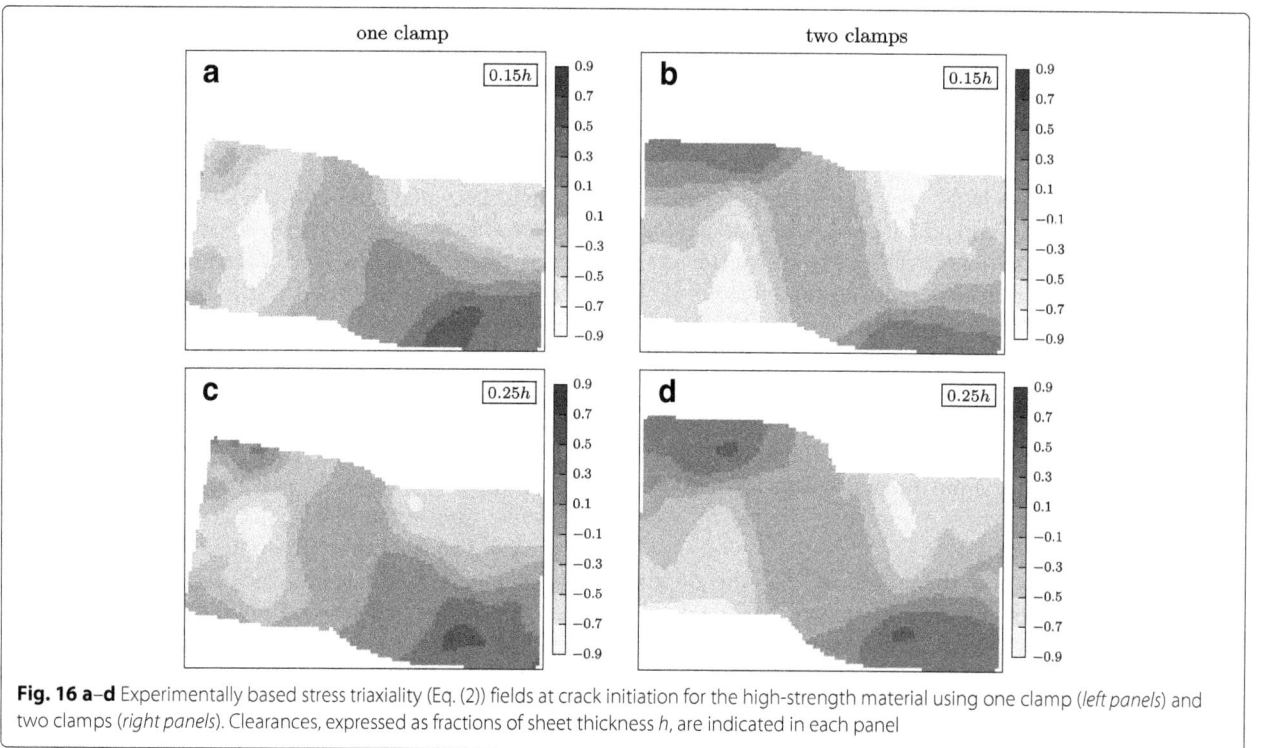

Fig. 16 a–d Experimentally based stress triaxiality (Eq. (2)) fields at crack initiation for the high-strength material using one clamp (*left panels*) and two clamps (*right panels*). Clearances, expressed as fractions of sheet thickness *h*, are indicated in each panel

Table 2 Summary of effective strains $\bar{\varepsilon}$ (Eq. (3)), von Mises stresses $\bar{\sigma}$ [MPa] (Eq. (1)) and stress triaxiality η (Eq. (2)) at crack initiation for both experiments and simulations

| | Medium strength | | | | | | High strength | | | |
| | 1 clamp | | | 2 clamps | | | 1 clamp | | 2 clamps | |
	0.05h	0.15h	0.25h	0.05h	0.15h	0.25h	0.15h	0.25h	0.15h	0.25h
$\bar{\varepsilon}_{I}^{exp}$	1.7	1.7	3.0	1.0	1.1	1.2	0.6	0.5	0.6	1.0
$\bar{\varepsilon}_{I}^{sim}$	2.4	2.2	3.5	2.0	1.9	1.9	1.5	1.7	1.1	1.2
$\bar{\varepsilon}_{II}^{sim}$	1.7	1.5	2.9	1.5	1.5	1.5	1.0	0.9	0.8	0.9
$\bar{\sigma}_{I}^{exp}$	850	840	870	820	820	840	1480	1480	1480	1480
$\bar{\sigma}_{I}^{sim}$	930	930	990	850	900	910	1520	1530	1530	1480
η_{I}^{exp}	0.2	0.05	0.2	0.1	0.1	0.1	0.2	0.2	0.05	0.1
η_{I}^{sim}	0.3	0.05	0.3	0.2	0.4	0.5	0.0	0.1	-0.2	0.3
η_{III}^{sim}	0.7	0.7	0.8	0.4	0.5	0.6	0.3	0.4	0.2	0.4

These parameters were evaluated in three regions denoted I, II and III and located as shown in Fig. (17). Steel sheets of two strength levels, two clamping configurations and three clearances, expressed as fractions of sheet thickness h, were used

- Effective strains and von Mises stresses were similar to the sheet surface and inside the bulk material. Mean stress, and therefore also triaxiality, was however not comparable on the surface and inside the material.
- There were large gradients in strain and stress around the curved tool profiles that were not captured by the experimental method. The results are still useful in combination with finite element simulations that are first validated with the experimental data and later used to provide missing data of stress and strain at the sheet edges.
- Slightly larger effective strains and von Mises stresses were in general seen when shearing with one clamp compared with two clamps, but no clear trend was seen for the stress triaxialities.

Acknowledgements
This work was performed at Luleå University and at Dalarna University within the Swedish Steel Industry Graduate School, with financial support from Dalarna University (Högskolan Dalarna), Swedish Steel Producers' Association (Jernkontoret), County Administrative Board of Gävleborg (Länsstyrelsen i Gävleborg), Regional Development Council of Dalarna (Region Dalarna), Regional Development Council of Gävleborg (Region Gävleborg), Municipality of Sandviken (Sandvikens kommun) and SSAB.

Authors' contributions
EG conducted the experiments and FE simulations, collected the data and wrote the major part of the paper. SM calculated the stress from experimental measurements. MO and LK are EG's PhD supervisors, who guided and supported this work and contributed with their expertise and advices. All authors read and approved the final manuscript.

Competing interests
The authors declare that they have no competing interests.

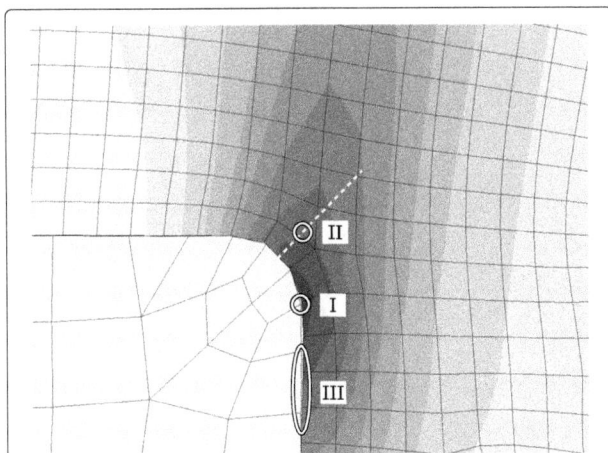

Fig. 17 Definition of locations, denoted I–III, close to the arc-shaped tool profile where the parameters in Table 2 were evaluated. *Point I*: transition from a vertical to an arc-shaped tool profile. *Point II*: 0.1 mm from middle of the arc profile along a direction 45° from the horizontal. *Region III*: close to the vertical tool profile just below point I. The image was taken from a simulation of shearing the medium-strength material with two clamps and 0.15h clearance. As length reference, the undeformed element size in the sheet mesh is 0.1 mm

References
Cockcroft, MG, & Latham, DJ (1968). Ductility and the workability of metals. *Journal of the Institute of metals, 96*(1), 33–39.
Gustafsson, E, Oldenburg, M, Jansson, A (2014). Design and validation of a sheet metal shearing experimental procedure. *Journal of Materials Processing Technology, 214*(11), 2468–2477.
Gustafsson, E, Karlsson, L, Oldenburg, M (2016a). Experimental study of forces and energies during shearing of steel sheet with angled tools. *International Journal of Mechanical and Materials Engineering, 11*(1), 10.
Gustafsson, E, Karlsson, L, Oldenburg, M (2016b). Experimental study of strain fields during shearing of medium- and high-strength steel sheet. *International Journal of Mechanical and Materials Engineering, 11*(1), 14.

Hild, F, & Roux, S (2006). Digital image correlation: from displacement measurement to identification of elastic properties—a review. *Strain*, *42*(2), 69–80.

Hollomon, JH (1945). Tensile deformation. *Transactions of the American institute of mining and metallurgical engineers*, *162*, 268–290.

Johnson, GR, & Cook, WH (1985). Fracture characteristics of three metals subjected to various strains, strain rates, temperatures and pressures. *Engineering Fracture Mechanics*, *21*(1), 31–48.

Kajberg, J, & Lindkvist, G (2004). Characterisation of materials subjected to large strains by inverse modelling based on in-plane displacement fields. *International Journal of Solids and Structures*, *41*(13), 3439–3459.

Marth, S, Häggblad, HÅ, Oldenburg, M, Östlund, R (2016). Post necking characterisation for sheet metal materials using full field measurement. *Journal of Materials Processing Technology*, *238*, 315–324.

McClintock, FA (1968). A criterion for ductile fracture by the growth of holes. *Journal of Applied Mechanics*, *35*(2), 363–371.

Rice, J, & Tracey, D (1969). On the ductile enlargement of voids in triaxial stress fields. *Journal of the Mechanics and Physics of Solids*, *17*(3), 201–217.

Sjödahl, M (1994). Electronic speckle photography: increased accuracy by nonintegral pixel shifting. *Applied Optics*, *33*(28), 6667–6673.

Sjödahl, M (1997). Accuracy in electronic speckle photography. *Applied Optics*, *36*(13), 2875–2885.

Stegeman, Y, Goijaerts, A, Brokken, D, Brekelmans, W, Govaert, L, Baaijens, F (1999). An experimental and numerical study of a planar blanking process. *Journal of Materials Processing Technology*, *87*, 266–276.

16

Numerical investigation on turbulent forced convection and heat transfer characteristic in spirally semicircle-grooved tube

Pitak Promthaisong[1], Amnart Boonloi[2] and Withada Jedsadaratanachai[1*]

Abstract

Turbulent forced convection and heat transfer structure in the spirally semicircle-grooved tube heat exchanger are numerically examined. The computational problem is solved by finite volume method (FVM) with the SIMPLE algorithm. The influences of groove depth and helical pitch on heat transfer, pressure loss, and thermal performance are investigated for turbulent regime, $Re = 5000\text{–}20{,}000$. As a result, the swirling flow is found through the test section due to the groove on the tube wall. The flow structure in the spirally semicircle-grooved tube can separate into two types: main and secondary swirling flows. The main swirling flow is found in all cases, while the secondary swirling flow is detected when $DR \geq 0.06$. The swirling flow disturbs the thermal boundary layer on the tube wall that is an important reason for heat transfer augmentation. In range studies, the enhancements on heat transfer and friction loss are around 1.16–1.96 and 1.2–10.8 time above the smooth tube, respectively. The optimum thermal performance is around 1.11, which detected at $DR = 0.06$, $PR = 1.4$, and $Re = 5000$. The correlations of the Nusselt number and friction factor for the spirally semicircle-grooved tube with $PR = 1.4$ are produced to help to design the tube heat exchanger.

Keywords: Turbulent flow, Heat transfer, Pressure drop, Spirally semicircle-grooved tube

Background

Many types of heat exchangers, shell and tube heat exchanger, fin and tube heat exchanger, plate fin heat exchanger, etc., are used in various industries such as heat treatment process, chemical industry, power plant, automotive industry, food industry, etc. The heat exchangers had been improved with the main aims to augment the thermal performance and to reduce the size of the heat exchanger. The improvement techniques of the heat exchanger are divided into two types: active and passive techniques. The thermal performance improvement with the passive method is widely used more than the active mode due to it does not require the additional power into the heating system. The passive method is to use the turbulator or vortex generator such as fin, rib, baffle, wing, winglet, and roughness surface to generate the vortex flow or swirling flow in the heat exchanger that helps to increase the heat transfer rate and performance. The selection of the turbulator depends on the application of the heat exchanger. The roughness tube is a popular type of the vortex generator, which always use to enhance heat transfer rate and performance, due to it gives lower friction loss than the other types of the turbulators.

The performance improvement in the heat exchangers with the roughness surface had been investigated by many researches. For example, Lu et al. (2013a; 2013b) presented the convective heat transfer in a spirally grooved tube and transversely grooved tube as Fig. 1a, b, respectively. The spirally grooved tube with $e/D = 0.0238$, 0.0306, 0.0381, and 0.0475 and the transversely grooved tube with $e/D = 0.038$, 0.046, and 0.092 were studied. Aroonrat et al. (2013) investigated the heat transfer in an internally grooved tube with constant wall heat flux. The internally grooved tube with the grooved depth, $e = 0.2$ mm, grooved width, $w = 0.2$ mm, and grooved pitch, $p = 12.7$, 203, 254, and 305 mm was investigated. They found that the internally grooved tube provides the Nusselt number and friction factor higher than the smooth tube. The effects of the helically grooved tube on heat transfer and pressure drop of R–134a were experimentally studied by Laohalertdecha and Wongwises (2010).

* Correspondence: kjwithad@kmitl.ac.th
[1]Department of Mechanical Engineering, Faculty of Engineering, King Mongkut's Institute of Technology Ladkrabang, Bangkok 10520, Thailand
Full list of author information is available at the end of the article

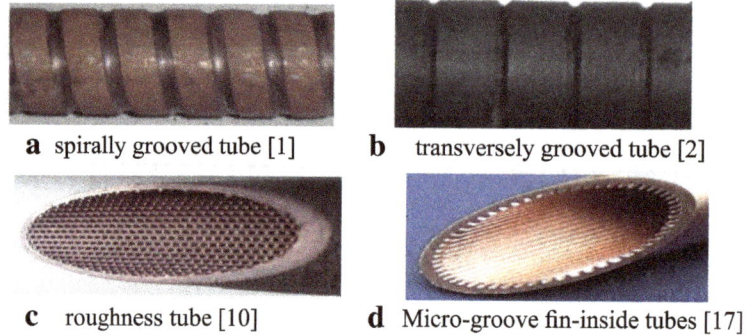

a spirally grooved tube [1]   **b** transversely grooved tube [2]

c   roughness tube [10]   **d** Micro-groove fin-inside tubes [17]

Fig. 1 Geometries of the roughness tube from the published papers. **a** Spirally grooved tube (Lu et al. 2013a); **b** transversely grooved tube (Lu et al. 2013b); **c** roughness tube (Webb 2009); and **d** Micro-groove fin-inside tubes (Tang et al. 2007)

They studied with the grooved depth, $e = 1.5$ mm and grooved pitch, $p = 5.08$, 6.35 and 8.46 mm. Their results show that the maximum heat transfer coefficient ratio and friction factor ratio up to 50 % and 70 % higher than the smooth tube, respectively. Khoeini et al. (2012) experimental investigated of R–134a on heat transfer in a corrugated tube with the depth of corrugation, $e = 1.5$ mm and pitch of corrugation, $p = 8$ mm. They found that the corrugated tube has a significant effect on condensation heat transfer characteristic and provides the condensation heat transfer coefficient higher than smooth tube. The numerical study on turbulent flow and heat transfer enhancement in a rectangular grooved tube was reported by Liu et al. (2015). They found that the swirling flow, which induced by the grooved wall, leads to enhanced heat transfer. Al-Shamani et al. (2015) presented the heat transfer enhancement in a channel with a trapezoidal rib–groove in a turbulent region. They claimed that the channel with trapezoidal rib–groove can enhance heat transfer higher than smooth tube. The friction loss in a spirally internally ribbed tube with a vapor liquid two-phase was investigated by Cheng and Chen (2007). They stated that the friction loss in the spirally internally ribbed tube is around 1.6–2.7 times over the smooth tube. Hwang et al. (2003) presented the heat transfer and pressure drop in an enhanced titanium tube on turbulent region. The four types of the roughness tube were studied. They found that the thermal performance for using the roughness tube is higher than the smooth tube. Webb (2009) studied the heat transfer and friction loss in a three-dimensional roughness tube (see Fig. 1c). Webb (2009) concluded that the roughness tube in the range studied provides the highest Nusselt number 3.74 times over the smooth tube. The comparison between the carbon steel smooth tube and spirally fluted tube on heat transfer characteristics was investigated by Wang et al. (2000). The experimental results show that the total heat transfer coefficient of the carbon steel spirally fluted tube is around 10–17 % higher than the carbon steel smooth tube. Dong et al. (2001) presented the experimental studied the

turbulent flow in a spirally corrugated tube on pressure drop and heat transfer. The roughness height ratio, $e/D = 0.0243$, 0.0199, 0.0302, and 0.0398 and spiral pitch ratio $p/D = 0.623$, 0.62, 0.438, 0.598 were investigated. The spirally corrugated tube gives the heat transfer coefficient and the friction factor around 30–120 % and 60–160 % when compared with the smooth tube. The heat transfer and pressure drop during condensation of refrigerant 134a were studied in an axially grooved tube by Graham et al. (1999). They summarized that the axially grooved tube performs heat transfer rate better than smooth tube. The experimental investigation of a vertical tube with internal helical ribs on heat transfer was presented by Zhang et al. (2015). They concluded that the heat transfer coefficient for using the vertical tube with internal helical ribs is around 1.41 times higher than the smooth tube. Dizaji et al. (2015) reported the experimental investigation on heat transfer and pressure drop on turbulent flow in a corrugated tube. The four types of the corrugated tube were investigated. They stated that the inner tube provides the Nusselt number and friction factor about 10–52 % and 150–190 %, respectively, over the smooth tube. Zhang et al. (2015) presented the effect of an internally grooved tube on heat transfer enhancement and pressure drop for using R417A. Their results indicated that the grooved tube in the range studied gave the maximum heat transfer performance and pressure drop about 2.8 and 2.6 times, respectively, above the smooth tube. Tang et al. (2007) used the roughness tube (see Fig. 1d) to study an oil-filled high-speed spinning process. Aroonrat and Wongwises (2011) studied the evaporation heat transfer characteristic and friction loss of R–134a in a vertical corrugated tube. They pointed out that the heat transfer and two-phase friction factor for using the corrugated tube are around 0–10 % and 70–140 %, respectively, when compared with the base case. Liu et al. (2013) investigated the flow behavior and heat transfer characteristic in shell and tube heat exchanger for using a helical baffled cooler with a spirally corrugated tube. The results show that the Nusselt number increases around 60–130 % over the smooth tube on the tube side. Chen et al. (2013) presented the experimental investigation on the effect of a transversally corrugated tube for heat transfer and pressure drop with molten salt. The corrugated tubes with $e/D = 0.034$, 0.047 and $p/D = 0.31$, 0.57, 1 were investigated. They showed that the transversally corrugated tube gives the heat transfer and pressure drop grater that the smooth tube.

Most of previous researches, the heat transfer enhancement in the spirally grooved tube had been studied with the experimental method. In the present work, the heat transfer and flow structure in the spirally semicircle-grooved tube are investigated numerically. The numerical analysis can describe the flow configuration and heat transfer characteristic in the tube heat exchanger. The understanding on the mechanisms of flow and heat transfer is an important way to improve the thermal performance of the heat exchanger. The effects of groove depth in term of depth ratio ($DR = e/D$) and helical pitch length ($PR = p/D$) are considered for the turbulent regime, $Re = 5000$–$20,000$.

Physical geometry and computational domain

The geometries and computational domain for the spirally semicircle-grooved tube are shown in Fig. 2. In the present study, the spirally semicircle-grooved tube is investigated with different helical pitch ratios, p/D, $PR = 0.6$, 0.8, 1.0, 1.2, 1.4, 1.6, 1.8, and 2.0 and depth ratios, d/D or $DR = 0.02$, 0.04, 0.06, 0.08, and 0.10. The characteristic diameter of tube, D, is fixed around 0.05 m, while the Reynolds numbers, Re, ranging from 5000 to $20,000$.

The tetrahedral mesh is selected for the computational domain of the spirally semicircle-grooved tube. At the near wall region, grids have high density based on the gradients of temperature, pressure, and velocity, which the first near-wall cell next to the wall which is determined by the Reynolds number and satisfies $y^+ \approx 1$.

Assumption and boundary condition

The inlet and outlet of the computational domain are set to be periodic boundary. No-slip wall condition is applied for the tube wall. The uniform heat flux around 600 W/m^2 is adopted for the heating wall. Air as the working fluid flows into the test section with the uniform mass flow rate at 300 K ($Pr = 0.707$). The thermodynamic properties of the air are assumed to be constant at mean bulk temperature. The forced convection heat transfer is considered, while the natural convection and radiation heat transfer are disregarded. The body force and viscous dissipation are ignored.

Mathematical foundation and numerical method

The incompressible turbulent flow with steady operation in three dimensions and heat transfer characteristic in the spirally semicircle-grooved tube are governed by the continuity equation, the Navier-Stokes equation, and energy equation. The realizable k-ε turbulent model (Launder and Spalding 1974) is used for the present investigation.

$$\frac{\partial(\rho k)}{\partial t} + \frac{\partial(\rho k u_j)}{\partial x_j} = \frac{\partial}{\partial x_j}\left[\left(\mu + \frac{\mu_t}{\sigma_k}\right)\frac{\partial k}{\partial x_j}\right] + G_k + G_b + \rho\varepsilon - Y_m + S_k \quad (1)$$

and

$$\frac{\partial(\rho\varepsilon)}{\partial t} + \frac{\partial(\rho\varepsilon u_j)}{\partial x_j} = \frac{\partial}{\partial x_j}\left[\left(\mu + \frac{\mu_t}{\sigma_\varepsilon}\right)\frac{\partial\varepsilon}{\partial x_j}\right] + \rho C_1 S\varepsilon - \rho C_2 \frac{\varepsilon^2}{k + \sqrt{v\varepsilon}} + C_{1\varepsilon}\frac{\varepsilon}{k}C_{3\varepsilon}G_b + S_\varepsilon \quad (2)$$

where

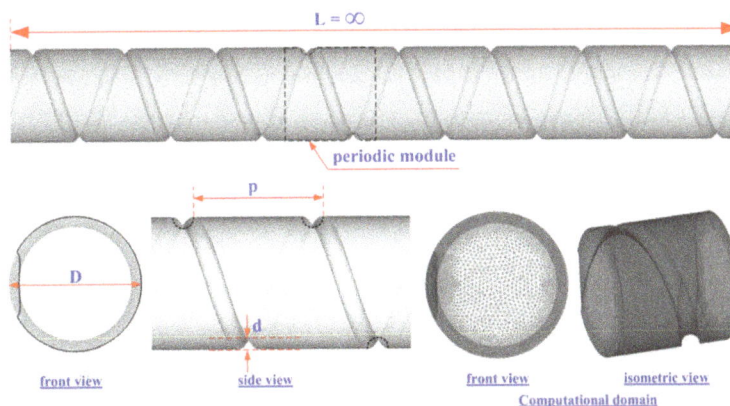

Fig. 2 Spirally semicircle-grooved tube geometries and computational domain

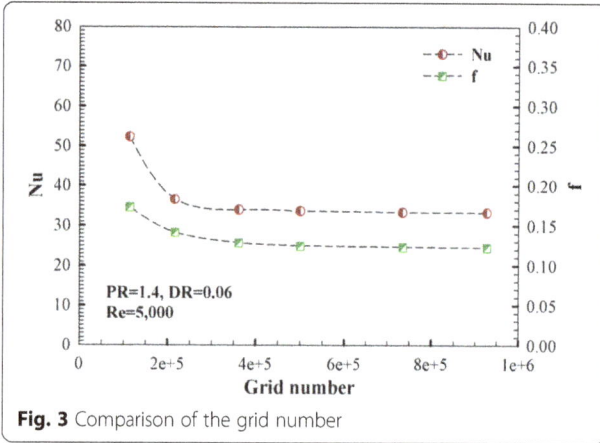

Fig. 3 Comparison of the grid number

$$Re = \frac{\rho u_0 D_h}{\mu} \tag{5}$$

The friction factor is defined as

$$f = \frac{(\Delta P / L) D_h}{2 \rho \bar{u}^2} \tag{6}$$

where ΔP is the pressure drop, D_h is the hydraulic diameter, and u is mean flow velocity.

The local Nusselt number is given by

$$Nu_x = \frac{h_x D_h}{k} \tag{7}$$

where h is the convective heat transfer coefficient and k the thermal conductivity.

The average Nusselt number can be obtained by

$$Nu = \frac{1}{A} \int Nu_x dA \tag{8}$$

Thermal performance enhancement factor (TEF) is defined as the ratio of the heat transfer coefficient of an augmented surface, h, to that of a smooth surface, h_0, under the constant pumping power condition. TEF can be calculated from

$$C_1 = \max \left[0.43, \frac{\eta}{\eta + 5} \right], \eta = S \frac{k}{\varepsilon}, S$$
$$= \sqrt{2 S_{ij} S_{ij}}, S_{ij} = \frac{1}{2} \left(\frac{\partial u_j}{\partial x_i} + \frac{\partial u_i}{\partial x_j} \right) \tag{3}$$

the constants in the model are given as follows:

$$\sigma_k = 1.0, \sigma_\varepsilon = 1.2, C_{1\varepsilon} = 1.44, C_2 = 1.9 \tag{4}$$

where S_k and S_ε are user-defined source terms, σ_k and σ_ε are the turbulent Prandtl numbers for k and ε, respectively.

For the convection terms, the SIMPLE algorithm for handling the pressure–velocity coupling and the QUICK scheme are solved using a finite volume approach (Patankar 1980). The energy equation is considered to be converged when the normalized residual values are less than 10^{-9} while the other variables are less than 10^{-5}.

In the present study, to analyze the flow behavior and heat transfer characteristics, the Reynolds number, friction factor, Nusselt number, and thermal enhancement factor are the most important parameters, which can be written as follows:

The Reynolds number is calculated as follows:

$$\text{TEF} = \left. \frac{h}{h_0} \right|_{pp} = \left. \frac{Nu}{Nu_0} \right|_{pp} = \left(\frac{Nu}{Nu_0} \right) \Big/ \left(\frac{f}{f_0} \right)^{1/3} \tag{9}$$

where, f_0 and Nu_0 are the friction factor and the Nusselt number of the smooth tube, respectively.

Numerical validation

The grid numbers around 113803, 216084, 361257, 500146, 735175, and 927547 for the spirally semicircle-grooved tube at $PR = 1.4$, $DR = 0.06$, and $Re = 5000$ are compared as shown in Fig. 3. Under similar conditions, the values of the Nusselt number and friction factor rarely change when increasing grid cell from 500146 to 927547. Therefore, the grid around 500146 is used in all

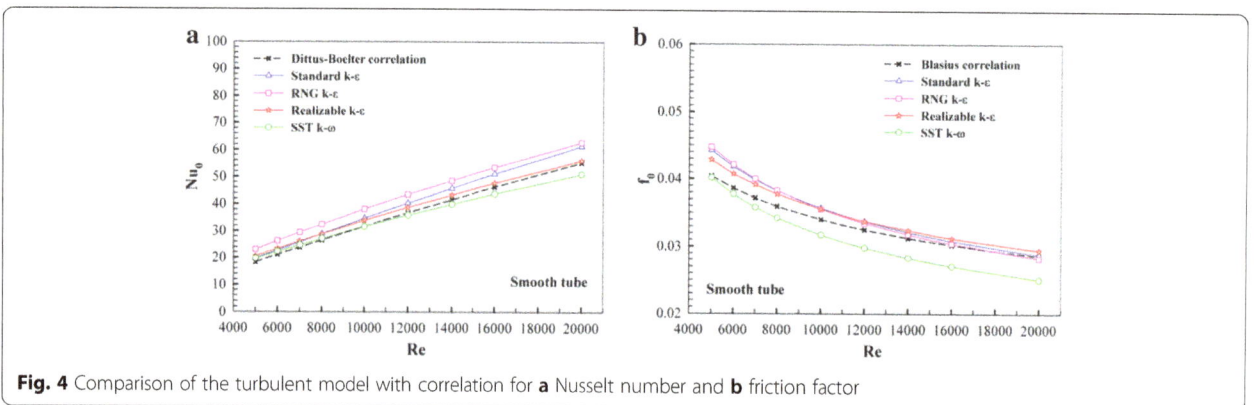

Fig. 4 Comparison of the turbulent model with correlation for **a** Nusselt number and **b** friction factor

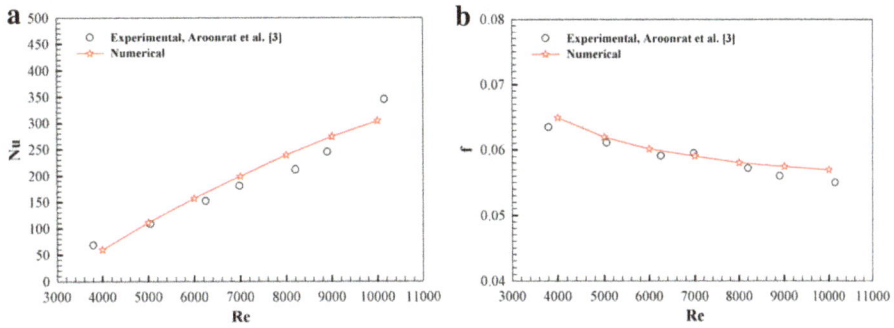

Fig. 5 Comparison of the numerical results with experimental results for **a** Nusselt number and **b** friction factor

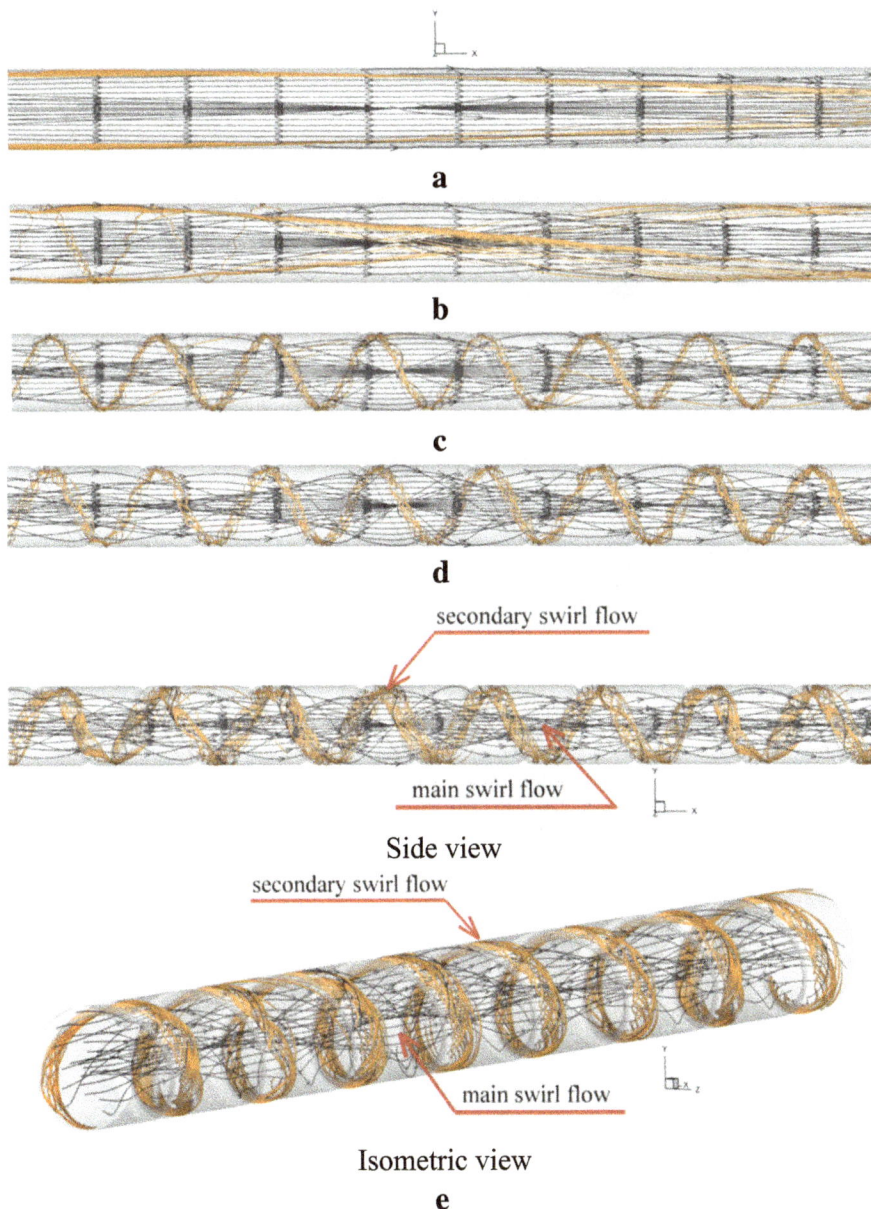

Side view

Isometric view

Fig. 6 Flow structure in axial flow for **a** DR = 0.02, **b** DR = 0.04, **c** DR = 0.06, **d** DR = 0.08, and **e** DR = 0.10 at PR = 1.4 and Re = 5000 of the spirally semicircle-grooved tube

cases of the computational domain for the spirally semicircle-grooved tube.

The comparisons of the turbulence models, the Standard k-ε turbulent model, the renormalized group (RNG) k-ε turbulent model, the realizable k-ε turbulent model, and the shear stress transport (SST) k-ω turbulent model and correlations (Incropera and Dewitt 2006) on the Nusselt number and friction factor of the smooth tube are shown in Fig. 4a, b, respectively. The numerical results show that the use of the realizable k-ε turbulent model provides the excellent solutions with the values from the correlations. The realizable k-ε turbulent model gives relative errors within 10.1 and 6.3 % for the Nusselt number and friction factor, respectively.

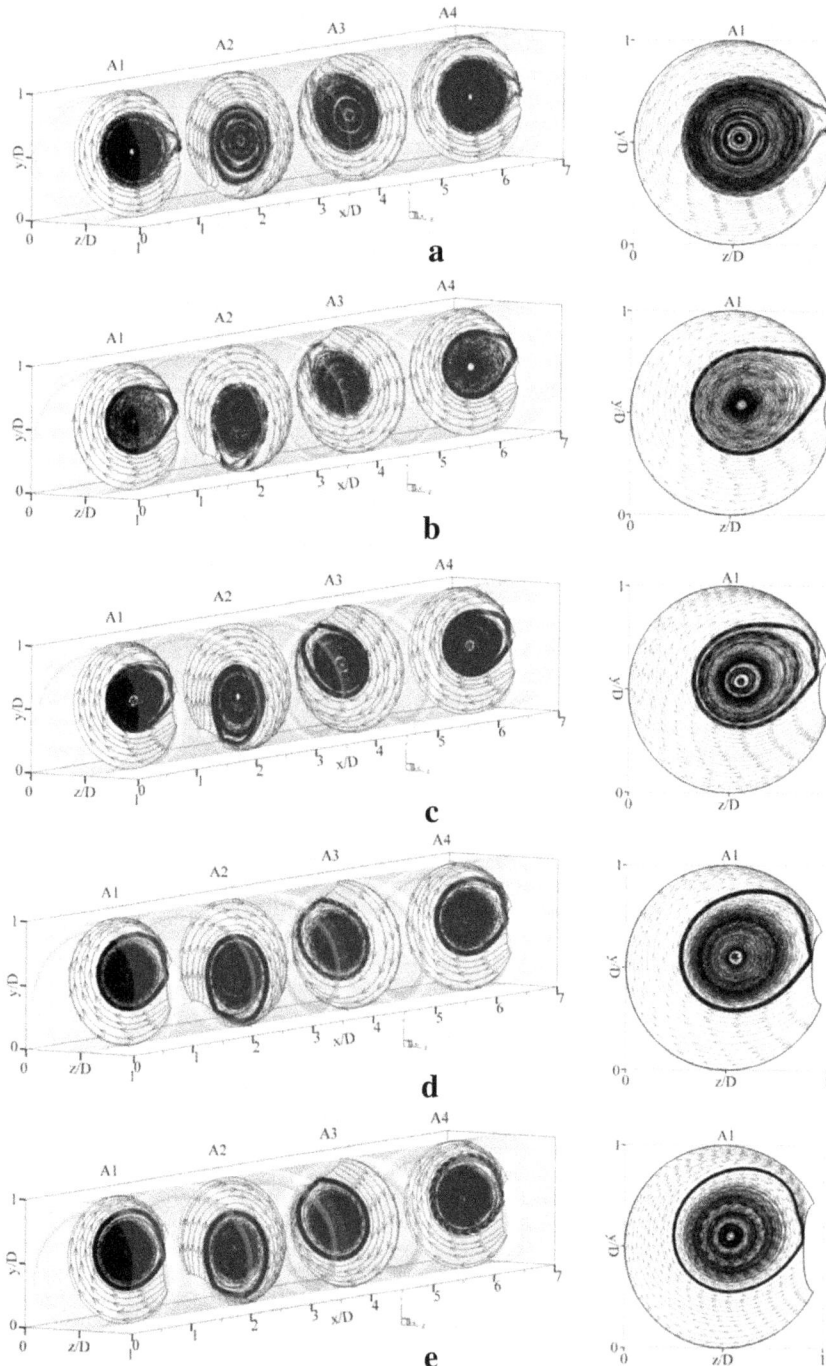

Fig. 7 Streamline in transverse planes for **a** $DR = 0.02$, **b** $DR = 0.04$, **c** $DR = 0.06$, **d** $DR = 0.08$, and **e** $DR = 0.10$ at $PR = 1.4$ and $Re = 5000$ of the spirally semicircle-grooved tube

Figure 5 presents the comparison of the numerical results with experimental results (Aroonrat et al. 2013) for using the realizable k-ε turbulent model in the case of the spirally grooved tube with $e/D_i = 0.028$, $w/D_i = 0.028$, and $p/D_i = 1.78$. In the figure, the deviations around 5.15 and 2.03 % for the Nusselt number and friction factor, respectively, are detected. As a result in this part, it can be concluded that the computation domain of the spirally semicircle-grooved tube has the reliability to predict the flow, heat transfer, and thermal performance.

Numerical result and discussion

The numerical results in the spirally semicircle-grooved tube with various DRs and PRs are separated into four parts: flow configuration, heat transfer characteristics, thermal performance evaluation, and correlation.

Flow configuration

The longitudinal vortex flow and streamlines in cross sectional plane are selected to describe the flow pattern in the spirally semicircle-grooved tube. Fig. 6a−e presents the longitudinal vortex flow through the test

section of the spirally semicircle-grooved tube at $PR = 1.4$ and $Re = 5000$ for $DR = 0.02$, 0.04, 0.06, 0.08, and 0.1, respectively. In general, the generation of the longitudinal vortex flow or swirling flow is due to the groove on the tube wall. As the figure shows, the flow structure can be divided into two types: main and secondary swirling flows. The main swirling flow is found at the core of the test tube, while the secondary swirling flow is detected on the groove surface (see Fig. 6e). The $DR = 0.02$ and 0.04 produce the main swirling flow through the test section but not generate the secondary swirling flow on the groove surface, while the $DR = 0.06$, 0.08 and 0.1 can induce on both main and secondary swirling flows through the tube heat exchanger. This is because the depth of the groove on the tube heat exchanger is an important factor to create the vortex flow. Moreover, the groove depth effects for the strength or intensity of the swirling flow. The $DR = 0.1$ provides the highest strength of the swirling flow, while the $DR = 0.02$ gives the opposite result. The creation of the swirling flow is a disturbance of the thermal boundary layer

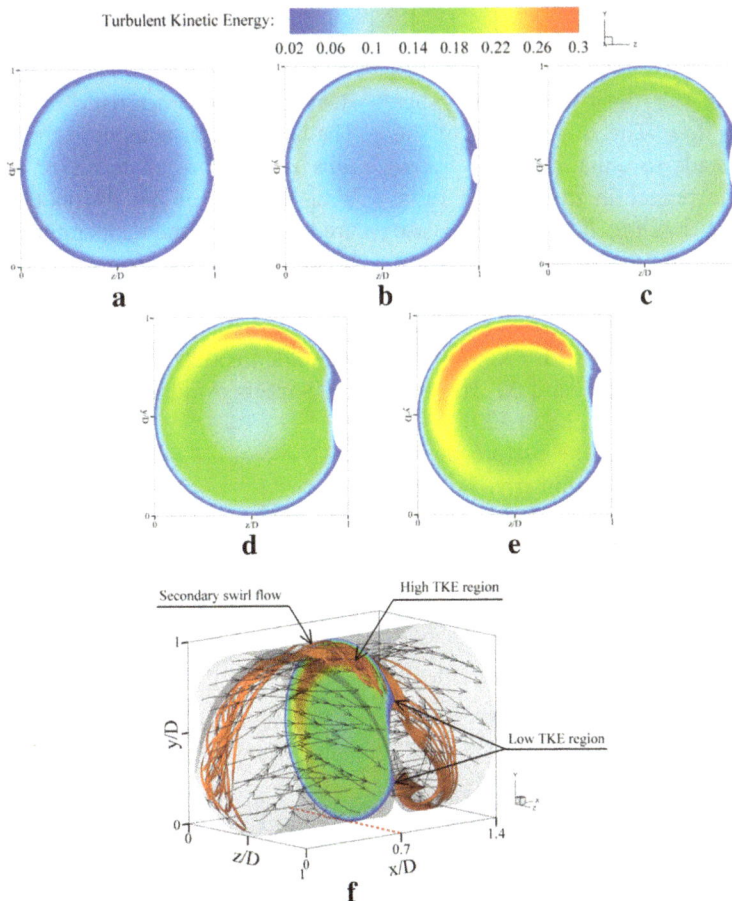

Fig. 8 Turbulent kinetic energy distribution in transverse planes for **a** $DR = 0.02$, **b** $DR = 0.04$, **c** $DR = 0.06$, **d** $DR = 0.08$, **e** $DR = 0.10$, and **f** isometric view of the $DR = 0.10$ at $PR = 1.4$ and $Re = 5000$ of the spirally semicircle-grooved tube

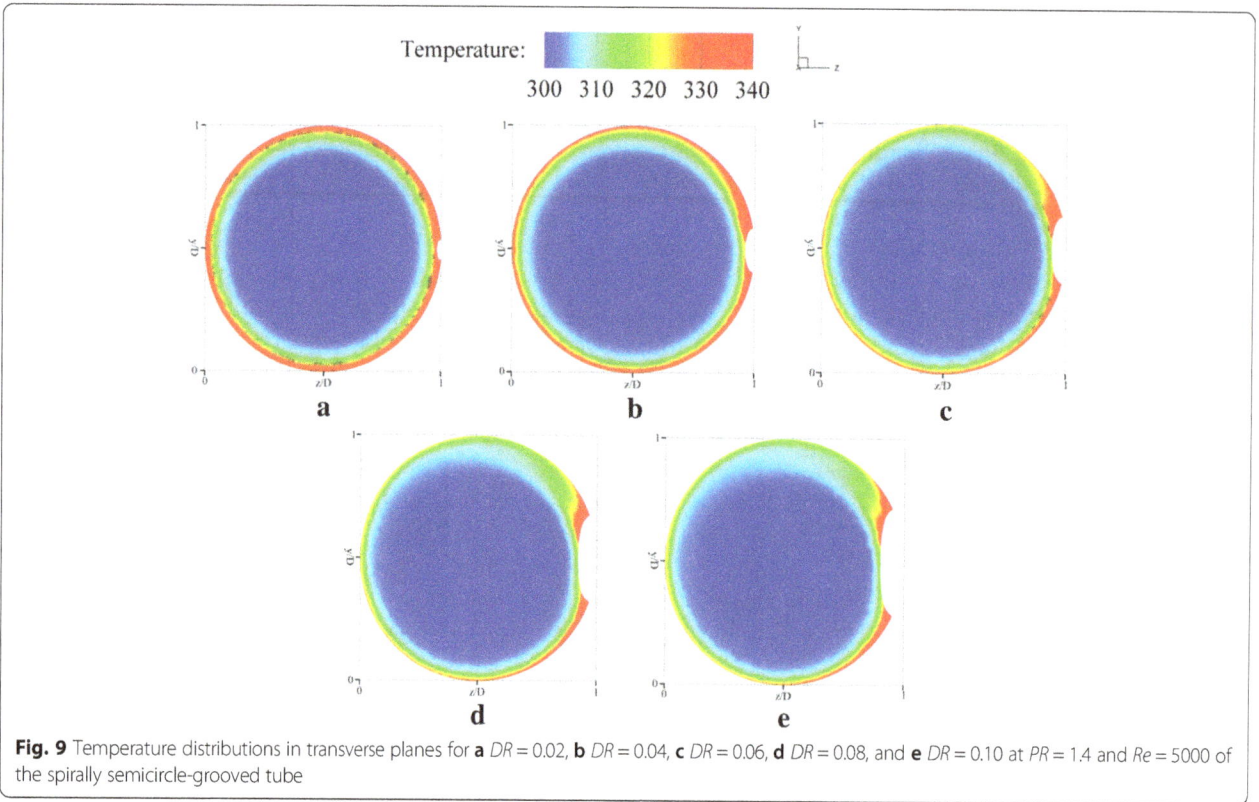

Fig. 9 Temperature distributions in transverse planes for **a** $DR = 0.02$, **b** $DR = 0.04$, **c** $DR = 0.06$, **d** $DR = 0.08$, and **e** $DR = 0.10$ at $PR = 1.4$ and $Re = 5000$ of the spirally semicircle-grooved tube

on the heating wall of the tube heat exchanger, especially, secondary swirling flow. The disturbance of the thermal boundary layer leads to enhanced heat transfer rate and thermal performance in the heating tube. The rising strength of the swirling flow may increase the intensity of the disturbance that a main reason for augmenting heat transfer rate.

Figure 7a–e shows the streamlines in transverse plane of the spirally semicircle-grooved tube with $PR = 1.4$ and $Re = 5000$ for $DR = 0.02, 0.04, 0.06, 0.08$, and 0.1, respectively. Generally, the vortex flows are found in all cases of the spirally semicircle-grooved tube. The pattern of the vortex flow at various DRs is found similarly, but the strength of the vortex flow is not equal. The strength of the vortex flow is considered from the turbulent kinetic energy (TKE) distributions in transverse plane as depicted in Fig. 8a–e, for $DR = 0.02, 0.04, 0.06, 0.08$, and 0.1, respectively, at $PR = 1.4$ and $Re = 5000$. The low TKE distribution is shown with blue contour, while the high TKE value is presented by red contour. The low TKE value is found at the edges of the groove, while the high TKE value is detected at the upper curve of the plane. The highest TKE distribution is found in case of $DR = 0.1$ due to the highest vortex strength. Moreover, it is found that the TKE distribution directly relates with the secondary swirling flow (see Fig. 8e); therefore, the high TKE is not found in case of $DR = 0.02$ and 0.04 due to the secondary swirling flow is not created.

Heat transfer characteristic

The heat transfer analysis in the spirally semicircle-grooved tube is described in terms of temperature contours and local Nusselt number distributions on the tube wall. Figure 9a–e shows the temperature distributions in y-z planes of the spirally semicircle-grooved tube at $PR = 1.4$ and $Re = 5000$ for $DR = 0.02, 0.04, 0.06, 0.08$, and 0.1, respectively. The high temperature is presented with red layer, while the opposite trend is depicted with blue contour. The high temperature is found in a large area at $DR = 0.02$ due to the

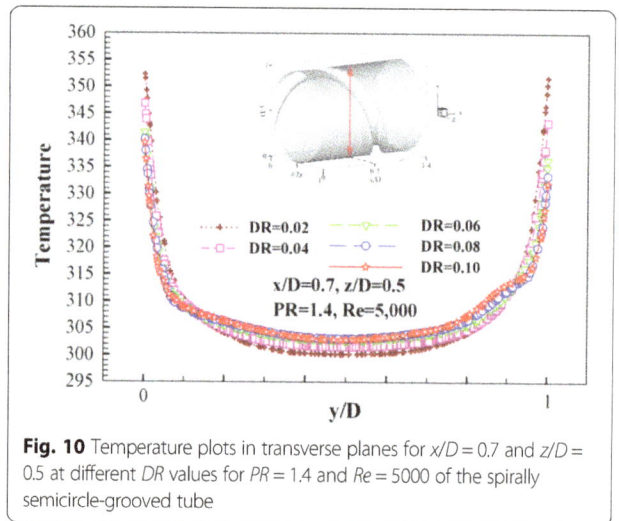

Fig. 10 Temperature plots in transverse planes for $x/D = 0.7$ and $z/D = 0.5$ at different DR values for $PR = 1.4$ and $Re = 5000$ of the spirally semicircle-grooved tube

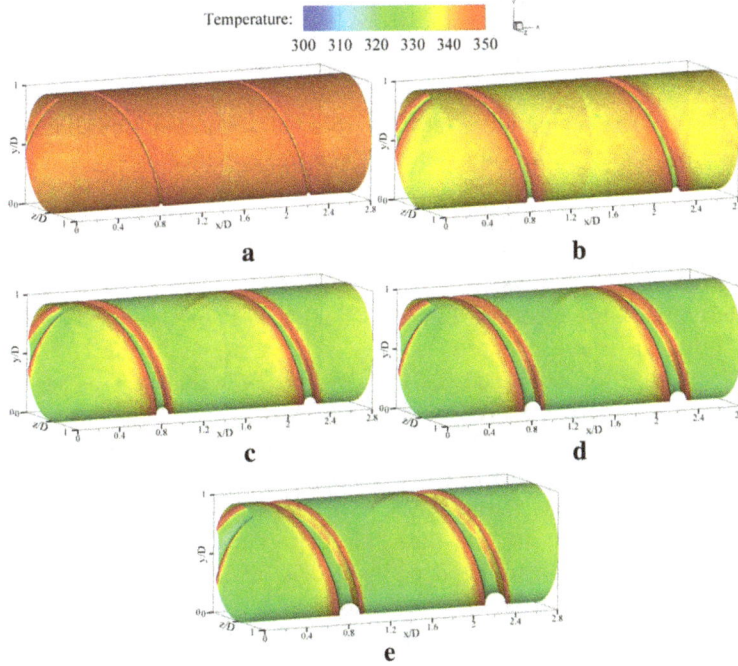

Fig. 11 Temperature distributions on the tube wall for **a** $DR = 0.02$, **b** $DR = 0.04$, **c** $DR = 0.06$, **d** $DR = 0.08$, and **e** $DR = 0.10$ at $PR = 1.4$ and $Re = 5000$ of the spirally semicircle-grooved tube

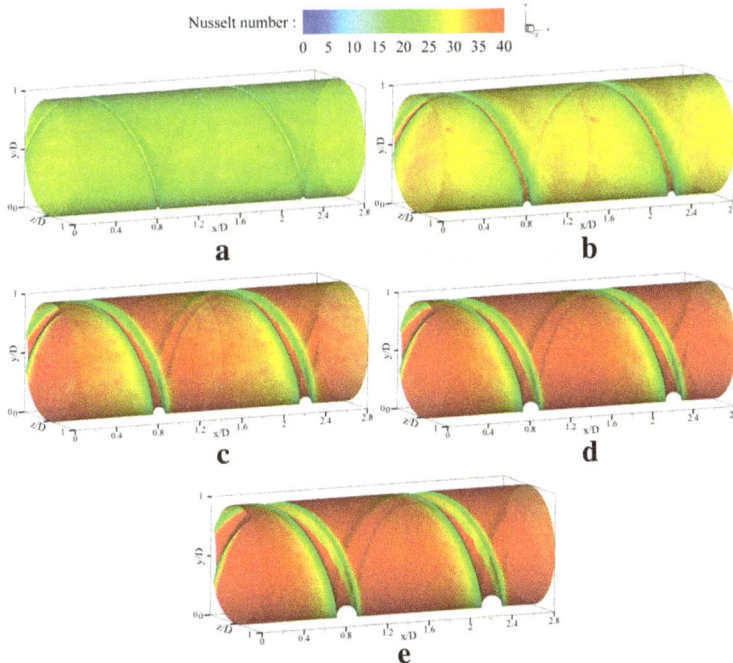

Fig. 12 Nu_x distributions on the tube wall for **a** $DR = 0.02$, **b** $DR = 0.04$, **c** $DR = 0.06$, **d** $DR = 0.08$, and **e** $DR = 0.10$ at $PR = 1.4$ and $Re = 5000$ of the spirally semicircle-grooved tube

poor heat transfer rate, while the reverse trend is detected at $DR = 0.1$. It can be concluded that the swirling flow and strength of the flow in the test tube has directly effect for heat transfer enhancement.

Figure 10 presents the plots of the temperature (at $x/D = 0.7$ and $z/D = 0.5$) with the y/D at various DRs of the spirally semicircle-grooved tube. As the figure, it is found that the $DR = 0.08$ and 0.1 give a better mixing of the fluid flow than the other cases when considered at the middle of the plane (in range $y/D = 0.2–0.8$).

The temperature distributions on the tube wall of the spirally semicircle-grooved tube at various DRs are illustrated in Fig. 11a–e, respectively, for $DR = 0.02$, 0.04, 0.06, 0.08, and 0.1 at $PR = 1.4$ and $Re = 5000$. A similar trend of the heat transfer characteristic, as depicted in Fig. 9, is found. It is interesting to note that the edges of the groove provide terrible heat transfer rate for all DRs.

The local Nusselt number distributions on the tube wall of the spirally semicircle-grooved tube are depicted in Fig. 12a–e for $DR = 0.02$, 0.04, 0.06, 0.08, and 0.1, respectively, at $Re = 5000$ and $PR = 1.4$. It is clearly seen in the figures that the $DR = 0.02$ gives the lowest heat transfer rate. The heat transfer rate increases when

increasing the depth of the semicircle groove on the tube wall due to the intensity of the swirling flow, especially, secondary swirling flow.

Thermal performance evaluation

The performance assessments of the spirally semicircle-grooved tube are reported in terms of the Nusselt number ratio (Nu/Nu_0), friction factor ratio (f/f_0), and thermal enhancement factor (TEF). The influences of the DR and PR on heat transfer, friction loss, and thermal performance are considered.

The variations of the Nu/Nu_0 with the Re at various PR and DR values are presented in Fig. 13. In general, the Nu/Nu_0 slightly decreases when increasing the Reynolds number for all cases. At similar PR, the $DR = 0.1$ provides the highest heat transfer rate, while the $DR = 0.02$ performs the lowest values. The increasing DR leads to enhancing the Nusselt number. In addition, the Nu/Nu_0 of $DR = 0.08$ and 0.1 is very close for all PRs. The reduction of the PR leads to increasing heat transfer rate. The spirally semicircle-grooved tube with $PR = 0.6$ gives the highest heat transfer rate. The Nu/Nu_0 has no change when increasing PR of the spirally semicircle-

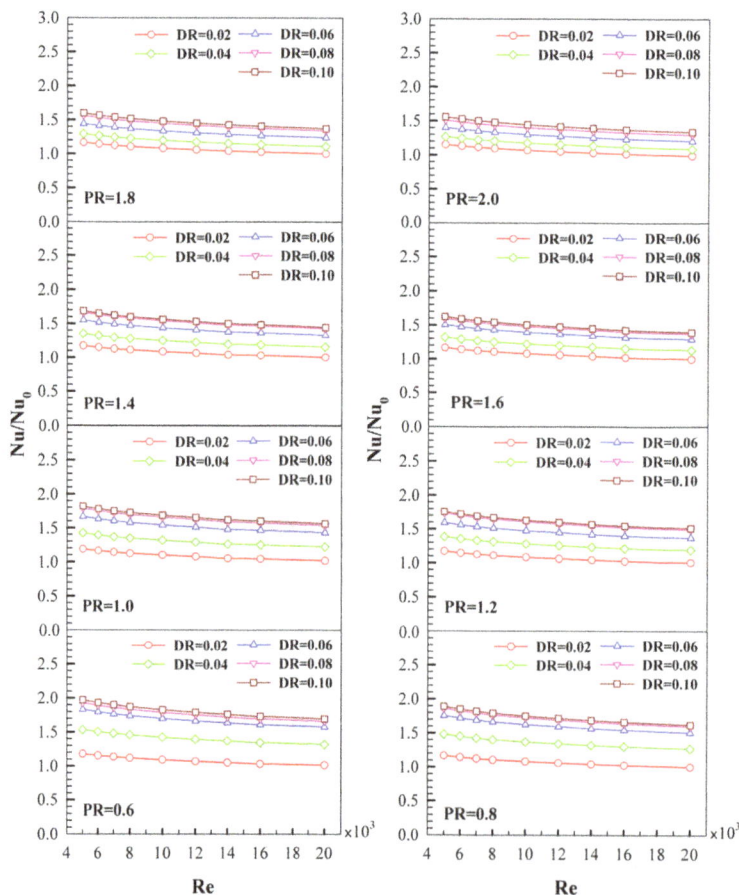

Fig. 13 The relation between Nu/Nu_0 and DR

grooved tube from 1.6 to 2.0. In range investigates, the Nu/Nu_0 for the spirally semicircle-grooved tube is around 1.16–1.96 depended on DR, PR, and Re.

The relations of the f/f_0 with the Reynolds number at various PR and DR values are reported in Fig. 14. Generally, the spirally semicircle-grooved tube has higher friction loss than the smooth circular tube in all cases. The f/f_0 slightly decreases when increasing the Reynolds number. The $DR = 0.1$ provides the highest friction loss, while the $DR = 0.02$ offers the lowest values of the pressure loss when considered at similar PR. The groove depth is the main cause for the augmentation of the pressure loss in the tube heat exchanger. The reduction of the PR can help to reduce the pressure loss in the test tube. The spirally semicircle-grooved tube with $PR = 1.8$ and 2.0 gives very close values of the friction factor ratio. The f/f_0 of the spirally semicircle-grooved tube is around 1.20–10.80 depending on DR, PR and Re.

The plots of the thermal enhancement factor with the Reynolds number with various PRs and DRs are plotted in Fig. 15. The TEF of the smooth circular tube is equal to 1; therefore, the advantage of the spirally semicircle-grooved

tube should give the TEF higher than 1. In general, the TEF decrease with increasing the Reynolds number in all cases. Almost all cases of the spirally semicircle-grooved tube with $PR = 0.6$ perform the TEF lower than 1 due to the pressure loss of the heating tube increases much more than the enhancement of the heat transfer rate when measured at similar pumping power. The maximum TEF is found at $DR = 0.04$ for $PR = 0.6$–1.2, while found at $DR = 0.06$ for $PR = 1.4$–2.0. The computational result reveals that the optimum TEF is around 1.11.

The variations of the Nu/Nu_0, f/f_0, and TEF with the DR at various PR values are created in Figs. 16, 17 and 18, respectively. The increasing DR and decreasing PR lead to augmentations on both heat transfer rate and friction loss in the test tube. The optimum case, which gives the highest thermal performance around 1.11, is found at $DR = 0.06$ and $PR = 1.4$ when considered at the lowest Reynolds number, $Re = 5000$.

Correlation of the Nusselt number and friction loss
The correlations of the Nusselt number and friction factor at $PR = 1.4$, $Re = 5000$–20,000, and $DR = 0.02$–0.10

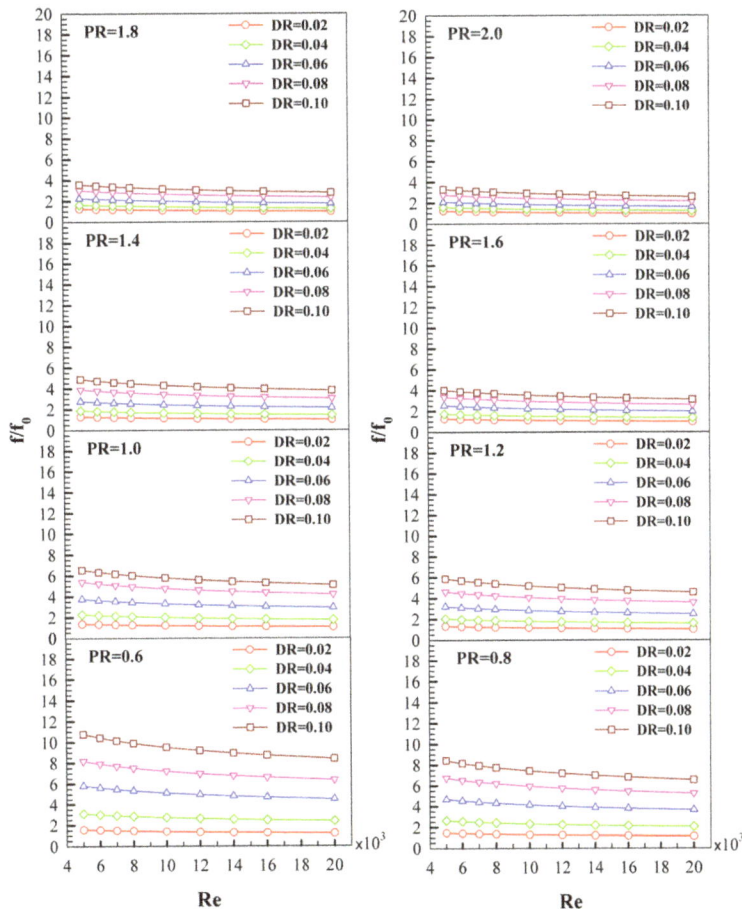

Fig. 14 The relation between f/f_0 and DR

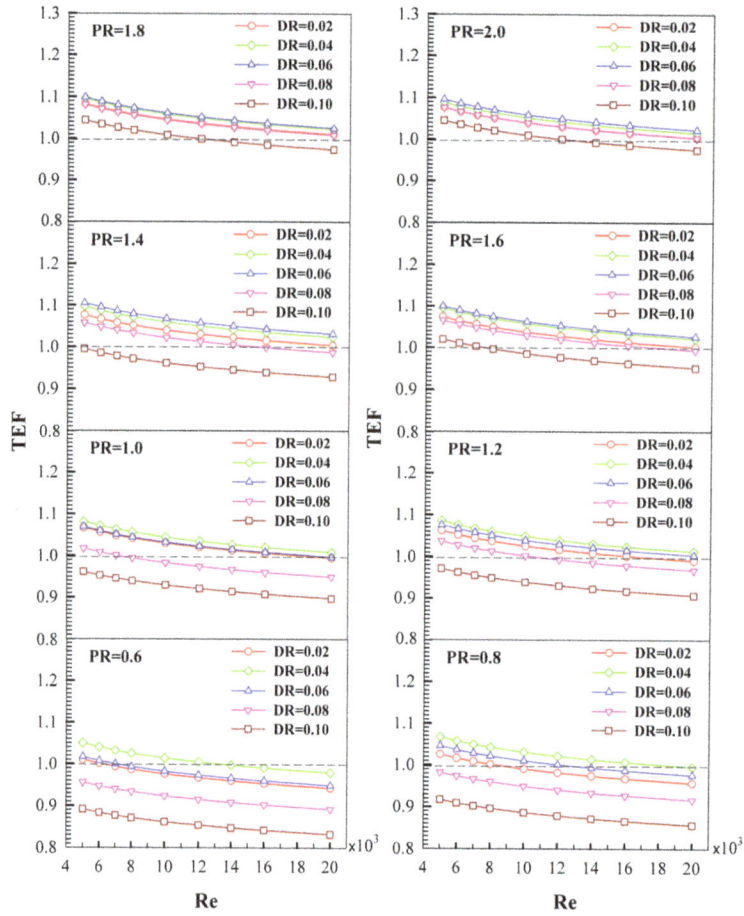

Fig. 15 The relation between TEF and *DR*

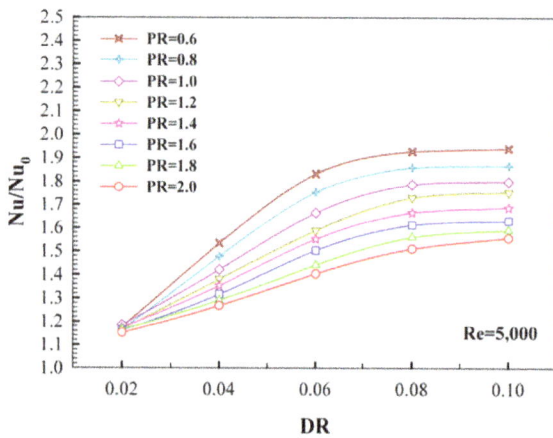

Fig. 16 The relation between *Nu/Nu₀* and *DR*

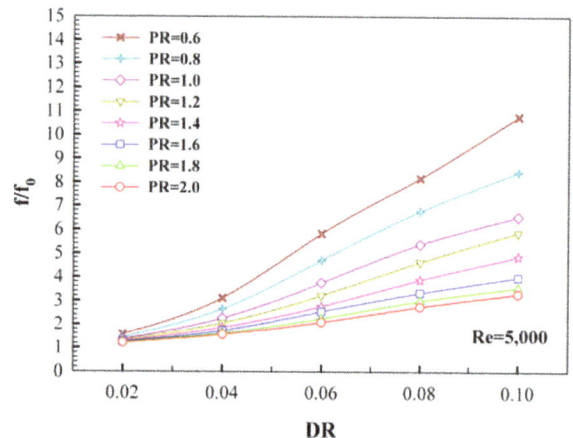

Fig. 17 The relation between *f/f₀* and *DR*

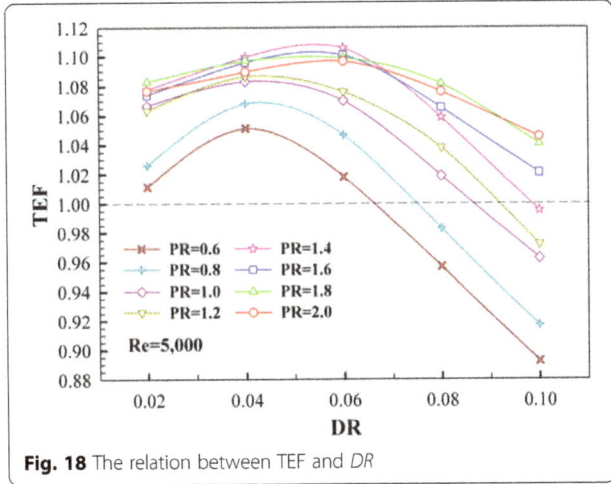

Fig. 18 The relation between TEF and *DR*

are created. The correlations are separated into two parts: $DR \leq 0.06$ (uptrend of TEF) and $DR \geq 0.06$ (downtrend of TEF) as shown in the Fig. 19.

For $DR \leq 0.06$,

$$Nu = 0.411 \mathrm{Re}^{0.614} \mathrm{Pr}^{0.4} DR^{0.249} \tag{10}$$

$$f = 30.568 \mathrm{Re}^{-0.43} DR^{0.674} \tag{11}$$

For $DR \geq 0.06$,

$$Nu = 0.333 \mathrm{Re}^{0.614} \mathrm{Pr}^{0.4} DR^{0.166} \tag{12}$$

$$f = 111.788 \mathrm{Re}^{-0.43} DR^{1.11} \tag{13}$$

The deviations of the Nu and f between numerical data and correlation data are within 1.8 and 6 %, respectively, for uptrend, and within 1.5 and 1.5 % for downtrend, respectively.

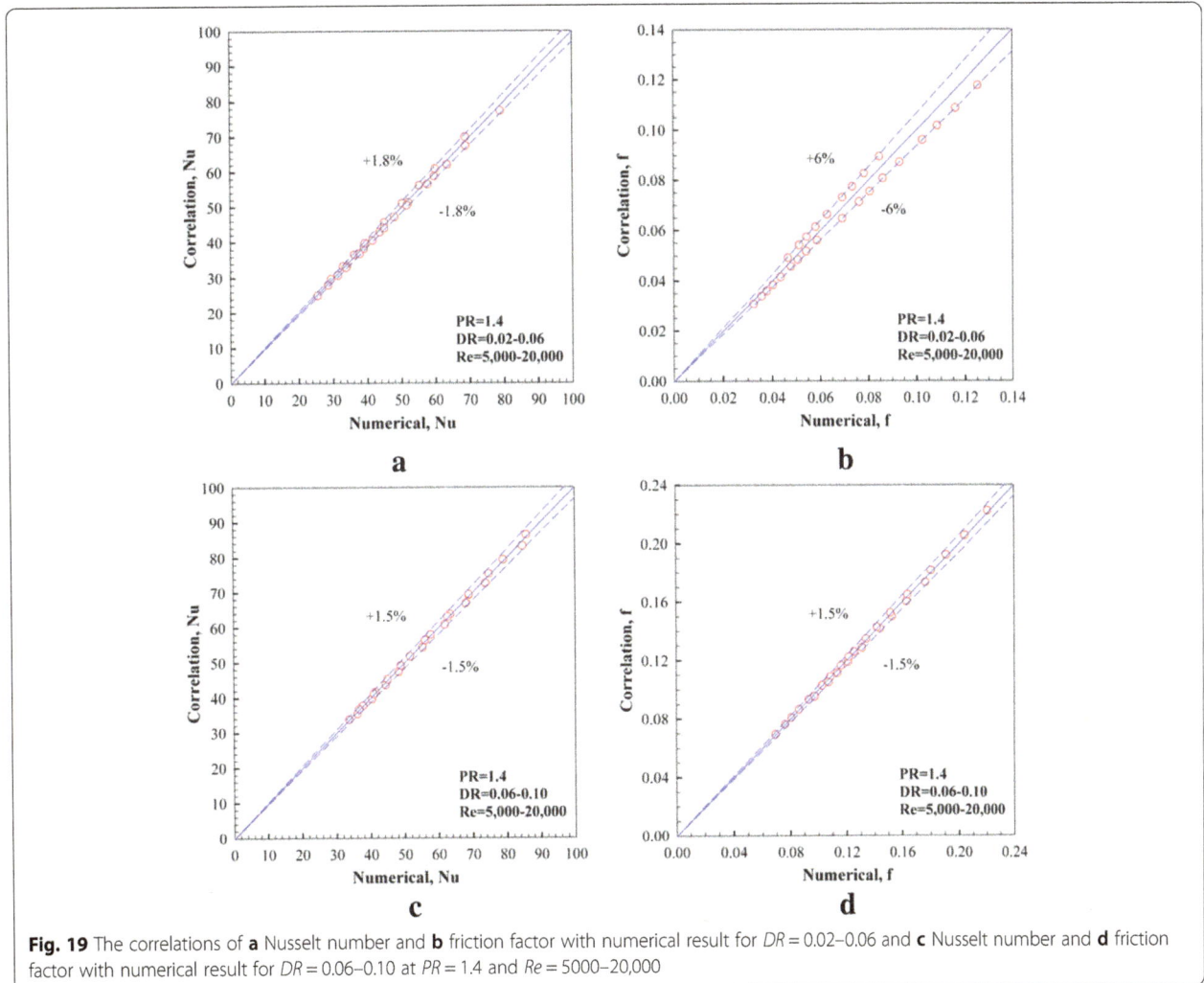

Fig. 19 The correlations of **a** Nusselt number and **b** friction factor with numerical result for $DR = 0.02$–0.06 and **c** Nusselt number and **d** friction factor with numerical result for $DR = 0.06$–0.10 at $PR = 1.4$ and $Re = 5000$–$20{,}000$

Conclusions

The numerical investigations on turbulent forced convection, heat transfer characteristic, and performance assessment in the spirally semicircle-grooved tube are examined. The influences of the groove depth and groove pitch for the spirally semicircle-grooved tube are investigated for the Reynolds number, Re = 5000–20,000. The flow visualization and heat transfer behavior in the spirally semicircle-grooved tube are described. The major findings for the present investigation are concluded as follows.

The semicircle groove on the tube surface can produce the swirling flow through the test section. The swirling flow can separate into two types: main and secondary swirling flows. The main swirling flow is found in all DRs, while the secondary swirling flow is detected at $DR \geq 0.06$. The swirling flow disturbs the thermal boundary layer near the tube wall that is reason for heat transfer augmentation. The intensity or strength of the swirling flow depended on DR and PR of the groove. The increasing DR and reducing PR lead to augmenting strength of the flow that helps to increase heat transfer rate in the test section.

In the range examined, the augmentations on the Nusselt number and friction loss are around 1.16–1.96 and 1.2–10.8 times above the smooth tube. The optimum TEF around 1.11 is detected at DR = 0.06 and PR = 1.4 and Re = 5000.

Nomenclature

D characteristic diameter of semicircle-grooved tube, m
DR semicircle-grooved depth ratio, (d/D)
PR semicircle-grooved pitch ratio, (p/D)
d semicircle-grooved depth, m
p semicircle-grooved pitch, m
f friction factor
h convective heat transfer coefficient, W m^{-2} K^{-1}
k turbulent kinetic energy, $\left(k = \frac{1}{2}\overline{u_i'u_j'}\right)$
k_a thermal conductivity of air, W m^{-1} K^{-1}
Nu Nusselt number
P static pressure, Pa
Pr Prandtl number
Re Reynolds number, $(\rho u_0 D/\mu)$
T temperature, K
TEF thermal performance enhancement factor, $(Nu/Nu_0)/(f/f_0)^{1/3}$
u_i velocity component in x$_i$-direction, m s^{-1}
$u_{i'}$ fluctuation velocity in x$_i$-direction, m s^{-1}
u_0 mean or uniform velocity in smooth tube, m s^{-1}
x coordinate direction

Greek letter

μ dynamic viscosity, kg s^{-1} m^{-1}
Γ thermal diffusivity
ε dissipation rate
ρ density, kg m^{-3}

Subscript

0 smooth tube
pp pumping power

Acknowledgements

The funding of this study is supported by King Mongkut's Institute of technology Ladkrabang research fund, Thailand. The researchers would like to thank Assoc. Prof. Dr. Pongjet Promvonge for suggestions.

Authors' contributions

PP conducted experiments and collected the data. AB and WJ wrote the paper and presented discussion. All authors read and approved the final manuscript.

Competing interests

The authors declare that they have no competing interests.

Author details

[1]Department of Mechanical Engineering, Faculty of Engineering, King Mongkut's Institute of Technology Ladkrabang, Bangkok 10520, Thailand. [2]Department of Mechanical Engineering Technology, College of Industrial Technology, King Mongkut's University of Technology North Bangkok, Bangkok 10800, Thailand.

References

Al-Shamani, A. N., Sopian, K., Mohammed, H. A., Mat, S., Ruslan, M. H., & Abed, A. M. (2015). Enhancement heat transfer characteristics in the channel with Trapezodial rib-groove using nanofluids. Case Studies in Thermal Engineering, 5, 48–58.
Aroonrat, K., & Wongwises, S. (2011). Evaporation heat transfer and friction characteristics of R-134a flowing downward in a vertical corrugated tube. Experimental Thermal and Fluid Science, 35, 20–28.
Aroonrat, K., Jumpholkul, C., Leelaprachakul, R., Dalkilic, A. S., Mahian, O., & Wongwises, S. (2013). Heat transfer and single-phase flow in internally grooved tubes. International Communications in Heat and Mass Transfer, 42, 62–68.
Chen, C., Wu, Y. T., Wang, S. T., & Ma, C. F. (2013). Experimental investigation on enhanced heat transfer in transversally corrugated tube with molten salt. Experimental Thermal and Fluid Science, 47, 108–116.
Cheng, L., & Chen, T. (2007). Study of vapor liquid two-phase frictional pressure drop in a vertical heated spirally internally ribbed tube. Chemical Engineering Science, 62, 783–792.
Dizaji, H. S., Jafarmader, S., & Mobadersani, F. (2015). Experimental studies on heat transfer and pressure drop characteristics for new arrangements of corrugated tubes in a double pipe heat exchanger. International Journal of Thermal Sciences, 96, 211–220.
Dong, Y., Huixiong, L., & Tingkuan, C. (2001). Pressure drop, heat transfer and performance of single-phase turbulent flow in spirally corrugated tubes. Experimental Thermal and Fluid Science, 24, 131–138.
Graham, D., Chato, J. C., & Newell, T. A. (1999). Heat transfer and pressure drop during condensation of refrigerant 134a in an axially grooved tube. International Journal of Heat and Mass Transfer, 42, 1935–1944.
Hwang, K., Jeong, J., Hyun, S., Saito, K., Kawai, S., Inagaki, K., & Ozawa, R. (2003). Heat transfer and pressure drop characteristics of enhanced titanium tubes. Desalination, 159, 33–41.
Incropera, F., & Dewitt, P. D. (2006). Introduction to heat transfer (3rd ed.). Hoboken: Wiley.
Khoeini, D., Akhavan-Behabadi, M. A., & Saboonchi, A. (2012). Experimental study of condensation heat transfer of R-134a flow in corrugated tubes with different inclinations. International Communications in Heat and Mass Transfer, 39, 138–143.
Laohalertdecha, S., & Wongwises, S. (2010). The effect of corrugation pitch and the condensation heat transfer coefficient and pressure drop of R-134a inside horizontal corrugated tube. International Journal of Heat and Mass Transfer, 53, 2924–2931.
Launder, B. E., & Spalding, D. B. (1974). The numerical computation of turbulent flows. Computer Methods in Applied Mechanics and Engineering, 3, 269–289.
Liu, L., Ling, X., & Peng, H. (2013). Analysis on flow and heat transfer characteristics of EGR helical baffled cooler with spiral corrugated tubes. Experimental Thermal and Fluid Science, 44, 275–284.

Liu, J., Xie, G., & Simon, T. W. (2015). Turbulent flow and heat transfer enhancement in rectangular channels with novel cylindrical grooves. *International Journal of Heat and Mass Transfer, 81*, 563–577.

Lu, J. F., Shen, X. Y., Ding, J., & Yang, J. P. (2013). Transition and turbulent convective heat transfer of molten salt in spirally grooved tube. *Experimental Thermal and Fluid Science, 47*, 180–185.

Lu, J. F., Shen, X. Y., Ding, J., Peng, Q., & Wen, Y. L. (2013). Convective heat transfer of high temperature molten salt in transversely grooved tube. *Applied Thermal Engineering, 61*, 157–162.

Patankar, S. V. (1980). *Numerical heat transfer and fluid flow*. New York: McGraw-Hill.

Tang, Y., Chi, Y., Chen, J. C., Deng, X. X., Liu, L., Liu, X. K., & Wan, Z. P. (2007). Experimental study of oil-filled high-speed spin forming micro-groove fin-inside tubes. *International Journal of Machine Tools and Manufacture, 47*, 1059–1068.

Wang, L., Suna, D. W., Liang, P., Zhuang, L., & Tan, Y. (2000). Heat transfer characteristics of carbon steel spirally fluted tube for high pressure preheaters. *Energy Conversion and Management, 41*, 993–1005.

Webb, R. L. (2009). Single-phase heat transfer, friction, and fouling characteristics of three-dimensional cone roughness in tube flow. *International Journal of Heat and Mass Transfer, 52*, 2624–2631.

Zhang, Q., Li, H., Zhang, W., Li, L., & Lei, X. (2015). Experimental study on heat transfer to the supercritical water upward flow in a vertical tube with internal helical ribs. *International Journal of Heat and Mass Transfer, 89*, 1044–1053.

Zhang, X., Zhang, J., Ji, H., & Zhao, D. (2015). Heat transfer enhancement and pressure drop performance for R417A flow boiling in internally grooved tubes. *Energy, 86*, 446–454.

17

Effect of torch process on the steels used for bucket, shovel handle, and other high-tonne mining equipment

H. Ochoa Medina[1,2], J. Leiva Yapur[1,3], O. Fornaro[4,5*] [iD] and Z. Cárdenas Quezada[1]

Abstract

Background: The gouging torch process using air carbon arc cutting (CAC) device is a standard maintenance procedure carrying out in high-tonne equipment used in the minery industry. The application of this process could locally affect the mechanical properties and the microstructure in the thermally affect zone (HAZ). The changes involve variation in the local carbon concentration and a tempering effect. In commonly used steels in the manufacture of buckets (SAE 5130) and shovel handles (ASTM 514 grade S), the processes influence negatively the work lifetime and the future maintenance works on the device.

Methods: Hardness, metallographic analysis trhough optical (OM) and scanning electron microscopy (SEM) were used to evaluate the affected zone.

Results: An increasing carbon content up to 2 wt% C was observed in the affected area of the sample, on the slag adhered to it. Presumably, the rest of carbon is lost by evaporation during the process.

Conclusions: The hardness measured on the surface of the cut zone shows an increased value for ASTM A 514 grade S, which does not present a notable change for SAE 5130. However, both steels showed a tempering effect. Microcracks of 20 to 40 μm appear, and in a few opportunities, a larger crack was found, reaching a total length of 1480 μm.

Keywords: Air carbon arc cutting, Heavy equipment, High-strength steels

Background

In the air carbon arc cutting (CAC) process, the molten metal generated by the electric arc is swept off by a strong dry airflow rate of 700 to 1000 l/min and a working pressure up to 6 kg/cm². The electric arc is generated by a graphite electrode (typically 90% C) with other metallic elements (Cu, Ni, Fe, among others). The composition is chosen to facilitate the passage of current and to avoid corrosion of the sample due to the pressurized air (American Society of Heating 2011; Victor Technologies, Inc. 2013; Varkey et al. 2013; Gutiérrez 2006; IS 1979; U.S. Department of Transportation 1994; AWS 1980; Das and Abarna 2015). The most common uses of this technique are (a)

preparations of welded joints, (b) removal of welding defects, and (c) removal of welding and joints of structures.

During the application of the CAC gouging torch process, it is possible to cause damage to the steel base substrate of the mining equipment, due to the high temperatures involved in the process, which can easily reach 2000 °C in the molten metal. At first, thermal fluctuations so as the addition of carbon concentration may cause effects on the thermally affected zone (HAZ) (Tingaev et al. 2016; Andrés et al. 2016), which may include residual stresses (Yilbas and Arif 2011; Michaleris 1999; Hu et al. 2007; Kyong-Ho and Lee 2007; Ma et al. 2016) and subsequently generate fatigue in the material (Goldberg 1978; Ramaswamy 1989; Cicero et al. 2017; Liu et al. 2016; Zhang et al. 2016; ASM 2013; Lara et al. 2013) or even spontaneous fracture, starting from micro-crack formation. It is also possible to suppose that solid-solid phase transformations so as micro-structure

* Correspondence: ofornaro@exa.unicen.edu.ar
[4]Instituto de Física de Materiales Tandil, IFIMAT (UNCPBA), Buenos Aires, Argentina
[5]Centro de Investigaciones en Física e Ingeniería del Centro de la Provincia de Buenos Aires, CIFICEN (UNCPBA-CICPBA-CONICET), Pinto 399, B700GHG Tandil, Argentina
Full list of author information is available at the end of the article

Table 1 Chemical composition of the used steels

Chemical component	ASTM 514 grade S	SAE 5130
	Wt% of component	
C	0.167	0.268
Si	0.358	0.249
Mn	1.166	0.975
P	< 0.003	< 0.003
S	< 0.003	< 0.003
Cr	0.0067	0.0067
Ni	0.0098	0.057
Mo	0.118	0.036
Al	0.0038	0.063
Ti	0.0012	0.0048
W	< 0.02	< 0.02
Fe	Balance	Balance

changes take place into the affected area during the cooling after the process end.

In maintenance work, a direct relationship between zones that have been treated through torching gauging processes and failure fractures due to fatigue or embrittlement has been found. For companies engaged in the minery industry, the recovery of abrasion-worn parts and the elimination of deteriorated areas in the mining equipment occupied in the extraction of high tonnage are extremely important.

The aim of this work is to study how the maintenance that uses CAC gouging torch process affects the microstructural behavior and mechanical properties in the most commonly used SAE 5130 and ASTM 514 S grade steels used in the manufacture of high-tonne mining equipment.

Methods

Used material

The samples were extracted from out-of-service equipment. The commonly used steels are ASTM 514 grade S for bucket and SAE 5130 in the case of shovel handles. The samples were initially cut with a Leco MSX255M saw equipment. After that, torch cuts were performed under ANSI/AWS C5.3-91 standard (Christensen 1973; Hause 1980; Panter 1977; Ridal 1977; Marshall 1980).

Determination of chemical composition

The steel samples were pickled with SiC 60 paper, washed with distilled water, cleaned with acetone, washed again with bi-distilled water, and dried at room temperature prior to analysis. The determination was performed according to ASTM E 415-E 1806, on a Bruker Q2 Ion spectrometer. The results obtained are presented in Table 1.

Determination of carbon in the slag

The slag is formed by small drops of molten and solidified steel in contact with the substrate material. Generally, it is joined with the exposed surface and could be easily removed by mechanics means. In other situations, it is expelled from the working zone by the high-pressure air flux. The carbon content in this residue was determined using a Primus Rigaku X-ray fluorescence device on milled slag particles. At least three determinations were made for each sample, taking the average of the obtained values.

Metallographic preparation and analysis

The metallographic analysis was performed according to ASTM E-3 standard. The specimens were prepared with a careful mechanical polishing, using graduated SiC paper of 240, 340, 400, and 600 grades, cooled in all cases with water. The polishing was finished with an alumina solution (Al_2O_3) in water on soft cloth. The microstructure was revealed by chemical etching by using nital (3%) for 2-s intervals. Optical micrographs were performed with an Optika Microscope model XD-3MET at 500× and 1000×. Electron microscopy images were taken with a scanning electron microscope JEOL model JSM-6360 LV, under the standard ASTM E 1508.

Results and discussion

As were previously said, during the cut process, part of the material is melted by the effect of an electric arc. Since the chemical composition of the electrode is rich in carbon, it

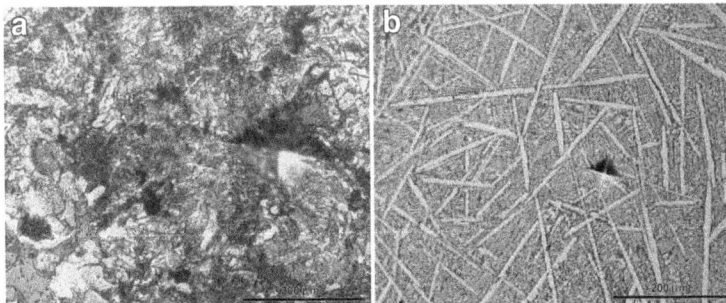

Fig. 1 Microstructure in different zones of the slag material of ASTM 514 grade S. **a** Austenitic zone. **b** Ferrite + carbides

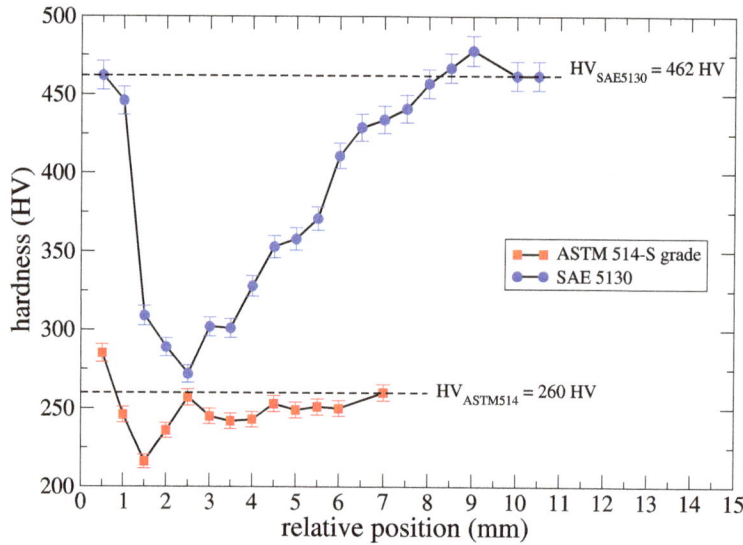

Fig. 2 Hardness and microhardness measured into the affected zone for both steels

can happen that the bath modifies its composition increasing the original solute carbon content. If this happens, a change in the microstructure would be more likely to occur, not only because of the high temperatures involved but also because of the diffusion of C in the liquid steel in the bath where the cutting takes place. For this reason, it is

interesting to know the composition of both the extracted material and the remnant in the piece, as part of the discussion related to the effects of the torch process.

As a first step, the composition of the material that comes out as slag from the cut is checked. This material is in principle liquid steel whose composition can be modified by the chemical elements that compose the electrode and that is solidified very quickly by the effect of the gas at high speed and pressure that extracts it from the bath generated once produced. The carbon content in these slag samples is higher than the original steel in both cases studied. For SAE 5130 steel, the C content determined in the slag was 1.69 wt% C, while for ASTM 514 grade S, a value of 1.93 wt% C was obtained. In Fig. 1, the microstructure obtained for ASTM

Table 2 Microhardness profile measured in treated steel SAE 5130 post-torching

Relative position (See text) (mm)	SAE 5130 (HV)
0.5	462
1	446
1.5	309
2	289
2.5	272
3.0	302
3.5	301
4.0	328
4.5	353
5.0	358
5.5	371
6.0	411
6.5	429
7.0	434
7.5	441
8.0	457
8.5	467
9.0	478
10	462
10.5	462

Table 3 Idem for and ASTM A 514 grade S

Relative position (see text) (mm)	ASTM 514 (HV)
0.5	285
1	246
1.5	216
2.0	236
2.5	257
3.0	245
3.5	242
4.0	243
4.5	253
5.0	249
5.5	251
6.0	250
7.0	260

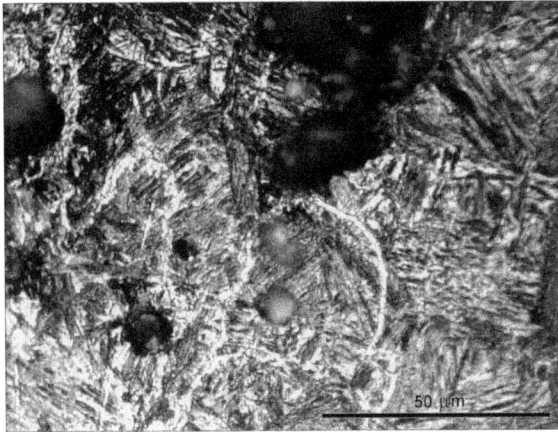

Fig. 3 Optical micrograph of SAE 5130 post-torching. Widmanstatten ferrite and martensite and some porosities could be observed

514 grade S ejected slag is shown in two different zones. Figure 1a corresponds to an austenitic structure, while Fig. 1b shows a microstructure of zone of high densities of needle carbides, consistent with a high-composition steel.

The microhardness is a reflect of the different structures. The obtained values fluctuate between 677 HV in the austenitic zone with dispersed carbides of Fig. 1a and 793 HV in the sample of Fig. 1b. The high dispersion in the structures could be derived from the different local cooling conditions, size of the slag drops, and other considerations during the application of the CAC procedure.

In the case of ASTM 514 grade S steel sample, rod-like and laminar carbides were found in the microstructure of the slag. Note that since the evaporation temperature of the graphite is 4827 °C, which is very close to the temperature that is reached at the electrode tip in the order of 5000–8000 °C, whereby the graphite of the electrode will evaporate, generating a reduction atmosphere near the cutting zone. While this gas is partially removed by pressurized air, carbon-rich areas could occur on the surface. In addition, the molten steel can facilitate the graphite entry since the temperature is higher enough to maintain the liquid bath for a short time. This allows the mixture of gas enriched with the steel to be produced by a diffusion mechanism through the gas + liquid interface, allowing the precipitation of carbides.

Figure 1b shows the typical dendritic microstructure of a foundry found in the cutting residue. The eutectic reaction L $\rightarrow$ y + MC, together with the austenite, is evidenced in the observed structure. The formation of MC nuclei is originally in the form of sticks, later branching. Growth is in conjunction with austenite as the temperature decreases (Bochnowski et al. 2003, Yang et al. 2006, Hetzner and Geertruyden 2008, Luan et al. 2010).

The behavior of bulk material is of course of great interest. As a consequence of the high temperature of the CAC gouging torch process, a softening effect can be observed in the HAZ. The dimension of this zone is quite well defined and is a consequence of the thermal and geometric characteristics of the treated elements. This behavior can be observed by microhardness on a profile that begins on the outer surface of the sample and is directed towards the center of the same. The microhardness of unmodified SAE 5130 steel was 462 HV and for ASTM 514 grade S steel 260 HV. These values were taken in areas far enough away from the cutting zone and will be used as a reference for the analysis that follows. In principle, the surface hardness taken on the cutting zone does not show a substantive difference with the reference value. A little increasing was noted on ASTM 514 grade S, prior to decrease, and shows a softening starting from the 500 μm of depth. This effect of tempering has a depth that varied in each case, as can be seen in more detail in Fig. 2. The obtained values of hardness and microhardness in the profile are detailed in Tables 2 and 3.

As can be seen in Fig. 2, the HAZ area in the case of SAE 5130 can reach approximately 9 mm in depth. In this zone, the observed microstructure corresponds to martensitic type with a large number of Widmanstatten-type needles, as can be seen in Fig. 3.

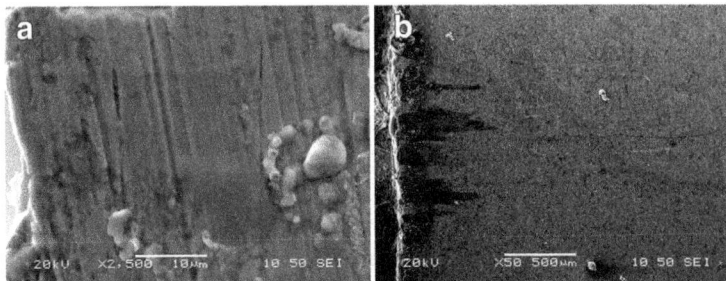

Fig. 4 SEM micrographs of SAE 5130 steel, adjacent to cutting zone, showing **a** the apparition of micro-cracks and **b** deep crack that reach the bulk material

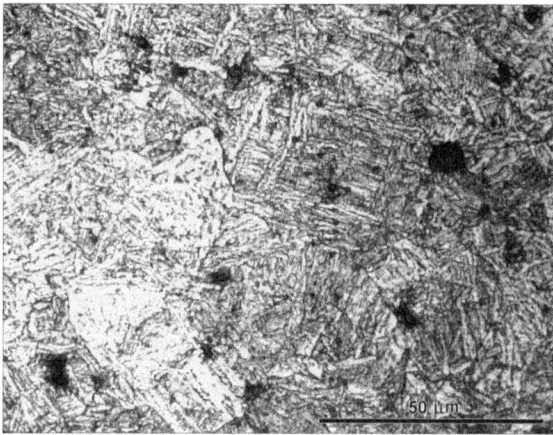

Fig. 5 Optical micrograph of ASTM 514 grade S post-torching

The reduction of the microhardness could be a consequence of a tempering caused by a high temperature delivered by the heat energy delivered by the torch. As can be seen in the SEM images of Fig. 4 for SAE 5130 steel, in the zone close to the surface, fissures of the order of 20 μm are observed in the micrographs, although some may agglomerate to form larger cracks, several mm in length (1500 μm).

On the other side, the ASTM 514 grade S steel shows a small increase in surface hardness after the torch process, reaching a surface hardness of 285 HV. This value is an increase of approximately 10% with respect to the original value of 260 HV determined far away from the cutting zone. This increase was found within the 500 μm closest to the cut. As in the previous case, after this zone, a softening occurs in the interior of the sample, although the magnitude of the zone affected by heat is rather narrower than for the other steel, since in approximately 3 mm, it recovers practically the original mechanical behavior.

In the affected zone, the observed microstructure is martensitic, with ferrite Widmanstatten as shown in Fig. 5. Also, a large amount of porosity and microcracks of approximately 50 μm in length are found as can be seen in the scanning electron micrographs of Fig. 6.

The observed behavior was similar to the reported in the bibliography for other thermal cutting methods. For example, after application of oxyfuel, laser, and plasma cutting, in the samples of S460 M steel, an increase in the superficial hardness followed by a decreasing in the mechanical properties was reported (Andrés et al. 2016). Also, for S345 and S390 steels treated by oxyfuel and plasma, cutting was found with similar results after being subjected to thermal cutting (Tingaev et al. 2016).

Conclusions

The analysis of the hardness and the observed microstructure shows that the torch process influences the behavior of the parts to be repaired, affecting the mechanical properties of the area. Formation of microcracks and porosities was also observed due to the same process.

The carbon content of the substrate material shows a small increment, apparently affected by the roasting process. Also, the C content is increased in the residue, reaching values close to 2 wt% C.

In the bulk material, the SAE 5130 steel shows microcracks less than 20 μm in length and also some greater fissures up to 1480 μm. In the ASTM 514 grade S steel, formation of microcracks of up to 40 μm in length was observed, but not a greater scale fissuration.

In all cases, a surface hardness increase tendency was observed both for SAE 5130 as for ASTM 514 steels, up to a 500 μm deep. A deeper observation reveals a softening beyond this 500-μm zone, product of a tempering in the hot affected zone. The affected zone varies between studied steels. In the case of SAE 5130, the tempering reach 8000 μm deep, while for steel ASTM 514 grade S, the affected zone reaches 5500 μm.

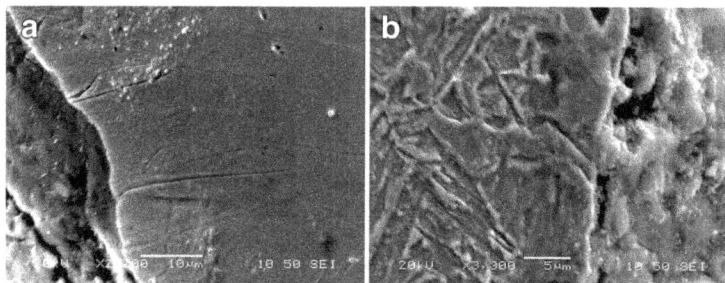

Fig. 6 SEM micrographs of ASTM 514 grade S post-torched steel. **a** Found defects are different sized micro-cracks into the HAZ, from 15 to 48 μm. **b** Microstructural change in the HAZ-bulk interface

Acknowledgements
The authors appreciate the collaboration of Minera Sierra Gorda for providing steels, personnel, and equipment and also Company Sorena for cutting by torch and personnel. The work was performed in the Centro de Ingeniería y Tecnología de los Materiales de la Universidad de Antofagasta.

Authors' contributions
JL and HO worked experimentally with the obtained samples. HO, OF, and ZC worked in metallography techniques. HO, JL, OF, and ZC had participated in the analysis and discussion of results. All authors read and approved the final manuscript.

Competing interests
The authors declare that they have no competing interests.

Author details
[1]Centro de Ingeniería y Tecnología de los Materiales Universidad de Antofagasta, Antofagasta, Chile. [2]Departamento de Ingeniería Mecánica Universidad de Antofagasta, Av. Angamos 601, Antofagasta, Chile. [3]Minera Sierra Gorda SCM, Antofagasta, Chile. [4]Instituto de Física de Materiales Tandil, IFIMAT (UNCPBA), Buenos Aires, Argentina. [5]Centro de Investigaciones en Física e Ingeniería del Centro de la Provincia de Buenos Aires, CIFICEN (UNCPBA-CICPBA-CONICET), Pinto 399, B700GHG Tandil, Argentina.

References
American Society of Heating (2011). Refrigerating and air-conditioning. Engineers. Environmental Health Committee (EHC) minutes. Winter Meeting January 31, 2011.

Andrés, D., García, T., Cicero, T., Lacalle, R., Álvarez, J., Martín-Meizoso, A., Aldazabal, J., Bannister, A., & Klimpel, A. (2016). Characterization of heat affected zones produced by thermal cutting processes by means of Small Punch tests. *Materials Characterization, 119*, 55–64.

ASM (2013). *Introduction to Surface Hardening of Steels.* ASM handbook, volume 4A, steel heat treating fundamentals and processes. Dossett and G.E. Totten, editors. Heat treating, Vol 4, ASM handbook, ASM International, 1991, p 259–267.

AWS. (1980). *Specification for low alloy steel electrodes for flux cored arc welding.* American Welting Society, Inc. Miami PL 33125. Appoved by AWS Board of Directors, August 15, 1980.

Bochnowski, W., Leitner, H., Major, L., Ebner, R., & Major, B. (2003). Primary and secondary carbides in high-speed after conventional heat treatment and laser modification. *Material Chemistry and Physics, 81*, 503–506.

Christensen, L. J. (1973). Air carbon arc cutting. *Welding Journal, 52*(12), 782–791.

Cicero, S., García, T., Álvarez, J. A., Klimpel, A., Bannister, A., & Martín-Meizosod, A. (2017). Fatigue behaviour and BS7608 fatigue classes of steels with thermally cut holes. *Journal of Constructional Steel Research, 128*, 74–83.

Das, A., & Abarna, R. (2015). Comparative study of air carbon arc gouging process on Sae 316 stainless steel. *International Research Journal of Engineering and Technology (IRJET), 02*(02), 1069–1074.

Goldberg F. (1978). Influence of thermal cutting and its quality in the fatigue strength of steel. Welding Research Supplement. *International Institute I-483-72:* 392-404.

Gutiérrez J. (2006) *Guía de instrucción en fabricación y reparación según curso modelo 7.04 de la omi oficial encargado de la guardia de máquinas,* Universidad Austral de Chile. Master Tesis.

Hause, W. O. (1980). What you should know about air carbon arc metal removal. *Welding Desing & Fabrication, 51*(1), 52–56.

Hetzner, D. W., & Van Geertruyden, W. (2008). Crystallography and metallography of carbides in high alloy Steels. *Materials Characterization, 59*, 825–841.

Hu, J.-f., Yang, J.-g., Hong-yuan, F., Guang-min, L., Yong, Z., & Wan, X. (2007). Temperature, stress and microstructure in 10Ni5CrMoV steel plate during air-arc cutting process. *Computational Materials Science, 38*, 631–641.

IS (1979). *Recommended practices for air carbon arc gouging and cutting.* Indian Standard Institution. WDC 621-791-948-054-4:006-76. IS 8987-1978.

Kyong-Ho, C., & Lee, C.-H. (2007). Residual stresses and fracture mechanics analysis of a crack in welds of high strength steels. *Engineering Fracture Mechanics, 74*, 980–994.

Lara, A., Picas, I., & Casellas, D. (2013). Effect of the cutting process on the fatigue behaviour of press hardened and high strength dual phase steels. *Journal of Materials Processing Technology, 213*, 1908–1919.

Liu, G., Huang, C., Zou, B., Wang, X., & Liu, Z. (2016). Surface integrity and fatigue performance of 17-4PH stainless steel after cutting operations. *Surface & Coatings Technology, 307*, 182–189.

Luan, Y., Song, N., Bai, Y., Kang, X., & Li, D. (2010). Effect of solidification rate on the morphology and distribution of eutectic carbides in centrifugal casting high-speed steel rolls. *Journal of Materials Processing Technology, 210*, 536–541.

Ma, Y., Pingfa, F., Jianfu, Z., Wu, Z., & Yu, D. (2016). Prediction of surface residual stress after end milling based on cutting force and temperature. *Journal of Materials Processing Technology, 235*, 41–48.

Marshall, W. J. (1980). Optical radiation levels produced by air carbon arc cutting processes. *Welding Journal, 59*(3), 43–46.

Michaleris P., Dantzig and Tortorelli D. (1999). Minimization of welding residual stress and distortion in large structures. Welding American Society. Supplement to the Welding Journal, November: 361-366.

Panter, D. (1977). Air carbon arc gouging. *Welding Journal, 56*(5), 32–37.

Ramaswamy, Murali T. (1989). *Heat affected zone studies of thermally cut structural steels.* Scholar Archive. Paper 130.

Ridal, E. J. (1977). Preparation for welding by air carbon arc gouging. *Welding & Metal Fabrication* 45 (6): 347–353 - 356–362.

Tingaev, A. K., Gubaydulin, R. G., & Ilin, R. G. (2016). Study of the effect of thermal cutting on the microstructure and chemical composition of the edges of workpieces made of steel brands S345, S390. International Conference on Industrial Engineering, ICIE 2016. *Procedia Engineering, 150*, 1783–1790.

U.S. Department of Transportation (1994). Federal Highway Administration, *Heat-affected zone studies of thermally cut structural steels.* Publication No. FHWA-RD-93-015 December. Research and Development Turner-Fairbank Highway Research Center 6300 Georgetown Pike McLean, Virginia, pp.2101–2296.

Varkey B., Balakrishnan K. and Chandran M. (2013). Experimental analysis of air carbon arc gouging process on SAE 316 stainless steel. Proceedings of International Conference on Materials.

Victor Technologies, Inc. (2013). *Air Carbon-Arc Guide Form* Number: 89–250-008.

Yang, J., Zheng, Q., Sun, X., Guan, H., & Hu, Z. (2006). Relative stability of carbides and their effects on the properties of K465 superalloy. *Material Science and Engineering A, 429*, 341–347.

Yilbas, B. S., & Arif, A. F. (2011). Laser cutting of steel and thermal stress development. *Optics & Laser Technology, 43*, 830–837.

Zhang, J., Wang, X., Paddea, S., & Zhang, X. (2016). Fatigue crack propagation behaviour in wire+arc additive manufactured Ti-6Al-4V: effects of microstructure and residual stress. *Materials and Design, 90*, 551–561.

18

A quasi-analytical solution of homogeneous extended surfaces heat diffusion equation

Ernest Léontin Lemoubou and Hervé Thierry Tagne Kamdem[*]

Abstract

Background: In this study, a quasi-analytical solution for longitudinal fin and pin heat conduction problems is investigated.

Methods: The differential transform method, which is based on the Taylor series expansion, is adapted for the development of the solution. The proposed differential transform solution uses a set of mathematical operations to transform the heat conduction equation together with the fin profile in order to yield a closeform series of homogeneous extended surface heat diffusion equation.

Results and conclusions: The application of the proposed differential transform method solution to longitudinal fins of rectangular and triangular profiles and pins of cylindrical and conical profiles heat conduction problems showed an excellent agreement on both fin temperature and efficiencies when compared to exact results. Therefore, the proposed differential transform method can be useful for optimal design of practical extended surfaces with suitable profile for temperature response.

Keywords: Analytical solution, Differential transform method, Heat conduction, Longitudinal fin, Pins

---

Background

Extended surfaces in the forms of longitudinal or radial fins or spines with various cross sections are widely used in industrial applications such as air-cooled craft engines, cooling of computer processors and electrical components, air-conditioning, refrigeration, heat exchangers, and solar collectors. These devices, which provide a considerable increase in the surface area for heat transfer between a heated source and a cooler ambient fluid, are most effective to enhance heat transfer between a surface and an adjacent fluid (Kraus et al. 2001). In designing extended surfaces, the first step consists of assuming that the heat transfer is governed by a one-dimensional homogeneous steady conduction along an extended surface and a uniform convection at the surface area (Arauzo et al. 2005; Kang 2009; Brestovic et al. 2015). Although an exact solution for this problem can be found in the literature, it may be either difficult or not possible to obtain when designing practical extended surfaces with profile matching suitable geometry for temperature response. Moreover, the exact solution of some problems

requires difficult manipulation of special functions such as Bessel functions.

Several efficient analytical or quasi-analytical methods have been developed over the past two decades to solve linear and nonlinear problems in science and engineering including the power series, the Adomian decomposition, the homotopy, and the differential transformation methods (Diez et al. 2009; Aziz and Bouaziz 2011; Torabi and Zhang 2013; Hayat et al. 2016). In general, exact solutions are mostly based on the tedious computation of special functions (Kraus et al. 2001; Diez et al. 2009). The power series method has been used by Arauzo et al. (2005) to approximate the solution of the one-dimensional steady heat conduction equation that governs the temperature variation in annular fins of hyperbolic profile. This method is a standard technique for solving linear ordinary differential equations with variable coefficients. The Adomian decomposition method has been used by Arslanturk (2005) and Bhowmik et al. (2013) to compute a closed-form solution for a straight convecting rectangular and hyperbolic profile annular fin with temperature-dependent thermal conductivity. This method represents the solution by an infinite series of the so-called Adomian polynomials and uses an iterative method such as the Newton–Raphson for the evaluation of the undetermined temperature at the fin tip.

---

[*] Correspondence: herve.kamdem@univ-dschang.org
Laboratory of Mechanics & Modeling of Physics Systems, Department of Physics/Faculty of Science, University of Dschang, P.O. Box 67, Dschang, Cameroon

The homotopy method is a quasi-analytical solution for solving nonlinear boundary value problem (Domairry and Fazeli 2009; Hayat et al. 2017a). This method does not require the calculation of Adomian polynomials as required for the Adomian decomposition method, but it requires an initial approximation (Roy and Mallick 2016; Hayat et al. 2017b; Waqas et al. 2016). Inc (2008) used the homotopy analysis method to evaluate the efficiency of straight fins with temperature-dependent thermal conductivity and determined the temperature distribution within the fin. The results show that the homotopy analysis method presents faster convergence and higher accuracy than the Adomian decomposition method and the homotopy perturbation method for nonlinear problems in science and engineering. The differential transform method (DTM) is based on the Taylor series expansion and constructs an analytical solution in the form of a polynomial. In general, the solution necessitates an iterative method such as the Newton–Raphson methods for the determination of the initial temperature transform function value. Joneidi et al. (2009) applied the differential transform method to predict temperature and efficiency of convective straight fins with temperature-dependent thermal conductivity. Their obtained results compared to exact and numerical results reveal that the differential transform method is an effective and accurate method for analyzing extended surfaces' nonlinear heat transfer problems. Kundu and Lee (2012) determined the performance of different fin geometries by analyzing heat transfer in rectangular, triangular, convex, and exponential geometric longitudinal fins using the differential transform method. These authors demonstrated that the differential transform is precise and cost efficient for analyzing nonlinear heat and mass transfer effects in extended surfaces.

The literature presents several quasi-analytical methods for analyzing heat transfer problems in extended surfaces. The different quasi-analytical methods, which used an iterative method such as the Newton–Raphson for the evaluation of an undetermined parameter, are applied mostly for nonlinear problems. On the other hand, the solution of linear problems is generally obtained as a particular case of nonlinear problems. The contribution of the present work is to present close-form series solution of the homogeneous extended surface heat diffusion equation using the differential transform method. This can be a useful strategy in developing an analytical solution when designing practical extended surfaces with suitable geometry for temperature response. The proposed differential transform solution uses a set of mathematical operations to transform the heat conduction equation together with the fin profile in order to yield a close-form series of homogeneous extended surface heat diffusion equation which avoid using an iterative method. The homogeneous extended surfaces in the forms of longitudinal fins of

rectangular and triangular profiles and pins of cylindrical and conical profiles are attached to a primary surface at constant temperature heat losses by convection to the surrounding medium and where the heat loss from the tip of the extended surfaces is assumed to be negligible. The temperature distribution and efficiency within extended surfaces are analyzed and compared against exact results.

Methods

Problem description

The problem under consideration consists of four homogeneous extended surfaces, longitudinal fins of rectangular and triangular profiles and pins of cylindrical and conical profiles as shown in Fig. 1. These extended surfaces are attached to a primary surface at constant temperature heat losses by convection to the surrounding medium and where the heat loss from the tip is assumed to be negligible.

For most engineering problems, the extended surface temperature must be maintained lower than the surrounding air to be cooled. Therefore, the extended surface which is attached to a primary surface at constant temperature T_b loses heat by convection to the surrounding medium of temperature T_∞. In this work, the heat loss from the tip of the extended surface is assumed to be negligible and the heat conduction is assumed to occur solely in the longitudinal direction. The governing of the conduction equations for this problem is well described in the literature (Kraus et al. 2001; Yaghoobi and Torabi 2011; Kundu and Lee 2012). For the problem under consideration, the governing equation for $0 \le x \le b$ and boundary conditions can be written as

$$\frac{d}{dx}\left(k_c \, p^\xi \frac{dT}{dx}\right) - \xi \, p^{(\xi-1)} \, h[T(x) - T_\infty] = 0 \qquad (1)$$

$$T|_{x=b} = T_b, \qquad A(x)\frac{dT}{dx}\bigg|_{x=0} = 0 \qquad (2)$$

where k_c is the constant thermal conductivity, h is the constant convective coefficient of the cooled air, A is the fin section area, the parameter $\xi = 1$ for longitudinal fins and $\xi = 2$ for pins and the extended surfaces profile, $p(x)$, is given by (Kraus et al. 2001)

$$p(x) = \begin{cases} \dfrac{w}{2}\left(\dfrac{x}{b}\right)^{(1-2\gamma)(1-\gamma)} & , if \ longitudinal \\[2em] \dfrac{w}{2}\left(\dfrac{x}{b}\right)^{(1-2\gamma)/(2-\gamma)} & , if \ pins \end{cases}$$

$$\qquad (3)$$

with $\gamma = 1/2$ for fins of rectangular and cylindrical profile and $\gamma = 0$ for triangular profile and $\gamma = -1$ for pins or

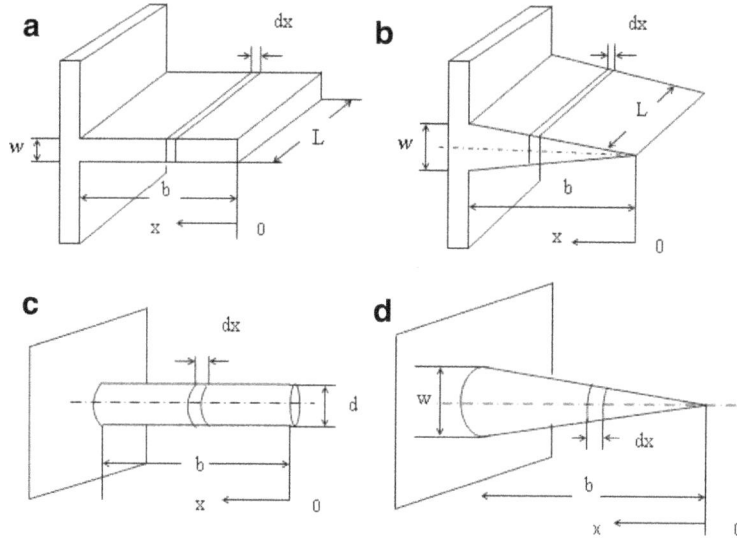

Fig. 1 Sketch diagram of longitudinal fins of rectangular (**a**) and triangular profile (**b**) and pins of cylindrical (**c**) and conical profile (**d**)

conical profile. Let us consider the following dimensionless temperature and coordinate, respectively, as

$$\theta = \frac{T(x)-T_\infty}{T_b-T_\infty}, \qquad \zeta = \frac{x}{b}, \tag{4}$$

The governing Eqs. (1) for and (2) can be written in dimensionless form as

$$\frac{d}{d\zeta}\left[p^\xi \frac{d\theta}{d\zeta}\right] - \frac{\xi h b^2}{k_c} p^{(\xi-1)}\theta = 0, \quad 0 \le \zeta \le 1 \tag{5}$$

$$\theta(\zeta = 1) = 1, \quad A(\zeta)\frac{d\theta}{d\zeta}(\zeta = 0) = 0 \tag{6}$$

Equation (6) can be further written as

$$p\frac{d^2\theta}{d\zeta^2} + \xi\frac{dp^\xi}{d\zeta}\frac{d\theta}{d\zeta} - \frac{\xi h b^2}{k_c}\theta = 0 \tag{7}$$

Differential transform operations

The differential transform method is generally used to derive a solution for a wide class of linear and nonlinear ordinary differential equations in the form of Taylor series (Chen and Ho, 1999; Hassan 2002). Consider an analytical function $f(x)$ defined in the domain D. The development of the function $f(x)$ around a point $x = x_i$ in Taylor series can be represented as (Jang et al. 2010; Chen and Ju 2004)

$$f(x) = \sum_{k=0}^{\infty} \frac{(x-x_i)^k}{k!}\left[\frac{d^k f(x)}{dx^k}\right]_{x=x_i} \tag{8}$$

It should be noted that for the development around point $x = 0$, Eq. (9) leads to Maclaurin series. The

function $F(k)$ which represents the transform form of the original function $f(x)$ can be expressed by the following relation (Jang et al. 2010; Chen and Ju 2004)

$$F(k) = \frac{(H)^k}{k!}\left[\frac{d^k f(x)}{dx^k}\right]_{x=x_i}, k = 0, 1, \cdots, \infty \tag{9}$$

where H is the space horizon of interest. Considering that the space horizon is $H = 1$, the transform function $F(k)$ is related to $f(x)$ by the differential inverse transform (Hassan 2002; Chen and Ju 2004)

$$f(x) = \sum_{k=0}^{\infty}(x-x_i)^k F(k) \tag{10}$$

If we consider that $f(x)$ is developed in the Taylor series around point $x = 0$, the different operators between $f(x)$ and the transform function $F(k)$ of Table 1 can be derived (Hassan 2002; Joneidi et al. 2009;

Table 1 One-dimensional DTM fundamental operations

Functions	Transform functions
$f(x) =$	$F(k)$
$af_1(x) + bf_2(x)$	$F(k) = aF_1(k) + bF_2(k)$
$\frac{df}{dx}$	$F(k) = (k+1)F(k+1)$
$\frac{d^2f}{dx^2}$	$F(k) = (k+1)(k+2)F(k+2)$
x^n	$F(k) = \delta(k-n) = \begin{cases} 1 & si \ k = n \\ 0 & si \ k \ne n \end{cases}$
$\exp(\lambda x)$	$F(k) = \frac{\lambda^k}{k!}$
$f_1(x)f_2(x)$	$F(k) = \sum_{l=0}^{k} F_1(l)F_2(k-l)$
$f_1(x)f_2(x)f_3(x)$	$F(k) = \sum_{h=0}^{k}\sum_{l=0}^{h} F_1(l)F_2(h-l)F_3(k-h)$

Yaghoobi and Torabi 2011). The key feature of the DTM is applied by using the operator of Table 1 to the ordinary differential equation of heat conduction through the extended surfaces. An easy way to derive the differential transform of Eqs. (6) and (7) is to find the differential transform of each term.

Considering Table 1 operators, the differential transform of Eq. (7) for longitudinal fins and pins gives, respectively

$$(k+1)(k+2)P(k)\Theta(k+2) + \sum_{l=0}^{k} P(l)\Theta(l)(k-l+2)$$

$$\times (k-l+1)\Theta(k-l+2) - \frac{h}{k_c}\Theta(k) = 0 \tag{11}$$

$$2\sum_{l=0}^{k}(l+1)(k-l+1)P(l+1)\Theta(k-l+1)$$

$$+ \sum_{h=0}^{k}\sum_{l=0}^{h}(k-h+1)(k-h+2)P(l)P(h-l)\,\Theta(k-h+2)$$

$$- \frac{2hb^2}{k_c}\Theta(k) = 0 \tag{12}$$

Taking the differential transform of the boundary conditions, Eq. (5), it can be obtained, respectively

$$\Theta(1) = 0 \tag{13}$$

$$\sum_{k=0}^{\infty}\Theta(k) = 1, \tag{14}$$

The transform condition (13) is applied only for rectangular and cylindrical fins whereas (14) is adapted for all the geometries.

Results and discussion

For stationary heat conduction through fins and pins with a dry surface and constant thermal conductivity cooled by air of constant heat transfer coefficient, exact analytical solution using a standard method of ordinary differential equations can be found in the literature (Kraus, et al., 2001; Kundu and Lee 2012). In the present work, the DTM is considered as alternative for obtaining analytical solution and the predictions are compared to the results from the standard method of ordinary differential equations.

Longitudinal fin with rectangular profile

For the first problem, the DTM is applied to a longitudinal fin of rectangular profile cooled by convection to the surrounding medium and with negligible heat loss from the tip. The terminology of this fin is described in

Fig. 1a, and the profile function is constant. The transform function of the profile is then

$$P(l) = \frac{w}{2}\delta(l) \tag{15}$$

The temperature transform function is reduced to

$$\Theta(k+2) = \frac{m^2\Theta(k)}{(k+1)(k+2)} \tag{16}$$

with $m = (2h/k_c w)^{\frac{1}{2}}$. Considering Eqs. (13) and (16), the temperature transform function can be written in a generalized form as

$$\Theta(2k) = \frac{m^{2k}}{(2k)!}\Theta(0), \quad k = 1,2,3,.... \tag{17}$$

$$\Theta(2k+1) = 0, \quad k = 0,1,2,3,.... \tag{18}$$

The discrete value $\Theta(0)$ is obtained by applying the second boundary condition Eq. (14) as

$$\Theta(0) = 1 \Big/ \sum_{k=0}^{n} \frac{m^{2k}}{(2k)!} \tag{19}$$

Considering Eq. (10), the temperature distribution for the fin of rectangular profile can then be expressed as

$$\theta(\zeta) = \sum_{k=0}^{n}\frac{(m\zeta)^{2k}}{(2k)!} \Big/ \sum_{k=0}^{n}\frac{m^{2k}}{(2k)!} \tag{20}$$

The temperature distribution given by Eq. (20) is computed and compared to the exact solution obtained using the standard method of ordinary differential equations (Kraus et al. 2001). The results were truncated to four decimal places since this precision is generally sufficient for engineering problems. The relative errors between the DTM and the exact results for the temperature are presented in Table 2. As expected, the DTM converges towards the exact solution as the order of the approximations is increasing. For both low and high value of the thermal length characteristic parameter, the DTM convergence in four significant figures at relatively low approximation orders: the first fourth terms of the Taylor series is sufficient.

The dimensionless temperature within the longitudinal fin of rectangular profile versus the dimensionless length for the different thermal length characteristic parameter $mb = (2h/k_c w)^{\frac{1}{2}} b$ can be seen in Fig. 2. Excellent agreement can be observed between DTM and the exact results. It is also noticed that the temperature increases with the height of the rectangular fin for a fixed value of the thermal length and tends to the unit when this fin characteristic is equal to one. For a dimensionless length, the magnitude of temperature

Table 2 Comparison of the DTM and exact results for different approximations order

ζ	mb = 0.6 Exact	DTM n = 1	2	mb = 1.4 Exact	DTM n = 1	2	3	4
0.0	0.8436	0.0039	0	0.4649	0.0401	0.0024	0.0001	0
0.1	0.8451	0.0039	0	0.4695	0.0405	0.0024	0.0001	0
0.2	0.8496	0.0039	0	0.4833	0.0416	0.0024	0.0001	0
0.3	0.8573	0.0039	0	0.5065	0.0431	0.0026	0.0001	0
0.4	0.8680	0.0039	0	0.5397	0.0445	0.0027	0.0001	0
0.5	0.8818	0.0038	0	0.5836	0.0452	0.0029	0.0001	0
0.6	0.8988	0.0036	0	0.6388	0.0444	0.0030	0.0001	0
0.7	0.9191	0.0032	0	0.7066	0.0409	0.0030	0.0001	0
0.8	0.9426	0.0025	0	0.7883	0.0335	0.0027	0.0001	0
0.9	0.9696	0.0015	0	0.8855	0.0205	0.0018	0.0001	0
1.0	1.0000	0.0000	0	1.0000	0.0000	0.0000	0.0000	0

decreases with increasing thermal length characteristic parameter. It should be noted that the thermal length characteristic parameter increases with the convection coefficient and decreases with conduction coefficient and fin thickness. In such case, the use of the longitudinal fin of rectangular profile will be optimal for a low convection coefficient of the cooling fluid or if the fin is designed with high thermal conductivity and thickness.

From the Fourier law, the heat transfer rate dissipated from the longitudinal fin to a neighboring fluid can be expressed in a dimensionless variable as

$$q(\zeta) = k_c A \frac{d\theta(\zeta)}{d\zeta} \qquad (21)$$

with the fin cross-sectional area $A = wL$ where δ and L are the fin thickness and length, respectively.

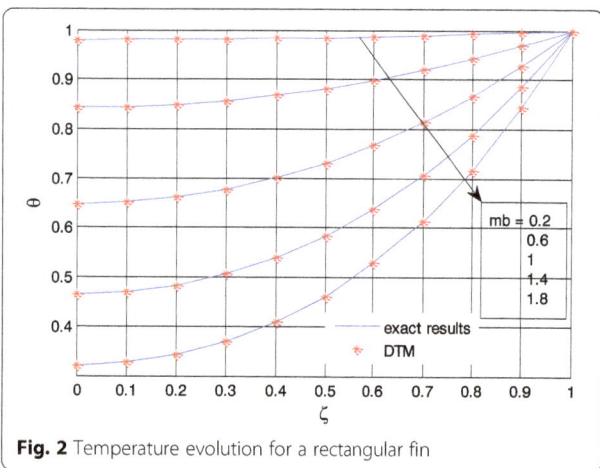

Fig. 2 Temperature evolution for a rectangular fin

Considering the derivative of Eq. (20), the heat transfer rate from Eq. (21) becomes

$$q(\zeta) = k_c w L m^2 (T_b - T_\infty) \sum_{k=1}^{n} \frac{m^{2k} \zeta^{2k-1}}{(2k-1)!} \Big/ \sum_{k=0}^{n} \frac{m^{2k}}{(2k)!} \qquad (22)$$

The fin efficiency can be evaluated as the ratio of heat transfer rate at the base of the fin to its ideal transfer rate if the entire fin were at the same temperature as its base

$$\eta = \frac{q(\zeta = 1)}{q_{ideal}} = \sum_{k=1}^{n} \frac{m^{2k}}{(2k-1)!} \Big/ \sum_{k=0}^{n} \frac{m^{2k}}{(2k)!} \qquad (23)$$

Figure 3 presents the behavior of a longitudinal fin of rectangular profile efficiency against thermal length characteristic parameter. Excellent agreement between the DTM efficiency prediction and exact results is observed. From Fig. 3, it is noted the decrease of efficiency is in accordance with increasing thermal length. This means that the performance of the rectangular fin decreases when the fin has high thickness and conduction coefficient or low convection coefficient of the cooling fluid.

Longitudinal fin of triangular profile
The second test problem considers longitudinal fins of triangular profile cooled by convection to the surrounding medium and with negligible heat loss from its tip. For this fin, shown in Fig. 1b, $\gamma = 0$ and the profile transform function gives

$$P(l) = \frac{w}{2} \delta(l-1) \qquad (24)$$

The temperature transform function for the longitudinal fin given by Eq. (11) reduces to

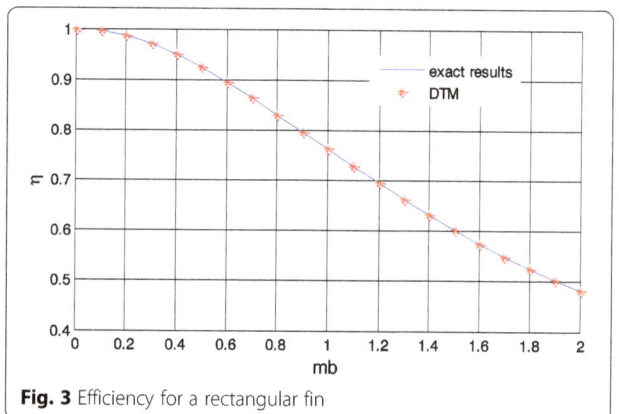

Fig. 3 Efficiency for a rectangular fin

$$\Theta(k+1) = \frac{1}{k+1} \sum_{l=0}^{k} \delta(l-1)(k-l+1)(k-l+2)$$

$$\Theta(k-l+2) + \frac{m^2}{k+1}\Theta(k)$$

$$(25)$$

with $m = \sqrt{2\,h/k_c\,w}$. The temperature transform function can be generalized as

$$\Theta(k) = \Theta(0)m^{2k}/(k!)^2, \quad k = 1, 2, 3, \dots. \quad (26)$$

From the boundary condition, Eq. (14), the value of the constant $\Theta(0)$ is given by

$$\Theta(0) = 1 \Big/ \sum_{k=0}^{n} \frac{m^{2k}}{(k!)^2} \quad (27)$$

Therefore, the temperature distribution for a longitudinal fin of triangular profile using Eq. (10) can be expressed as

$$\theta(\zeta) = \sum_{k=0}^{n} \frac{m^{2k}\zeta^k}{(k!)^2} \Big/ \sum_{k=0}^{n} \frac{m^{2k}}{(k!)^2} \quad (28)$$

The temperature computed using Eq. (28) is compared to the exact solutions from the standard method of ordinary differential equations. The DTM solutions converge to the exact solutions for the five terms of the Taylor series, and for all the significant figures of the exact solution. On Fig. 4, it can be seen that the evolution of the dimensionless temperature of the longitudinal fin of triangular profile with the dimensionless length increased as thermal length characteristic parameter increased. The decrease of the fin temperature with the thermal length characteristic parameter indicates that the loss of heat fin is more significant for longitudinal fins of triangular profiles with high thickness and conduction coefficient or for low convection coefficient of the cooling fluid.

Considering the heat transfer rate dissipated from the fin to a neighboring fluid, Eq. (21), the heat transfer rate dissipated from the longitudinal fin of triangular profile having cross-sectional area $A = wL$ is

$$q(\zeta) = k_c wLm^2(T_b-T_\infty) \sum_{k=1}^{n} \frac{m^{2k}\zeta^{k-1}}{(k-1)!k!} \Big/ \sum_{k=0}^{n} \frac{m^{2k}}{(k!)^2}$$

$$(29)$$

Consequently, the fin efficiency can be expressed as

$$\eta = \frac{q(\zeta = 1)}{2hL(T_b-T_\infty)} = \sum_{k=1}^{n} \frac{m^{2k}}{(k-1)!k!} \Big/ \sum_{k=0}^{n} \frac{m^{2k}}{(k!)^2} \quad (30)$$

The decrease of the longitudinal fin of triangular profile efficiency with increased thermal length characteristic parameter can be seen in Fig. 5. An excellent agreement between present DTM and exact results is achieved.

Pin of cylindrical profile

A pin of cylindrical profile with the heat loss from the tip which is assumed to be negligible and cooled by convection to the surrounding medium is now considered. Since for this pin, the profile function has the same expression than that for longitudinal fin of rectangular profile, the profile transform function is given by Eq. (15). It can also be deduced from Eq. (12) that the temperature transform function is identical to Eq. (15) where the parameter m is now given by $m = (4h/k_c\,d)^{\frac{1}{2}}$. Therefore, the temperature distribution through the pin of cylindrical profile is given by Eq. (20). The fin cross-sectional area of this pin is $A = \pi d^2/4$ and therefore the heat transfer rate dissipated from the pin of cylindrical profile to the surrounding fluid is

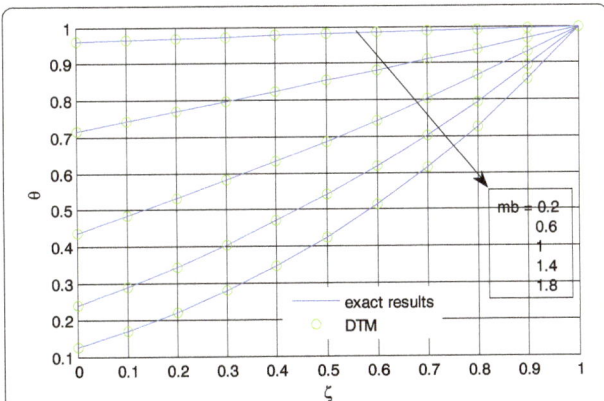

Fig. 4 Temperature evolution for a triangular fin

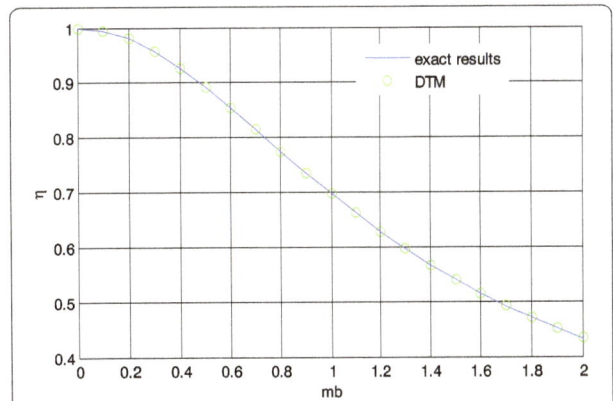

Fig. 5 Efficiency for a triangular fin

$$q(\zeta) = \pi d^2\, k_c m^2 (T_b - T_\infty) \sum_{k=1}^{n} \frac{m^{2k}\zeta^{2k-1}}{(2k-1)!} \Big/ 4\sum_{k=0}^{n} \frac{m^{2k}}{(2k)!}$$

(31)

The efficiency of this fin can be evaluated as

$$\eta = \frac{q(\zeta = 1)}{\pi d h (T_b - T_\infty)} = \sum_{k=1}^{n} \frac{m^{2k}}{(2k-1)!} \Big/ \sum_{k=0}^{n} \frac{m^{2k}}{(2k)!}$$

(32)

Figures 6 and 7 compare the pins' dimensionless temperature efficiency predicted using DTM and exact results. An excellent agreement between the two methods is observed for DTM with four terms of the Taylor series. It can be seen in Fig. 6 that the dimensionless temperature increases with increasing pin dimensionless length, while the pin efficiency decreased with increasing thermal length characteristic parameter. This indicates that a pin of cylindrical profile dissipates more heat transfer for high values of the pin diameter and thermal conductivity or for low convection coefficient of the cooling fluid.

Pin of conical profile

The last problem deals with a pin of conical profile cooled by convection to the surrounding medium cooled and where the heat loss from the tip is assumed to be negligible. For this pin, shown in Fig. 1d, the profile transform function is expressed as

$$P(l) = \frac{w}{2}\delta(l-1)$$

(33)

For the general discrete expression for pin, Eq. (12) is reduced to

$$\Theta(k+1) = \frac{1}{2(k+1)}\left\{ \sum_{l=0}^{k} \delta(l-1)(k-l+1)(k-l+2) \right.$$
$$\left. \Theta(k-l+2) - 2m^2\Theta(k)\right\}$$

(34)

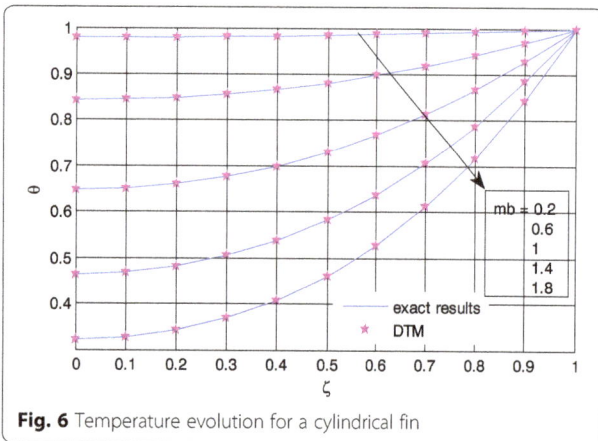

Fig. 7 Efficiency for a cylindrical fin

with $m = (4h/k_c\, w)^{\frac{1}{2}}$. The temperature transform function can be generalized as

$$\Theta(k) = \frac{2^k\, m^{2k}}{k!(k+1)!}\, \Theta(0), \quad k = 1, 2\dots$$

(35)

Application of boundary transform Eq. (14) gives

$$\Theta(0) = 1\Big/ \sum_{k=0}^{n} \frac{2^k\, m^{2k}}{k!(k+1)!}$$

(36)

The temperature distribution within the pin is obtained by substituting Eqs. (35) and (36) in Eq. (10) as

$$\theta(\zeta) = \sum_{k=0}^{n} \frac{2^k\, m^{2k}}{k!(k+1)!}\zeta^k \Big/ \sum_{k=0}^{n} \frac{2^k\, m^{2k}}{k!(k+1)!}$$

(37)

Figure 8 shows the variation of the dimensionless temperature according to the pin of the conical profile dimensionless length. Excellent agreement between present DTM using five terms of the Taylor series terms and exact results is observed. It can be seen that the temperature increases with increasing pin length, while

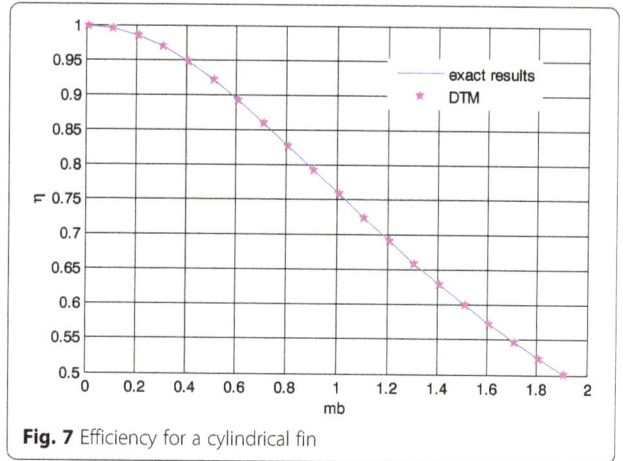

Fig. 6 Temperature evolution for a cylindrical fin

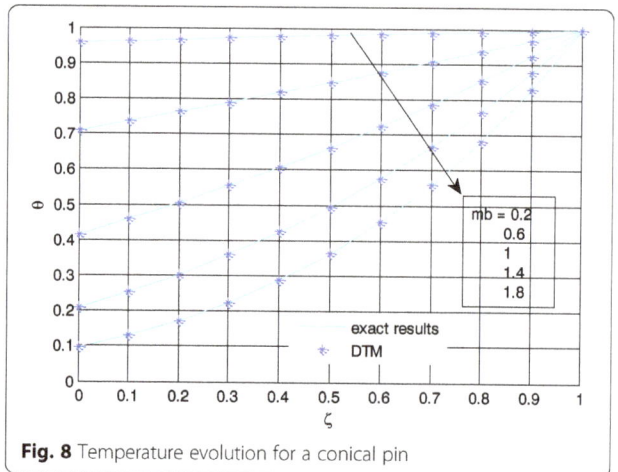

Fig. 8 Temperature evolution for a conical pin

it decreases with increasing thermal length characteristic parameter.

The heat transfer dissipated by the pin having cross-sectional area $A = \pi w^2/4$ can be evaluated as

$$q_b = k_c \frac{\pi w^2}{4} m^2 (T_b - T_\infty) \sum_{k=1}^{n} \frac{2^k \, m^{2k-2}}{(k-1)!(k+1)!} \bigg/ \sum_{k=0}^{n} \frac{2^k \, m^{2k}}{k!(k+1)!}$$

(38)

The pin of conical profile efficiency is given by

$$\eta = \frac{q(\zeta = 1)}{\pi w h (T_b - T_\infty)/2} = \sum_{k=1}^{n} \frac{2^{k+1} \, m^{2k-2}}{(k-1)!(k+1)!} \bigg/ \sum_{k=0}^{n} \frac{2^k \, m^{2k}}{k!(k+1)!}$$

(39)

The decrease of pin efficiency with thermal length characteristic parameter is illustrated in Fig. 9, where an excellent agreement between present DTM and exact results is observed. The variation of temperature and efficiencies with the thermal length characteristic parameter confirms that the performance of pins of conical profile will be optimal for the combination of high conduction coefficient and fin thickness or for low convection coefficient of the cooling fluid.

Conclusions

A differential transform method to analyze stationary heat conduction through homogeneous extended surfaces with negligible heat loss from the tip has been presented. The major conclusion of this work is that for stationary heat conduction through extended surfaces, the DTM solution can be obtained in a closed series solution which does not necessitate an iterative method such as the Newton–Raphson methods for the determination of the initial value of temperature transform function. The proposed method is shown to converge with a few Taylor series for both low and high values of the thermal length characteristic parameter. Application

of the present DTM solution to longitudinal fins of rectangular and triangular profiles and pins of cylindrical and conical profiles show an excellent agreement with exact results. For all cases studied, the magnitude of temperature decreases with increasing thermal length characteristic parameter. This indicates that the loss of heat from extended surfaces is more significant for low convection coefficient of the cooling fluid than that for extended surfaces with high thickness and conduction coefficient.

Nomenclature

A fin section area, m^2
b height of the fin, m
h heat convectivity coefficient $W \cdot m^{-2} \cdot K^{-1}$
k_c thermal conductivity heat coefficient, $W \cdot m^{-1} \cdot K^{-1}$
q heat flux, W/m^2
T temperature, K
x coordinate, m
p profile function
P profile transform function
W thickness of the fin, m
Greek symbols
β dimensionless temperature
θ dimensionless temperature
Θ dimensionless temperature transform function
η fin efficiency
ζ dimensionless coordinate
Subscripts
b fin base
i ideal parameter
∞ ambient environment

Authors' contributions
ELL implemented the quasi-analytic differential transform method and HTTK defined the research problem. All authors contributed to the problem formulation, drafted the manuscript, and read and approved the final manuscript.

Competing interests
The authors declare that they have no competing interests.

References
Arauzo, I., Campo, A., & Cortes, C. (2005). Quick estimate of the heat transfer characteristics of annular fins of hyperbolic profile with the power series method. *Applied Thermal Engineering, 25*, 623–634.
Arslanturk, C. (2005). A decomposition method for fin efficiency of convective straight fins with temperature-dependent thermal conductivity. *International Communications in Heat and Mass Transfer, 32*, 831–841.
Aziz, A., & Bouaziz, M. N. (2011). A least squares method for a longitudinal fin with temperature dependent internal heat generation and thermal conductivity. *Energy Conversion and Management, 52*, 2876–2882.
Bhowmik, A., Singla, R. K., Roy, P. K., Prasad, D. K., Das, R., & Repaka, R. (2013). Predicting geometry of rectangular and hyperbolic fin profiles with temperature-dependent thermal properties using decomposition and evolutionary methods. *Energy Conversion and Management, 74*, 535–547.
Brestovič, T., Jasminská, N., & Lázár, M. (2015). Application of analytical solution for extended surfaces on curved and squared ribs. *Acta Mechanica et Automatica, 9*(2), 75–83.

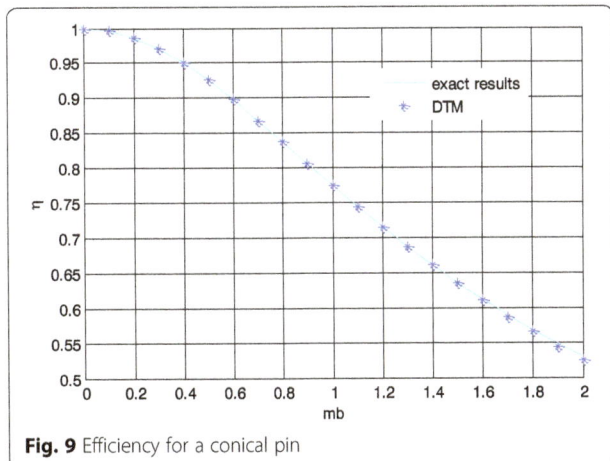

Fig. 9 Efficiency for a conical pin

Chen, C. K., & Ho, S. H. (1999). Solving partial differential equations by two-dimensional differential transform method. *Applied Mathematics and Computation, 106*, 171–179.

Chen, C. K., & Ju, S. P. (2004). Application of differential transformation to transient advective–dispersive transport equation. *Applied Mathematics and Computation, 155*, 25–38.

Diez, L. I., Campo, A., & Cortés, C. (2009). Quick design of truncated pin fins of hyperbolic profile for heat-sink applications by using shortened power series. *Applied Thermal Engineering, 29*, 815–821.

Domairry, G., & Fazeli, M. (2009). Homotopy analysis method to determine the fin efficiency of convective straight fins with temperature-dependent thermal conductivity. *Communications in Nonlinear Science and Numerical Simulation, 14*, 489–499.

Hassan, I. H. A. (2002). Different applications for the differential transformation in the differential equations. *Applied Mathematics and Computation, 129*, 183–201.

Hayat, T., Waqas, M., Ijaz Khan, M., & Alsaedi, A. (2016). Analysis of thixotropic nanomaterial in a doubly stratified medium considering magnetic field effects. *International Journal of Heat and Mass Transfer, 102*, 1123–1129.

Hayat, T., Ijaz Khan, M., Waqas, M., & Alsaedi, A. (2017a). Newtonian heating effect in nanofluid flow by a permeable cylinder. *Results in Physics, 7*, 256–262.

Hayat, T., Ijaz Khan, M., Farooq, M., Alsaedi, A., & Khand, M. I. (2017b). Thermally stratified stretching flow with Cattaneo-Christov heat flux. *International Journal of Heat and Mass Transfer, 106*, 289–294.

Inc, M. (2008). Application of homotopy analysis method for fin efficiency of convective straight fins with temperature-dependent thermal conductivity. *Mathematics and Computers in Simulation, 79*, 189–200.

Jang, M.-J., Yeh, Y.-L., Chen, C.-L., & Yeh, W.-C. (2010). Differential transformation approach to thermal conductive problems with discontinuous boundary condition. *Applied Mathematics and Computation, 216*, 2339–2350.

Joneidi, A. A., Ganji, D. D., & Babaelahi, M. (2009). Differential transformation method to determine fin efficiency of convective straight fins with temperature dependent thermal conductivity. *International Communications in Heat and Mass Transfer, 36*, 757–762.

Kang, H.-S. (2009). Optimization of a rectangular profile annular fin based on fixed fin height. *Journal of Mechanical Science and Technology, 23*, 3124–3131.

Kraus, A. D., Aziz, A., & Welty, J. (2001). *Extended surface heat transfer*. Hoboken: John Wiley & Sons, Inc.

Kundu, B., & Lee, K.-S. (2012). Analytic solution for heat transfer of wet fins on account of all nonlinearity effects. *Energy, 41*, 354e367.

Roy, P. K., & Mallick, A. (2016). Thermal analysis of straight rectangular fin using homotopy perturbation method. *Alexandria Engineering Journal, 55*, 2269–2277.

Torabi, M., & Zhang, Q. B. (2013). Analytical solution for evaluating the thermal performance and efficiency of convective–radiative straight fins with various profiles and considering all non-linearities. *Energy Conversion and Management, 66*, 199–210.

Waqas, M., Farooq, M., Khan, M. I., Ahmed Alsaedi, A., Hayat, T., & Yasmeen, T. (2016). Magnetohydrodynamic (MHD) mixed convection flow of micropolar liquid due to nonlinear stretched sheet with convective condition. *International Journal of Heat and Mass Transfer, 102*, 766–772.

Yaghoobi, H., & Torabi, M. (2011). The application of differential transformation method to nonlinear equation arising in heat transfer. *International Communications in Heat and Mass Transfer, 38*, 815–820.

Permissions

All chapters in this book were first published in IJMME, by Springer International Publishing AG.; hereby published with permission under the Creative Commons Attribution License or equivalent. Every chapter published in this book has been scrutinized by our experts. Their significance has been extensively debated. The topics covered herein carry significant findings which will fuel the growth of the discipline. They may even be implemented as practical applications or may be referred to as a beginning point for another development.

The contributors of this book come from diverse backgrounds, making this book a truly international effort. This book will bring forth new frontiers with its revolutionizing research information and detailed analysis of the nascent developments around the world.

We would like to thank all the contributing authors for lending their expertise to make the book truly unique. They have played a crucial role in the development of this book. Without their invaluable contributions this book wouldn't have been possible. They have made vital efforts to compile up to date information on the varied aspects of this subject to make this book a valuable addition to the collection of many professionals and students.

This book was conceptualized with the vision of imparting up-to-date information and advanced data in this field. To ensure the same, a matchless editorial board was set up. Every individual on the board went through rigorous rounds of assessment to prove their worth. After which they invested a large part of their time researching and compiling the most relevant data for our readers.

The editorial board has been involved in producing this book since its inception. They have spent rigorous hours researching and exploring the diverse topics which have resulted in the successful publishing of this book. They have passed on their knowledge of decades through this book. To expedite this challenging task, the publisher supported the team at every step. A small team of assistant editors was also appointed to further simplify the editing procedure and attain best results for the readers.

Apart from the editorial board, the designing team has also invested a significant amount of their time in understanding the subject and creating the most relevant covers. They scrutinized every image to scout for the most suitable representation of the subject and create an appropriate cover for the book.

The publishing team has been an ardent support to the editorial, designing and production team. Their endless efforts to recruit the best for this project, has resulted in the accomplishment of this book. They are a veteran in the field of academics and their pool of knowledge is as vast as their experience in printing. Their expertise and guidance has proved useful at every step. Their uncompromising quality standards have made this book an exceptional effort. Their encouragement from time to time has been an inspiration for everyone.

The publisher and the editorial board hope that this book will prove to be a valuable piece of knowledge for researchers, students, practitioners and scholars across the globe.

List of Contributors

Pavel Layus, Paul Kah and Markku Pirinen
Welding Technology Laboratory, Lappeenranta University of Technology, PL 20, 53851 Lappeenranta, Finland

Alexander Zisman and Sergey Golosienko
Central Research Institute of Structural Materials Prometey, Saint-Petersburg, Russia

K. Vajravelu
Department of Mathematics, University of Central Florida, Orlando, FL 32816, USA

K. V. Prasad and Hanumesh Vaidya
Department of Mathematics, VSK University, Vinayaka Nagar, Ballari 583 105, Karnataka, India

Chiu-On Ng
Department of Mechanical Engineering, The University of Hong Kong, Pokfulam, Hong Kong

E. Gustafsson and L. Karlsson
Dalarna University, SE-791 88 Falun, Sweden

M. Oldenburg
Luleå University of Technology, SE-971 87 Luleå, Sweden

Sambit Kumar Mohapatra and Kalipada Maity
Department of Mechanical Engineering, National Institute of Technology, Rourkela, Odisha 769008, India

Nian Zhou
Department of Material Science, Dalarna University, SE-79188 Falun, Sweden
KTH, SE-10044 Stockholm, Sweden

Rachel Pettersson
KTH, SE-10044 Stockholm, Sweden
Jernkontoret, SE-11187 Stockholm, Sweden

Ru Lin Peng
Department of Management and Engineering, Linköping University, SE-58183 Linköping, Sweden

J. Bala Bhaskara Rao
Department of Mechanical Engineering, SISTAM College, Kakinada, India

V. Ramachandra Raju
Department of Mechanical Engineering, JNTU Kakinada, Kakinada, India

Adel Benchabane and Amar Rouag
Laboratoire de Génie Energétique et Matériaux, LGEM, Université de Biskra, B.P. 145 R.P., Biskra 07000, Algeria

Noureddine Cherrad
Laboratoire de Génie Energétique et Matériaux, LGEM, Université de Biskra, B.P. 145 R.P., Biskra 07000, Algeria

Département de Génie Mécanique, Université de Ouargla, Faculté des Sciences Appliquées, Ouargla 30 000, Algeria

Lakhdar Sedira
Laboratoire de Génie Mécanique, LGM, Université de Biskra, B.P. 145 R.P., Biskra 07000, Algeria

Hoda Nasr El-Din and Hoda Refaiy
Plastic Deformation Department, Central Metallurgical R& D institute (CMRDI), Cairo, Egypt

Ezzat A. Showaib and Nader Zaafarani
Production and Mechanical Design Department, Faculty of Engineering, Tanta University, Tanta, Egypt

Vinoadh Kumar Krishnan and Kumaran Sinnaeruvadi
Department of Metallurgical and Materials Engineering, National Institute of Technology, Tiruchirappalli, Tamil Nadu 620015, India

T. C. Lim
School of Science and Technology, SIM University, Singapore, Singapore

S. Sree Sabari, S. Malarvizhi and V. Balasubramanian
Centre for Materials Joining and Research (CEMAJOR), Department of Manufacturing Engineering, Annamalai University, Annamalai Nagar, Chidambaram 608 002, Tamil Nadu, India

K. Ganesh Kumar, B. J. Gireesha and N. G. Rudraswamy
Department of Studies and Research in Mathematics, Kuvempu University Shankaraghatta, Shimoga, Karnataka 577 451, India

S. Manjunatha
Faculty of Engineering, Department of Engineering Mathematics, Christ University, Mysore Road, Bengaluru 560074, India

E. Gustafsson and L. Karlsson
Dalarna University, SE-791 88 Falun, Sweden

M. Oldenburg
Luleå University of Technology, SE-971 87 Luleå, Sweden

M. Manjaiah and Rudolph F. Laubscher
Department of Mechanical Engineering Science, University of Johannesburg, Kingsway Campus, Johannesburg 2006, South Africa

Anil Kumar
Department of Studies i0n Mechanical Engineering, University B.D.T. College of Engineering, Davangere 577 004, Karnataka, India

S. Basavarajappa
Indian Institute of Information Technology, Dharwad 580029, Karnataka, India

Gustafsson, L. Karlsson, S. Marth and M. Oldenburg
Dalarna University, SE-791 88 Falun, Sweden

Pitak Promthaisong and Withada Jedsadaratanachai
Department of Mechanical Engineering, Faculty of Engineering, King Mongkut's Institute of Technology Ladkrabang, Bangkok 10520, Thailand

Amnart Boonloi
Department of Mechanical Engineering Technology, College of Industrial Technology, King Mongkut's University of Technology North Bangkok, Bangkok 10800, Thailand

H. Ochoa Medina
Centro de Ingeniería y Tecnología de los Materiales Universidad de Antofagasta, Antofagasta, Chile Departamento de Ingeniería Mecánica Universidad de Antofagasta, Av. Angamos 601, Antofagasta, Chile

J. Leiva Yapur
Centro de Ingeniería y Tecnología de los Materiales Universidad de Antofagasta, Antofagasta, Chile Minera Sierra Gorda SCM, Antofagasta, Chile

O. Fornaro
Instituto de Física de Materiales Tandil,IFIMAT (UNCPBA), Buenos Aires, Argentina

Centro de Investigaciones en Física e Ingeniería del Centro de la Provincia de Buenos Aires, CIFICEN (UNCPBA-CICPBA-CONICET), Pinto 399, B700GHG Tandil, Argentina

Z. Cárdenas Quezada
Centro de Ingeniería y Tecnología de los Materiales Universidad de Antofagasta, Antofagasta, Chile

Ernest Léontin Lemoubou and Hervé Thierry Tagne Kamdem
Laboratory of Mechanics & Modeling of Physics Systems, Department of Physics/Faculty of Science, University of Dschang, P.O. Box 67, Dschang, Cameroon

Index

www.ingramcontent.com/pod-product-compliance
Lightning Source LLC
Chambersburg PA
CBHW082034190326
41458CB00010B/3367

9 781632 386304